Ethics and Sustainability in Digital Cultures

Digital technologies, now ubiquitous around the world, can promote positive values, as well as support those that are less socially acceptable. To better understand such technologies' impact on ethics and sustainability, this book situates digital technologies within a cultural context, arguing that the technology is received differently in different cultural contexts. The book contains chapters on state-of-the-art digital technologies such as artificial intelligence from various countries including Japan and Sweden to highlight the multifarious ways in how ethical and sustainability issues are being manifested in certain cultural contexts.

The book contributes to furthering understandings on the similarities and differences between digital technology implementations in different cultures, promoting a cross-cultural dialogue on desired values and how they are promoted or downplayed by such technologies. The book is divided into two parts: the former focuses on how individuals relate to new digital technologies, and the latter focuses on those who develop digital technologies.

The book targets scholars, businesspeople and policymakers interested in the connection between digital technologies, ethics and sustainability from various cultural viewpoints. It provides new case studies on a range of digital technologies and discussions about digital technology implementations in cultural contexts.

Thomas Taro Lennerfors is Professor and Head of the Division of Industrial Engineering and Management at Uppsala University, Sweden.

Kiyoshi Murata is Director of the Centre for Business Information Ethics and Professor of MIS at the School of Commerce, Meiji University, Tokyo, Japan.

Routledge Series on Digital Spaces
Series Edited by Chris Lukinbeal and Barney Warf

Digital and physical worlds have become so intertwined they are inseparable. The digital revolution has had enormous impacts on the people, economies, politics, cultures, and places. This book series engages with cutting edge research on the effect of digital technologies on the world(s) we live in, from a variety of perspectives and scales. It welcomes contributions from across the social sciences and humanities that seek to stretch disciplinary boundaries by encompassing new ways of seeing and dealing with digital technologies.

Bodies, Technologies and Methods
Phil Jones

Algorithms and the Assault on Critical Thought
Digitalized Dilemmas of Automated Governance and Communitarian Practice
Nancy Ettlinger

Ethics and Sustainability in Digital Cultures
Edited by Thomas Taro Lennerfors and Kiyoshi Murata

For more information about this series, please visit: https://www.routledge.com/geography/series/RSDS

Ethics and Sustainability in Digital Cultures

Edited by Thomas Taro Lennerfors and Kiyoshi Murata

Routledge
Taylor & Francis Group
LONDON AND NEW YORK

First published 2024
by Routledge
4 Park Square, Milton Park, Abingdon, Oxon OX14 4RN

and by Routledge
605 Third Avenue, New York, NY 10158

Routledge is an imprint of the Taylor & Francis Group, an informa business

British Library Cataloguing-in-Publication Data
A catalogue record for this book is available from the British Library

ISBN: 9781032434643 (hbk)
ISBN: 9781032434667 (pbk)
ISBN: 9781003367451 (ebk)

DOI: 10.4324/9781003367451

Typeset in Times New Roman
by codeMantra

Contents

Figures

Tables

Contributors

Editors

Thomas Taro Lennerfors is Professor and Head of the Division of Industrial Engineering and Management at Uppsala University, Sweden. He is interested in exploring the connections between ethics and philosophy on the one hand and business and technology on the other. His most recent ethics-related books are *Ethics in Engineering* (2019), and *Tetsugaku Companion to Japanese Ethics and Technology*, co-edited with Kiyoshi Murata.

Kiyoshi Murata is Director of the Centre for Business Information Ethics and Professor of MIS at the School of Commerce, Meiji University, Tokyo, Japan. He has studied information ethics since 1997 following his career of research on economics, operational research, business administration and management information systems and established the center in April 2006, which is the only research institute to study information ethics in Japan.

Contributors

Andrew A. Adams is Deputy Director of the Centre for Business Information Ethics and Adjunct Professor in the Graduate School of Business Administration at Meiji University, Tokyo, Japan. He has a PhD in computer science, a master's degree in law and has been researching Information Ethics since 2003 after earlier studying theorem proving and computer mathematics. He is interested in too many things.

Mario Arias-Oliva is Professor of the Marketing Department at the Faculty of Economics and Business, Complutense University of Madrid, Madrid, Spain. His research is related to technology acceptance, including social and ethical consideration of disruptive technologies such as cryptocurrencies, wearables and implantables. He collaborates with the Centre for Business Information Ethics led by Professor Kiyoshi Murata and the Centre for Computing and Social Responsibility. He is Director of the Telefonica Chair on Smartcities in collaboration with Universitat Rovira i Virgili and Universitat de Barcelona in Spain.

Anisa Aini Arifin has a background in Computer Science and is a former student in the Master's program of Industrial Management and Innovation at Uppsala University, Sweden. She is now product manager in an e-commerce company called Tokopedia, part of GoTo Group, managing the KYC service across the group. Anisa has interest in computer science, engineering ethics, artificial intelligence, voice assistants and data privacy.

Ryoko Asai is Research Fellow at the Faculty of Social Sciences, Ruhr University Bochum, Germany. She is also Research Fellow at the Centre of Business Information Ethics, Meiji University in Japan and Guest Researcher in the IT Department, Uppsala University, Sweden. She has a PhD in Political Science and her area of expertise is information ethics, gender and care in computing.

Vanessa Bracamonte is core researcher in the Usable Trust Group at KDDI Research, Inc., Japan. Her research topics include usable privacy and security, the issue of trust in relation to technology and the social impact of technologies. She obtained her PhD in Informatics from SOKENDAI (The Graduate University for Advanced Studies, Japan) in 2016.

Matthew Davis is doctoral candidate in the Division of Industrial Engineering and Management at Uppsala University. He has a keen interest in exploring the critical implications of disruptive technologies on society, with a particular focus on additive manufacturing, artificial intelligence and blockchain technology. Matthew is fascinated by how these technologies can address or exacerbate the challenges of our time, and his research seeks to uncover insights that will shape future discussions on their development.

Per Fors is Assistant Professor at Uppsala University's Division of Industrial Engineering and Management. His research interests center on the intersection of ICT and sustainability, including the P2P economy, e-waste and energy consumption of ICTs, among other topics. Additionally, he serves as chair of the IFIP TC9 Working Group on ICT and Sustainable Development.

Yasunori Fukuta is Professor of Marketing Research in the School of Commerce, Meiji University, Tokyo, Japan. He was a visiting scholar of Cardiff Business School, Cardiff University, Cardiff, UK and a member of board of directors of the Japan Management Diagnosis Association. His research interest is in statistical modeling of dynamics of market, influence of social practice on value co-creation process and ethics in marketing research and data collection through IoT.

Rickard Grassman is Senior Lecturer and Researcher in the Management and Organisation section of Stockholm University. He is interested in technology and social movements from an ethical and philosophical point of view, as well as in a wide range of ideological tropes and reverberations that this intersection brings to bear. Blockchain, AI and social media are among the technologies he is particularly interested in, insofar as they pertain to social organization and social mobilization in a plethora of ways.

Akira Ide is Professor at the Institute of Liberal Arts and Science, Kanazawa University, Japan. He is Researcher of tourism, with particular interest in dark tourism and tourism informatics. He worked as a visiting scholar at the Reischauer Institute of Japanese Studies, Harvard University from 2019 to 2020.

Iordanis Kavathatzopoulos is Professor of Human-Computer Interaction in the Department of Information Technology, Uppsala University, Sweden. His research interests cover a wide range of fields between philosophy, psychology and new technologies. They focus on the areas of measuring and training ethical skills, ethics of business, organizations and politics, and ethics of autonomous systems and AI.

Hiroshi Koga is Professor in the Faculty of Informatics, Kansai University, Osaka, Japan. He studies information system management. His major research achievements include "On the Lost Property of Telework during the COVID19 Pandemic in Japan: From the Perspective of Sociomateriality and Organizational Citizenship Behaviour", *the Review of Socionetwork Strategies*, 17(1), 2023.

Mikael Laaksoharju is Associate Professor at Uppsala University, Sweden. His research interests are in the intersection of interaction/UX design and ethics, specifically in how ethical considerations can be incorporated into interaction/ UX design processes and how technology can be understood in light of ethical theories. Furthermore, he is interested in how skills in handling ethical issues in design can be effectively taught.

Takashi Majima is Professor in the School of Business Administration, Senshu University, Kawasaki, Japan. He has a PhD in Business Administration. He is interested in studying the process of generating and changing organizational ethics and the connection between ethics and technology (e.g., clean technology, food technology, human resource technology).

Makoto Nakada is former head of the International and Advanced Japanese Studies in the Graduate School of Humanities and Social Sciences, and Professor Emeritus at the University of Tsukuba in Japan. He serves as the vice-editor of the *International Review of Information Ethics* published by the International Center for Information Ethics in Germany. He has been trying to make the basement of studies on "existential-hermeneutical perspectives on disasters, information ethics, social robotics, privacy" in Japan and in East Asia.

Lars Oestreicher is Associate Professor in the IT Department at Uppsala University, lecturing in Non-Excluding (Universal) Design and User Interface Design. He has been the Swedish representative in the IFIP TC13 on Human Computer Interaction between 1990 and 2008, receiving two IFIP Awards. He is involved in a collaboration project with the Department of Art History, studying the potential for AI agents to detect the agency within historic images, and has a large interest in AI methods for image generation.

Yohko Orito is Associate Professor of MIS at the Faculty of Collaborative Regional Innovation, Ehime University, Matsuyama, Japan. She has been Research Associate at the Centre for Business Information Ethics, Meiji University since 2007 and was a visiting scholar at the Centre for Computing and Social Responsibility, De Montfort University, Leicester, UK in 2010. Her research interest is in information ethics in business organizations, particularly use of personal information in businesses and protection of the right to privacy and freedom.

Paweł Pachciarek is Postdoctoral Researcher at Tama Art University, independent curator, art critic and occasional performer. He received his MA in Japanese Studies from Adam Mickiewicz University, Poznan, Poland (2012), and obtained his doctorate at Osaka University (2021). He is currently JSPS Postdoctoral Fellow for Foreign Researchers at the Japan Society for the Promotion of Science (JSPS).

Norberto Patrignani is teaching Computer Ethics at Politecnico of Torino and Italian national representative at the TC9-Technology and Society of the IFIP. From 1974 to 1999, he worked at Olivetti Research. He holds a degree in Computer Science from the University of Torino and a PhD in Computer Ethics from Uppsala University. His research interests are social, ethical and environmental impacts of digital technologies.

Anders Persson is PhD student in the Division for Human-Computer Interaction at Uppsala University. His thesis title is "Thinking with Technology", where he explores how thinking and reasoning humans interact and are affected by modern computer technology such as artificial intelligence.

Tina Ringenson is a senior administrative official at the Swedish Energy Agency. At the time of contribution to this book, she was a postdoctoral researcher in the Division of Industrial Engineering and Management at Uppsala University, Sweden, as well as a member of the Uppsala Smart Energy Research group. Her research interest lies in the relationship between humans and technology in transition to resilient, sustainable societies.

Akio Sato is Professor in the Department of Sociology at Ryukoku University, Otsu, Shiga, Japan. He studies in information society and has been engaged mainly in research on teleworking. He has published several papers on teleworking in English, including "Telework and its Effects in Japan", in Messenger, J. C. (ed.), *Telework in the 21st Century (The ILO Future of Work series)*, 2019.

Maki Sato is Fellow at The New Institute based in Hamburg, Germany. She also serves as Lecturer at Rikkyo University and Fellow at the University of Tokyo East Asian Academy for New Liberal Arts (EAA). Her research field is environmental philosophy. Her current research interest expands to the philosophy of religion (Shinto and Buddhism) and cognitive philosophy to question what it is to be human.

Sachiko Yanagihara is Professor of MIS in the School of Economics, University of Toyama, Toyama, Japan. She is interested in exploring the connections between information ethics and communication in virtual space on telework. Her most recent work is "On the Work Autonomy Supported by Information and Communication Technology: A Case Study of Work from Home in Japan before the COVID-19 Epidemic", *The Review of Socionetwork Strategies*, 17(1), 2023.

Acknowledgments

This study was supported by the JSPS (Japan Society for the Promotion of Science)/STINT (Swedish Foundation for International Cooperation in Research and Higher Education) Bilateral Joint Research Program "Information and Communication Technology for Sustainability and Ethics: Cross-national Studies between Japan and Sweden" (JPJSBP120185411 and JA2017-6999).

1 Ethics and sustainability in digital cultures

A prolegomena

Thomas Taro Lennerfors and Kiyoshi Murata

1 Why ethics and sustainability in digital cultures: Research background and objectives

At the end of 2022, an article called "The dangerous universalist grounds of 'AI ethics'" was published, where Emmanuel R. Goffi discussed "the UNESCO Recommendations on the Ethics of Artificial Intelligence" as an example of the tendency to try to establish global norms for the development and use of artificial intelligence (AI). Goffi argued that beneath these global norms are often hidden foundations that are not shared by all and everyone, such as the Universal Declaration of Human Rights. Rather, what is framed as universal is indeed Western stakeholders promoting their own vested interests. Ending the piece, Goffi argued that what is needed is to broaden the foundations beyond Western ethical foundations, including other traditions of ethics (Goffi, 2022).

That there is a need to continue to problematise how ethical frameworks, values, and norms related to AI, or other digital technologies, are framed in a universalistic and a-cultural manner is obvious from the aforementioned article. This is merely the tip of the iceberg of a long-standing debate about how culture should be integrated into discussions about ethics and information and communication technologies (ICTs) or digital technologies (Collste, 2016; Kelly & Bielby, 2016)—or, more broadly, multipurpose technologies in general (Imamichi, 2011). Recently, the rapid advances of AI have led to a new special issue in the field (Aggarwal, 2020), which presents a range of cultural perspectives on AI ethics, similarly arguing that discussions on this topic have been dominated by Western ethical perspectives, values, and interests, with limited consideration for broader ethical and sociocultural viewpoints.

Although this special issue is highly welcome, this book addresses the broader debates on ethics of a wider range of digital technologies. It is undoubtedly the case that AI has received unparalleled attention within discussions about ethics and new technologies. Similarly to how Stephen Toulmin (2020) argued that medicine saved the life of ethics, one could argue that AI once again saved the life of ethics. For example, we see the revival of the trolley problem for accident scenarios of self-driving cars as indicative of this tendency. In a world where sustainability has become popularised and it encompasses what was earlier seen as ethical issues,

DOI: 10.4324/9781003367451-1

AI has contributed to bringing ethics back on the agenda (see discussion about these concepts later in the introductory chapter). Although the number of AI ethics discussions, projects, and experts has increased significantly, there are other forms of technologies that have received somewhat less attention, which we also include in the book. We previously published a special issue on cross-national studies in Sweden and Japan, covering AI, robots, blockchain, digitally powered sharing economy, and extended reality (Lennerfors et al., 2021). Problematising that Japan is an Eastern country to be compared with the Western country of Sweden, we argued that Japanese society has undergone Americanisation to some extent after World War II while retaining its distinctive cultural characteristics and that Swedish culture and values cannot be easily categorised alongside those of the Western countries (Lennerfors et al., 2021). This book is both a deepening of the same topics and an extension to other countries and other technologies.

As will most probably be apparent to the reader, it is not an exaggeration that digital technologies of different kinds are an inevitable part of human lives in many parts of the globe with the ubiquity and pervasiveness of the Internet, social media, and the burgeoning impact of AI, augmented reality, etc., in our lives. Peters (2015) mentioned that these technologies are today the deep background for life on earth. Lagerkvist (2018), drawing on Peters, argues that digital culture is deeply existential. In other words, being human means to be entangled with digital technologies, a situation in which humanity and digital technologies are co-constituting each other (see also Verbeek, 2011). These digital technologies, similar to other modern technologies, both appear to us as a *tool* and as something which connects us to a sociotechnical *system*. For example, as Lagerkvist (2018) mentions, an iPhone is not only a tool for our own purposes, but it also connects us to the system of surveillance capitalism.

As is obvious when discussing any kind of innovation (Lennerfors & Murata, 2023), it is necessary to not fall into only highlighting the positive sides or the negative sides of digital technologies, but to have a reflexive discussion about the pros and cons of the digital technologies in particular implementations. This idea informed the discussion about information ethics, computer ethics, and digital ethics, as well as discussions within the digital culture domain. For example, Mosco (2005) unpacked what he saw as the prevalent myths of information technologies—that they will end both politics and distance, and end the division between social groups. Rather than contributing to creating myths, Mosco holds that we should elucidate the applications of the technologies, to avoid too positive views about the promises of new technology. However, one should avoid "moral panic" (Ess, 2020)—and technophobia or neophobia—in the face of new digital technologies, and not see certain forms of technology inevitably leading to the demise of the world as we know it. Still, this reflexive perspective which is highlighted is not straightforward, since technologies, not the least digital technologies, not only are evaluated in the current manifestations but also harbour expectations and ideas about future applications. Related to many of the technologies surveyed in this book such as voice assistant (VA) technologies, AI, and blockchain, one cannot merely separate their current use and ethical implications from the myths, expectations, and future desires that these technologies embody and their ethical implications. However, despite the deep difficulties, we hold that it is desirable to nurture a reflexive attitude towards ideas about the future, particularly when they are too black-and-white, in other words what Ess (2020) describes as technology-good and technology-bad. Furthermore, it is important to realise that we will most probably never reach a harmonious consensus about neither current uses nor future desires, but merely temporary discursive configuration which still harbours many contradictions and mixed feelings.

In other words, digital technologies are imbued with values and are far from value neutral. Perhaps they are more neutral than other technologies. ICT is by Moor (1985) seen as a perfectly logically malleable, in other words multi-purpose technology, able to be used in different ways, in the best cases being used to promote positive values (democracy, equal treatment, etc.), but also always running the risk of supporting less socially accepted values (crime, discrimination, environmental pollution, etc.). Rather than seeing such technologies as either neutral or determining, an ethics of ambiguity has been proposed (Lagerkvist, 2018; citing de Beauvoir, 1947), where digital technologies can be described as both beneficial and burdensome at once. In line with that mentioned argument, and also the suggestion of Coeckelbergh (2020), we try to eschew polarised debates between those who are for and against particular forms of digital technologies, but rather try to nuance and look at particular implementations, or cases, of digital technologies and the ethical problems and potential they raise. This goes much in line with an ethics of pluralism (Ess, 2020; Lennerfors, 2019), where the point is not to propose one normative set of values for the digital realm, but rather support our ethical reflection by juxtaposing different ideas of ethics, not with the aim to reach a reflective equilibrium, but as a way to continue thinking (see Kavathatzopoulos' contribution to this book: Chapter 2). These different implementations of digital technologies and the different ethical perspectives taken on them contribute to something more akin to diffraction, a more boundary-crossing exercise, where different perspectives, rather than being reified, are read through one another to spawn creative and unexpected outcomes (Barad, 1996).

The discussions about a-cultural AI ethics are thus merely the tip of the iceberg of discussions that have been ongoing within the academic realm for decades. While in the 1980s and 1990s, person-to-person issues dominated the debates, since the 2000s the focus is increasingly on our relationship to nature, often conceptualised in terms of environmental ethics or sustainability. In the remaining sections of this chapter, we will discuss how the ethical and sustainability aspects of digital technologies have many points in common and show the merits in not separating between these realms, but rather combining them. Then, we will discuss earlier attempts to take cultural perspectives into the ICT/digital ethics debate and present the concept of culture as suture. Finally, we go into the different chapters and situate them within the theme of the book.

2   Ethics and sustainability of digital technologies

This book is connecting ethics and sustainability as their premises are shared. This is certainly a provocative statement, as it approaches ethics from a pluralistic and pragmatic standpoint. As we have previously argued (Lennerfors & Murata, 2023), ethics involves critical thinking in relation to the core values that the development and use of digital technologies promote or hinder through the determination of what core values are and what each core value means may vary from situation to situation. Within discussions on ethics, we hold that it is central to recognise the various value conflicts that arise throughout the lifespan of digital technologies. Ethics is not limited to pivotal "design" moments but is an ongoing process that is ever-present. Despite the emphasis on critical thinking, the embodied, emotional, and relational aspects of ethics are still considered essential, as they often highlight salient ethical issues. However, such ethical impulses should not be acted upon blindly but subjected to critical examination. This approach to ethics could seem to imply that the entities involved in ethical examinations, or the ethical subjects, should distance themselves completely from the situation and view it "from nowhere", which is of course not possible. However, such an ethical subject can draw on concepts and experiences to reflect upon, diffract, and act in their immediate environment.

Both in ethics and sustainability in relation to technology, we maintain that what is at stake is a concern with the impact of actions mediated by technologies. Impact can be related to people, such as infringing on the freedom of a certain group of stakeholders, or also relate to the environment, for example where a certain way of using a technology can have a negative impact on the climate. However, within the public discourse and also in informal academic conversations, in the countries we are working in, there is a tendency to, when one needs to separate the concepts, equate sustainability with environmental issues and ethics with social issues. However, we hold that this separation is not tenable. Discussions about sustainability are often framed in terms of environmental, social, and economic sustainability, where some claim that a balance needs to be struck between these three value spheres and some hold that the environmental realm is the base for the other two realms, so it needs to be prioritised (Purvis et al., 2019). Within original definitions of sustainable development, there are also intrinsic concerns with distributive justice, given that we need to think about the needs of future generations: meeting the needs of the present without compromising the ability of future generations to meet their own needs (Brundtland, 1987). Ethics has for a long time been concerned with human-to-human matters, but for several decades there has been a discussion about environmental ethics, in other words our obligations to the natural environment (Jonas, 1985). Similarly, discussion about the relationships between human and non-human animals has also been intensified within ethics during the past several decades (Singer, 1975). Ethics is therefore similar to sustainability, concerned with impact on humans, animals, plants and the environment or ecosystem, and even with artefacts like machines when it comes to the ethics of computing (Bynum, 2006). A common concern of these two domains is thus to research impacts on human and non-human stakeholders relating to the implementation of digital technologies.

In fact, we are now placed in an environment where nature and technologies are interwoven, and the development and dissemination of ICTs or digital technologies, in particular, have made the boundaries between the natural and the artificial blurred. Our living environment is being filled with ambient intelligence technologies, and a lot of people have outsourced a significant portion of their intellectual activities such as memorisation, reasoning, and judgement to ICT applications. We have already been a human-machine hybrid, whether we realise this or not. Many of an individual's, and an organisation's, actions, decisions, and experiences are mediated by digital technologies. Through interacting with those technologies as well as with others and the external world via the technologies, we consciously or unconsciously define and redefine our own and others' existence, interpret and reinterpret technologies, and construct and reconstruct the world. This inevitably affects our way of thinking and how we perceive truth, reality, good, and evil. We are thus in the midst of digital cultures.

Within the field of sustainable ICT, a variety of frameworks are presented to classify the different impacts that ICT can have. A commonly used framework is the three-level framework which distinguishes between direct environmental effects of the production and use of ICTs, indirect environmental effects stemming from changes of production processes, products, and distribution systems, and indirect environmental impacts through the change of lifestyles and value systems (cf. Hilty & Aebischer, 2015). Within the ethics of technology, there are similar frameworks. For example, van de Poel (2001) argued that a number of ethical issues can occur within technical design processes. He mentioned that the design process involves several key components, including defining goals, establishing design criteria and requirements, selecting alternatives to explore, evaluating and comparing these alternatives, weighing trade-offs between design criteria, and assessing and managing risks and unintended consequences. Additionally, designers must consider the implicit political and social implications of their designs and decide whether these align with their values and desired outcomes. Ultimately, the design process requires a thoughtful and deliberate approach to ensure the best possible outcomes for all stakeholders. This framework can also be seen to concern different forms of impact, ranging from direct impacts, unforeseen or unintended impact, and the macro-level impact in terms of the scripts that the technology is seen to embody. Both ethics and sustainability are thus acknowledging impact on different scales, both in space and time, as well as intended and unintended impacts.

Given the multiplicity of different impacts and both positive and negative effects of the implementation of different ICTs, there is an explicit view within ethics and sustainability that there are no perfect solutions, in other words, solutions that only have positive impact on all potential human and non-human stakeholders. However, in practice, we sometimes tend to focus on some particular impact, for example CO_2 emissions related to two different implementations of the same technology, or how different forms of social media impact on democracy. If reducing what is "sustainable" or "ethically good" into one, more narrow, value, then it is easier to get an answer. However, if broadening the scope, some form of trade-off needs to be struck between the different impacts a new technology might have. As mentioned previously, this decision is of course not something that is made with perfect information and the decision itself is influenced by historical patterns of action or path dependencies. Moreover, although it might seem in some cases that no trade-off between values was struck, it might have been an implicit decision or action that caused a technology to be implemented in the particular way it was implemented. As mentioned above, the way digital technologies are implemented is most probably a mix of intended use of different stakeholders, as well as unintended effects. However, with this core idea of conflicts between different desirable and undesirable impacts, the domains of ethics and sustainability could be integrated. As mentioned, this represents a particular understanding of ethics and sustainability which is pluralistic and where there is no a priori hierarchy between different values at stake. This view stands in stark contrast to perspectives that take a particular value as having an elevated status, for example democracy or justice, and subordinating all other concerns to that value. We also see the value of such approaches, particularly when values have not been paid sufficient attention to throughout history, for example the neglect of the concept of care within the ethical canon up until the latter half of the 20th century (Gilligan, 1993). However, the pluralistic approach to ethics recognises that there always need to be a balance between different values and that the core concern of ethics is to both reason about this balance and acting for the maintenance or revision of current balances within society. This means that different forms of implementation of digital technologies could be seen as a trade-off between different values. One could then empirically examine implementations of ICTs and try to unpack what trade-offs have been struck and how. This opens up a range of interpretations, regarding the forces behind the particular forms of ICT implementation: Is it based on political visions of different kinds (neoliberalism, capitalism, and anarchism)? What preconceptions did designers have about users? How can the values guiding the development of this ICT be unmasked? This inevitably leads to questions of normative character, such as: Is this really the best way in which we can use this technology at the moment? How can we live with the technology in a better way than we are doing now? How can we steer the development of the technology to harmonise better with a set of values that we stand for?

Theoretically, the concepts of ethics and sustainability bring some differences. While sustainability studies could be seen as somewhat more open-ended theoretically (Purvis et al., 2019), ethics often brings with it a set of theories, for example normative ethical theories, such as utilitarianism and deontology, which are not seen as common within discussions about sustainability. However, within practice, sustainability discussions are often related to the Sustainable Development Goals (SDGs) (Sachs, 2012); while in matters related to the ethical realm, there are no clear and universal values (although, as mentioned in the introductory paragraph, such values are sometimes proposed in relation to particular technologies). Furthermore, some methodology that is quite commonplace in studies of sustainability, for example life cycle analysis, is not seen as one of the core methodologies within ethics. Within both domains, there is also a range of approaches ranging from impact assessment to collaborative approaches to working for ethics and sustainability together with stakeholders, such as Responsible Research and Innovation (RRI) (von Schomberg, 2013) and guidance ethics (Verbeek & Tijink, 2020) within ethics, and stakeholder collaborations (Fadeeva, 2005) and living labs (von Wirth et al., 2019) within sustainability studies. Perhaps on an institutional, research group, and individual basis, there are a range of differences relating to what happens when one positions oneself as being related to ethics or sustainability. A range of connotations that are often quite local and partial crop up. Furthermore, it is important to remember that neither ethics nor sustainability is a complete and harmonious body of knowledge but in themselves harbour disagreements and contestations. In this book, we will have a quite open-ended approach to ethics and sustainability, arguing that both are trying to describe and prescribe the "good", but do so in slightly different ways. By relating these concepts, we incidentally try to join scholars and debates that are linked but sometimes separated because of the disciplinary delimitations into the two mentioned domains.

3   Cultural sutures

Despite the ubiquity of discussions on the ethics and sustainability of the development and use of digital technologies, there is still, as Goffi (2022) mentioned, some tendency of universalism and cultural insensitivity in the debates. Much research about ICT ethics has not paid sufficient attention to the context in which ICT is introduced. This research rather treats ICT reception in all parts of the globe as more or less similar to how it would be received in a "Western" country.

However, as mentioned above, in the realm of intercultural information ethics (IIE), cultural and national differences are considered, with the belief that "culture matters" (Ess, 2017). IIE emphasises the importance of not overlooking culture—the values, beliefs, and practices of people—when examining how ICTs are adopted and adapted in various cultural contexts. Similarly, cultural differences must not be ignored when attempting to establish a normative ethical foundation for the acceptance or rejection of particular ICTs. Instead, as Rafael Capurro suggests, it is critical to explore the boundaries and diversity of moral values and ethical reasoning across various societies, historical periods, philosophical traditions, and how these impact the present-day adoption of information technology in social contexts (Bielby, 2015, p. 236).

Over time, IIE has progressed, resulting in a variety of empirical studies that examine the practical application of ICT in different countries with seemingly distinct national cultures. In addition to these empirical studies, there have been theoretical discussions about the notion of culture. Despite years of scholarship, ethnocentrism remains "intransigent, if not predominant" (Ess, 2017, p. 2), as evidenced by the article referenced in the introductory paragraph.

Furthermore, the idea of a shared national culture has been problematised. We have identified a tendency towards separating the world into cultures, each of which is constituted by cultural values that should be paid attention to when describing and analysing the spread of ICT in a particular context. These differences should also be paid respect to when constructing a normative base for evaluating ICT development and implementation. While it is straightforward to problematise ethnocentrism, the practice of intellectually dividing the world into different cultures constituted by their own common values and practices, separating them from others, is also problematic. Palm (2016) argues that IIE must be better at reflecting on its theoretical assumptions and that it is important to find better understandings of the key concepts of "culture" and "cultural". By separating the world into different cultures, the field can run the risk of producing stereotypical understanding of, for example, Western and Eastern cultures, while reality is more complex. A more nuanced understanding of culture is needed. Dominant theories in the field seem to be somewhat essentialist, arguing that people who belong to a certain country tend to share certain characteristics, as if they had a collective programming of the mind distinguishing them from other groups of people. In contrast, and much more marginalised within IIE, we have presented non-essentialist frameworks that do not argue that culture is related to essences, but rather discuss how "culture", as a concept, is used and mobilised, and what role "culture" plays in how one identifies oneself or identifies others. We argue that such understandings are a good counterweight to the predominant understanding of culture, which are held to overemphasise similarities and downplay the complexities of cultures.

In the adjacent discussions about digital cultures, we find literature that describes a co-constitution of digital technology and culture. Charlie Gere, in his book *Digital Culture* (2009), discusses the development of digital culture, where culture is seen as providing the impetus for the development of technology, and where technology in turn influences culture. What we find ourselves in today is predominantly a digital culture, as the digital is penetrating our lives. Digital culture is not only created by digitalisation, but our increasingly digital thinking also contributes to digital culture. Taking the perspective from humanities and social sciences, digital culture is produced not only by technology but also by techno-scientific discourses about information and systems, avant-garde art, counter-culture, and philosophy. Gere writes about the development of this digital culture, through a genealogical expose. In a review of Gere's book, Leggett (2003), although very positive to the book in general, wrote that the book is quite Europe- and US-centric, although there are important but neglected developments elsewhere, for example in Australia. Also within digital cultures, there have been calls and efforts to internationalise the discussion (see e.g. Goggin, 2016). This highlights the importance of problematising the notion of digital culture and rather thinking about it in terms of digital cultures. In general, the field of digital cultures discusses how cultural practices influence and are influenced by developments in digital technologies. Although the explicit discussion about ethics and sustainability is not as obvious in digital culture literature, it is obvious that a core part of the research concerns what Mihelj et al. (2019) described as an unanswered question: What kind of values and expectations are expressed in this "digital culture" (Gere, 2009)? The answer to the question Mihelj et al. (2019) provided is that the principal components of digital culture can be caught in three concepts, which should be seen as articulated with each other: *participation, remediation,* and *bricolage.* Another interesting strand within the digital culture discussions concerns culture in the form of the arts and how digital technologies have impacted on music and the graphic arts (Mihelj et al., 2019). In an influential article within the field, Deuze (2006) described that he sees culture as more or less a set of values, norms, practices, and expectations shared (and constantly renegotiated) of a group of people. Hand (2016) argued that cultures are "forms of life", encompassing all aspects of our lived experience whether they be deemed "ordinary" or "sublime".

In Lennerfors and Murata (2021), drawing on the concept of suture, we presented a conceptual argument relating to how we as researchers and practitioners use the concept of "culture". The first way is that we suture (connect, tie) someone to a national culture. We draw on Holliday (2011) to argue that although the problems of essentialism are accepted and critiqued, an essentialist perspective stems from a persistent yearning to establish a definitive understanding of culture and its variances (Långstedt, 2018). In line with the first understanding of suture, the nothing (the incomplete, undecided cultural identity) is replaced by something essentialist. Perhaps one knows that the theory is not correct, but one identifies with the fiction to get closure. Although the tone in our presentation of this kind of suture is somewhat sceptical, we are ourselves in our understanding of responses to ICT development and implementation in different countries, suturing the analysis to an essentialist notion of culture. Indeed, there are plenty of learnings that one can have from understanding basic philosophical tenets of different regions as well as their practices in order to understand why a particular form of technology was received in a certain way. Our point is merely, by pointing out that this is one form of suture, one should be wary to not fix the explanations to a certain conception of culture.

In the second understanding of culture as suture, we highlight that the seemingly neutral, and objective, point of view is seen as the point of view of some external, ominous agent. The procedure of this second kind of suture is thus not that individuals, such as researchers, make truth-claims about culture, but rather that the seemingly neutral and objective view provided by such truth-claims, particularly with the most seemingly objective and neutral, such as Hofstede's, suddenly are reframed and become sutured to something evil. This movement creates a connection between culture and the conditions of its production. For example, Goffi (2022) argues that what is hidden beneath a universalistic, shared human culture claim is a disguised Western take on ethics. Similarly, Ailon (2008) critiques Hofstede's cultural framework for devaluing and misrepresenting other than Western cultures. Wong (2020) holds that there is a risk of seeing cultural differences as excuses. This suture throws light on the biased or strategic nature of cultural representations. Following this line of thought, it seems as if all concepts, including cultural representations, are informed by a hidden agenda, but it is also important to not take this argument too far. We argue that practitioners and researchers are trying to make sense of the world and do not consciously inscribe their agendas into their worldview. This second sense of suture rather points to the importance to think about the conditions of production for cultural concepts.

Drawing on Rushton (2017), we have earlier argued that there needs to be a third interpretation of suture, in addition to tying the subject to a fixed position and exposing the gaze that guides the suture. We develop an idea of "sutures as always temporary, always containing the conditions of production, and always serving as an invitation to de-suturing" (Lennerfors & Murata, 2021, p. 81). In this third form, we know that culture is a suture and is also open to the temporary and fragile nature of cultural explanations, being always open to reinterpretation. This could lead to a more creative debate about how we use the concept of culture, "one in which we always expose the conditions of its production, suture again, and recognize that we constantly fail in describing culture or cultural differences" (Lennerfors & Murata, 2021, p. 81). With this understanding, it becomes interesting to study how culture is mobilised in different relationships between people and technologies and how cultural identification at different levels becomes a central concept for understanding our relationships to digital technologies. Within this book, the authors use the concept of culture in different ways, sometimes suturing it to larger cultural entities, sometimes to national culture, sometimes to the culture and values that are imbricated with a particular form of technology, and sometimes to local cultures or cultural differences at a more granular level than the nation-state. Culture is also sometimes used as referring to cultural practices, which are altered or promoted by digital technologies. Similarly to the discussion about ethics and sustainability, we are inspired by a flexible understanding of culture, as the third form of suture.

4   Positionality statement

As we will proceed to describe, this book discusses ethical and sustainability aspects of digital technologies in a range of cultural settings. The reason is to promote an international dialogue that includes contributions from different countries in the same edited volume, rather than zooming in on one particular geographical area. Leading up to this volume, we have conducted a series of workshops in both real life and online in order to share perspectives and discuss ethics and sustainability in digital cultures. In a previous publication, we focused on cross-national comparisons of ICT implementation between Sweden and Japan (Lennerfors et al., 2021), and the focus of this book was to broaden the scope and discuss different ways in which the culture manifests in relation to digital technologies.

The core part of the book concerns Japan and Sweden. This is due to the project in which the majority of the texts have been generated, but these two countries also bear interesting characteristics, sharing both support of democracy and capitalism, and being advanced countries within ICT. The countries also differ in their social characteristics and cultural traits, whereas Sweden is more culturally diverse than Japan (Lennerfors et al., 2021). In addition to this, we have included discussions about Poland, Italy, UK, Spain, China, and Indonesia. In relation to the above discussion of the biased nature of studies, one could critique the book as too concerned with countries in the Global North. We acknowledge that this is an important critique and invite future research that either makes a more balanced inclusion of perspectives from the Global South or solely focuses on the Global South.

The authors of this volume have engaged significantly with discussions, papers, conferences, within the ethics and sustainability of ICTs. Within that field, a part of the authors have previously been very active within discussions on cultural aspects of the ethics of such technologies. There is thus a predominance of experience and knowledge from those discussions, although a few authors have engaged in discussions on digital cultures as it is discussed within cultural studies. An implication of this author setup is that in relation to digital culture discussions, we put the technologies underlying digital cultures more in focus. Many of the authors also work within either departments of business and technology, rather than in humanities departments. This incidentally shows how discussions about ethics and sustainability in digital cultures are spreading across the whole university.

5   Overview of chapters

The book is divided into two sections. The first section focuses on users' experiences with digital technologies, which we refer to as "Practicing Ethics and Sustainability in Digital Cultures". Although these technologies are in this part primarily approached and viewed as tools, they are certainly also intertwined with sociotechnical systems. The second section focuses on the production side of digital systems, with an emphasis on developers, entrepreneurs, and policymakers. We refer to this section as "Creating Ethical and Sustainable Digital Cultures".

5.1   Practicing ethics and sustainability in digital cultures

In Chapter 2, Kavathatzopoulos engages with the discussions about the potential and threats of AI to humanity. Based on classical philosophy, he draws on his long-standing research to argue that thinking is a fundamental practice that defines human beings, irrespective of culture. What is in focus for the Kavathatzopoulos piece is the process of thinking rather than the production of results from the thought process, although these are of course facets of the same process. With this idea of the human in mind, and caring about the longevity or sustainability of this core human characteristic, Kavathatzopoulos discusses what role AI can play for promoting or downplaying thinking. Throwing light on how AI can support the dual processes of Aristotelian habit and phronesis, a hopeful view of the future use of AI arises—but whether this hope will be fulfilled is in the hands of our and future generations.

In Chapter 3, Laaksoharju et al. discuss ChatGPT and other AI chatbots and their ethical implications, through the lens of Don Ihde's embodiment, hermeneutic, alterity, and background relationships. This shows the different ways in which humans can and might relate to this new technology. Within embodiment relationships to chatbots, they can make your life more efficient by taking care of chores, pleasantries, and the like in order to free up time for you to focus on what matters. As a part of hermeneutic relationships, chatbots help you to interpret the world in a way which surpasses your own abilities, but can you trust the interpretations of the chatbot? Within alterity relationships, the chatbot becomes a speaking partner that you can form a meaningful relationship with, where you can exchange thoughts that you could perhaps not exchange with a human being, but what are the consequence for human-to-human relationships? As a background, and part of the social fabric, the argument is made that chatbots could contribute with introducing more doubt and suspicion, which could have detrimental effects on society.

In Chapter 4, Arifin and Lennerfors discuss ethical aspects of VA technologies, in other words applications such as Google Assistant, Siri, and Alexa. What has featured in earlier discussions about the topic is how people relate to that these applications are "always listening". Inspired by mediation theory, which explores different relationships between human, technology, and world, the authors embark on an empirical study of Indonesian users of VA technologies, exploring their current use of VAs, the ethical issues they perceive, as well as the dreams of what the future VA would be for these users. Although many of the users are wary of the privacy risks of the technology, they still keep using it and dream about VAs that are even more proactive, assisting the users to lead better lives.

In Chapter 5, as the basis for exploring the impact of digital technologies on ethics and sustainability in today's digital—or mathematically founded—world, Nakada, Kavathatzopoulos, and Asai examine the nature of truth and reality in mathematics and computing, which tend to be considered fixed and stable. Through a series of elaborate discussions on mathematics as a tool for machine learning-based AI and robotics based on Japanese and Western philosophers' insights and empirical findings of the authors' previous work, it is clarified that truth in mathematics, thus in today's world, is neither fixed nor stable and mathematics is rather a plural matter, and the importance of overcoming techno-determinism and reductionism is emphasised. How mathematics and technology enter into existence is also discussed.

Chapter 6 describes the recent situation in Japan regarding telework. Despite the great promise of telework as an enabler of diverse work arrangements and thus organisational and social sustainability, its adoption was slow in Japan. However, the COVID-19 pandemic started in early 2020 changed such a situation; a temporary telework boom was caused, yet the boom is leaving as the infectious disease is tamed. Koga, Sato, and Yanagihara hypothesise that this shows Japanese people's remaining blind to the ethical and social significance of telework and examine whether the hypothesis is correct based on empirical data they collected during the COVID-19 epidemic in Japan. They conclude that the hypothesis is almost correct and point out a necessity to reaffirm the roles expected of telework such as realising the ideal way of working and digital transformation.

In Chapter 7, Murata, Orito, Adams, Arias-Oliva, and Fukuta deal with the ethical and social aspects of human body modification or the cyborgisation of human beings through digital technologies such as wearables and implantables. Based on a cultural study on how the Japanese have related to body modification including piercing, tattoos, and organ transplants and interviews with ten experts in the research fields relevant to body modification ethics including life science, health science, and robotics, this chapter ends up in a research agenda about what ethical issues need to be studied before these technologies are rolled out on a large scale. In addition, comparative perspectives from the UK and Spain are provided to relativise and clarify the Japanese sociocultural circumstances surrounding human cyborgisation.

5.2   *Creating ethical and sustainable digital cultures*

In Chapter 8, Grassman, Asai, and Davis discuss the rise of memetic movements in social media. Originally seen as a potential for democracy, social media can be as subversive to democratic as to authoritarian governments. They examine the emergence of social movements online and their repercussions outside social media. The point of the chapter is to explore the medium and how it seemingly takes precedence over content generation, focusing on the conspiracy theory of the Q-anon movement and how it was received within Sweden and Japan. Taking a Levinasian perspective on ethics, the authors of the chapter argue against the reification of other people in order to create an identity, rather than appreciating alterity and seeing its potential.

In Chapter 9, Ide and Pachchiarek focus on the ethical aspects related to augmented reality, more specifically as they appear in the location-based game Pokémon Go. This widely popular game has been received well but has also spurred some ethical questions. The authors position their work in this literature and specifically intend to discuss the moral outrage it has spurred when it is placed in sites of genocide. By engaging in a comparative discussion between Poland and Japan, more specifically playing in the Auschwitz concentration camp or near the peace memorial park in Hiroshima, they throw light on how different cultures approach this issue.

In Chapter 10, Fors and Ringenson concern the ICT-based digital platforms that constitute the backbone of the sharing economy, in other words, where we can share resources, which could lead to positive effects relating to environmental sustainability. However, in their piece, the authors argue that such platforms could have positive effects on local community building, since they stimulate meetings between strangers. Taking the case of the Swedish sharing platform Hygglo as a point of departure, they discuss the potentials and downsides of such platforms in restoring community ties that have gradually been deteriorating over the years.

In Chapter 11, Orito and Majima discuss the role of ICT-based digital platforms to support the recreation of cultural sustainability, more specifically to the playing as well as crafting and maintaining of musical instruments. In a time where the role of cultural practices, such as music playing, is weakened, not the least due to the advances in digital technologies and the availability of streamed unlimited music, we hear the story of a start-up whose business model and aim are to share musical instruments.

In Chapter 12, Davis, Grassman, Bracamonte, and Sato focus on blockchain technologies that have been described as both something ethically positive, as a way to liberate economies from the legacies of old capitalistic structures, and as a potential environmental threat as they consume large amounts of electricity. In response to a fragmented discussion of sustainability aspects of blockchain technologies, the authors aim to take a comprehensive view and discuss blockchain from economic, environmental, and social aspects while at the same time acknowledging that these divisions are artificial and need to be thought together. They connect this discussion on sustainability to the cultural movement that is related to blockchain.

In Chapter 13, Patrignani focuses on the environmental impact both positive and negative that ICT technologies can have. He separates his discussions into two major strands of thought—one concerning what positive environmental sustainability effects that the use of digital technologies can have and one concerning how to reduce the negative environmental sustainability effect that such digital technologies have. Patrignani describes a European approach to these sustainability efforts, based on EU policies and directives. Such an approach takes a proactive stance against free market forces in order to promote values that are seen as desirable within the European context.

Acknowledgement

The studies in this book were supported by the JSPS (Japan Society for the Promotion of Science)/STINT (Swedish Foundation for International Cooperation in Research and Higher Education) Bilateral Joint Research Program "Information and Communication Technology for Sustainability and Ethics: Cross-national Studies between Japan and Sweden" (JPJSBP120185411 and JA2017-6999).

References

Aggarwal, N. (2020). Introduction to the special issue on intercultural digital ethics. *Philosophy & Technology, 33*, 547–550. https://doi.org/10.1007/s13347-020-00428-1

Ailon, G. (2008). Mirror, mirror on the wall: Culture's consequences in a value test of its own design. *Academy of Management Review, 33*(4), 885–904. https://doi.org/10.2307/20159451

Barad, K. (1996). Meeting the universe halfway: Realism and social constructivism without contradiction. In L. H. Nelson, & J. Nelson (Eds.), *Feminism, science, and the philosophy of science* (pp. 161–194). Springer. https://doi.org/10.1007/978-94-009-1742-2_9

Bielby, J. (2015). Comparative philosophies in intercultural information ethics. *Confluence: Online Journal of World Philosophies, 2*, 233–253.

Brundtland, G. H. (1987). Our common future—Call for action. *Environmental Conservation, 14*(4), 291–294. https://doi.org/10.1017/S0376892900016805

Bynum, T. W. (2006). Flourishing ethics. *Ethics and Information Technology, 8*, 157–173. https://doi.org/10.1007/s10676-006-9107-1

Coeckelbergh, M. (2020). *AI ethics*. MIT Press.

Collste, G. (Ed.). (2016). *Ethics and communication: Global perspectives*. Rowman and Littlefield.

de Beauvoir, S. (1947). *The ethics of ambiguity*. Gallimard.

Deuze, M. (2006). Participation, remediation, bricolage: Considering principal components of a digital culture. *The Information Society, 22*(2), 63–75. https://doi.org/10.1080/01972240600567170

Ess, C. (2017). What's "culture" got to do with it? A (personal) review of CATaC (Cultural Attitudes towards Technology and Communication), 1998–2014. In G. Goggin, & M. McLelland (Eds.), *Routledge companion to global internet histories* (pp. 34–48). Routledge. https://doi.org/10.4324/9781315748962

Ess, C. (2020). *Digital media ethics*. John Wiley & Sons.

Fadeeva, Z. (2005). Promise of sustainability collaboration—Potential fulfilled? *Journal of Cleaner Production, 13*(2), 165–174. https://doi.org/10.1016%2FS0959-6526(03)00125-2

Gere, C. (2009). *Digital culture.* Reaktion Books.

Gilligan, C. (1993). *In a different voice: Psychological theory and women's development.* Harvard University Press.

Goffi, E. R. (2022, October 18). The dangerous universalist grounds of 'AI ethics'. *The Yuan.* https://www.the-yuan.com/419/The-dangerous-universalist-grounds-of-AI-ethics.html

Goggin, G. (2016). Re-orienting global digital cultures. In S. S. Lim, & C. R. Soriano (Eds.), *Asian perspectives on digital culture* (pp. 191–198). Routledge.

Hand, M. (2016). *Making digital cultures: Access, interactivity, and authenticity.* Routledge.

Hilty, L. M., & Aebischer, B. (2015). ICT for sustainability: An emerging research field. In L. Hilty, & B. Aebischer (Eds.), *ICT innovations for sustainability* (pp. 3–36). Springer. https://doi.org/10.1007/978-3-319-09228-7_1

Holliday, A. (2011). *Intercultural communication and ideology.* Sage.

Imamichi, M. (2011). 未来を創る倫理学 エコエティカ [*Eco-ethica: Ethics for creating the future*]. Showado.

Jonas, H. (1985). *The imperative of responsibility: In search of an ethics for the technological age.* University of Chicago Press.

Kelly, M., & Bielby, J. (Eds.). (2016). *Information cultures in the digital age.* Springer.

Lagerkvist, A. (Ed.). (2018). *Digital existence: Ontology, ethics and transcendence in digital culture.* Routledge.

Långstedt, J. (2018). Culture, an excuse?—A critical analysis of essentialist assumptions in cross-cultural management research and practice. *International Journal of Cross Cultural Management, 18*(3), 293–308. https://doi.org/10.1177/1470595818793449

Leggett, M. (2003). Charlie Gere's digital culture. *Digital Creativity, 14*(4), 255–256. https://doi.org/10.1076/digc.14.4.255.27883

Lennerfors, T. T. (2019). *Ethics in engineering.* Studentlitteratur.

Lennerfors, T. T., & Murata, K. (2021). Culture as suture: On the use of "culture" in cross-cultural studies in and beyond intercultural information ethics. *The Review of Socionetwork Strategies, 15*(1), 71–85. https://doi.org/10.1007/s12626-021-00080-x

Lennerfors, T. T., & Murata, K. (2023). Innovation ethics. In A. Rehn, & A. Örtenblad (Eds.), *Debating innovation: Perspectives and paradoxes of an idealized concept* (pp. 33–53). Palgrave Macmillan. https://doi.org/10.1007/978-3-031-16666-2_3

Lennerfors, T. T., Murata, K., & Koga, H. (2021). Editorial preface: Why cross-national studies between Japan and Sweden? *The Review of Socionetwork Strategies, 15*(1), 67–70. https://doi.org/10.1007/s12626-021-00073-w

Mihelj, S., Leguina, A., & Downey, J. (2019). Culture is digital: Cultural participation, diversity and the digital divide. *New Media & Society, 21*(7), 1465–1485. https://doi.org/10.1177/1461444818822816

Moor, J. H. (1985). What is computer ethics? *Metaphilosophy, 16*(4), 266–275. https://doi.org/10.1111/j.1467-9973.1985.tb00173.x

Mosco, V. (2005). *The digital sublime: Myth, power, and cyberspace.* MIT Press.

Palm, E. (2016). What is the critical role of intercultural information ethics? In G. Collste (Ed.), *Ethics and communication: Global perspectives* (pp. 181–195). Rowman & Littlefield International.

Peters, J. D. (2015). *The Marvelous Clouds: Towards a Philosophy of Elemental Media.* University of Chicago Press.

Purvis, B., Mao, Y., & Robinson, D. (2019). Three pillars of sustainability: In search of conceptual origins. *Sustainability Science*, *14*, 681–695. https://doi.org/10.1007/s11625-018-0627-5

Rushton, R. (2017). Suture and Gus Van Sant's Milk. In A. Piotrowska, & B. Tyrer (Eds.), *Psychoanalysis and the unrepresentable: From culture to the clinic* (pp. 203–216). Taylor and Francis.

Sachs, J. D. (2012). From millennium development goals to sustainable development goals. *The Lancet*, *379*(9832), 2206–2211. https://doi.org/10.1016/S0140-6736(12)60685-0

Singer, P. (1975). *Animal liberation: The definitive classic of the animal movement.* HarperCollins.

Toulmin, S. (2020). How medicine saved the life of ethics. In J. P. DeMarco, & R. d. M. Fox (Eds.), *New directions in ethics: The challenges in applied ethics* (pp. 265–281). Routledge.

van de Poel, I. (2001). Investigating ethical issues in engineering design. *Science and Engineering Ethics*, *7*(3), 429–446. https://doi.org/10.1007/s11948-001-0064-0

Verbeek, P.-P. (2011). *Moralizing technology: Understanding and designing the morality of things.* University of Chicago Press.

Verbeek, P.-P., & Tijink, D. (2020). *Guidance ethics approach: An ethical dialogue about technology with perspective on actions.* Platform voor de InformatieSamenleving. https://ppverbeek.org/guidance-ethics

von Schomberg, R. (2013). A vision of responsible research and innovation. In R. Owen, J. Bessant, & M. Heintz (Eds.), *Responsible innovation: Managing the responsible emergence of science and innovation in society* (pp. 51–74). John Wiley & Sons. https://doi.org/10.1002/9781118551424.ch3

von Wirth, T., Fuenfschilling, L., Frantzeskaki, N., & Coenen, L. (2019). Impacts of urban living labs on sustainability transitions: Mechanisms and strategies for systemic change through experimentation. *European Planning Studies*, *27*(2), 229–257. https://doi.org/10.1080/09654313.2018.1504895

Wong, P. H. (2020). Cultural differences as excuses? Human rights and cultural values in global ethics and governance of AI. *Philosophy & Technology*, *33*(4), 705–715. https://doi.org/10.1007/s13347-020-00413-8

Part I

Practicing Ethics and Sustainability in Digital Cultures

2 Artificial intelligence and the sustainability of thinking

How AI may destroy us, or help us

Iordanis Kavathatzopoulos

1 Powerful technology, ethics and sustainability

Technology helps us with many things, and we expect artificial intelligence (AI), or machine learning (ML), to give us much more in the future. However, there are certain risks involved, even the ultimate risk of exterminating humanity. The discussions focus on issues like the probability of AI becoming exponentially more powerful than all human intelligence combined, acquiring an independent existence of itself, transforming us into something we do not want to be, direct evolution in a radically different way, even affect the whole universe, etc. (Bostrom, 2014; Harrari, 2015; Kurzweil, 2006; O'Neil, 2016; Reese, 2018; Tegmark, 2017; see also Future of Life Institute, 2023). Although not everyone agrees on whether any of these things will happen, or when they may happen, all these writers focus their interest mainly on the technical aspects of the issue or on AI itself and its potential. It is however another interesting angle to look at this phenomenon: human nature, or life generally, with its strengths and weaknesses, and the dynamics of its interaction with powerful technology.

In order to answer these very important questions, we should probably start by asking what is our relationship to technology in general. Let us for a moment think of a simple technological tool: a calculator. We use it for adding, multiplying, etc. It provides answers. Right answers. This is exactly what we expect from it and we have designed it exactly for this purpose. The answers delivered by the calculator would have no meaning unless they were the right answers. So, this tool has its meaning as a provider of the answer we need when we use it to perform micro-calculations as a small part of a bigger problem-solving process. However, some interesting things happen here. Before the advent of mini-calculators, we were using our cognitive skills sometimes in combination with simple tools to perform subtractions, multiplications, etc. In the beginning, when calculators first became available, some people did not trust them and they felt the need to calculate by hand in order to confirm the solution presented by the calculator. Today, it is mostly the other way around. If forced to calculate by using our own mental skills, we feel the need to repeat the calculation using a calculator in order to confirm the answer. Indeed, we really trust the machine's answer.

DOI: 10.4324/9781003367451-3

Loss of skill is presented most often as the main risk of using computer tools. It is. But there is another aspect that might be important too: We judge or trust the machine thinking to be better than our thinking. Rightly so, since a mini-calculator always gives us the correct answer, unless it is broken somehow. So, what happens here is that we feel we do not need to use our thinking for these micro-calculations anymore. And, actually, we do not.

2   Ethics, sustainability and culture as processes

Today, having access to sophisticated AI, which is capable of much more, we expect it to provide answers, products, and services to us, just as the mini-calculator does regarding micro-calculations (Persson & Kavathatzopoulos, 2017). Accordingly, the discussion about AI's blessing or curse is about whether its deliveries will be good or bad to humans, animals, environment, or the whole universe. This is a significant issue, and since we can probably not avoid it, we have to handle it somehow.

However, this is not easily done. One difficulty is foreseeing the future, generally of course, but here it is especially difficult because of the complexity and opacity of ML processes. In addition to that, there is an inherent contradiction in building AI systems. Its value is on doing things by itself. If we tell it what to do, then it is not so independent, and it loses a lot of its value. If it makes its own decisions, and therefore it is valuable to us, it will of course do things out of our control. So, if we do not close it down, we will let it act independently and we will not be able to predict what it will do. It is very difficult for us to foresee what will happen in the future: the more distant future the more difficult. On top of that, AI or ML processes are too complex for us to describe and control (see, e.g., Burrell, 2016). So, we cannot prepare ourselves for what AI will do. We cannot foresee what it will do in order to plan our actions. Another and probably more profound difficulty is the definition of what is a good development and future and, following this, the difficulty of defining for ourselves what it is that we really want to strive for and achieve by using AI. For example, the ideas of evolution and eternal change are very difficult to combine with the idea of sustainability.

Still, we have to do something because AI is crucial for humanity if not fateful. Being aware of this we turn our focus on what AI will do and, as we have discussed above, try to foresee and control somehow its impact on our lives and our societies. An alternative way to do this could be to try not to make ourselves blind for the role it can play for our thinking, like for example the role mini-calculators play for our micro-calculating thinking. Of course, we can focus our efforts on judging the impacts of AI on what we think is valuable for us or for the world, and try to guide AI toward the right direction. But still, as an alternative way or in combination to these efforts, the issue of the impact of AI, and especially its future impact, could be handled by considering the tension between what leads to something and what its product is, i.e., thinking and knowledge, choice and decision, judging and judgment, process and product, and the role they play for humans and for life generally.

Ethics cannot be defined through the normative content of a statement or a behavior. This varies depending on the circumstances and the conditions of the persons and the situation involved; see, for example, the dialog *Protagoras* (Platon, 1986). It is the process of investigation, argumentation, and dialog and its quality that gives ethics its meaning.

In a similar way, sustainability cannot be ethically evaluated with reference to a state of things, for example, nature or environment as they are during a certain period of time. Affecting the process of change is though empirically and theoretically possible and therefore ethically meaningful. We can try to hasten, slow, or redirect processes of change, but we cannot freeze a state of things.

Even in our understanding of culture, we can see a similarity regarding the issue of content and the process behind it. Although we can identify differences between groups and societies with reference to habits, we can see that these same groups do not differ in the ways they adopt deliberative or less deliberative ways of thinking (Kavathatzopoulos et al., 2017; Nakada et al., 2021).

In everyday life, this is what we usually do: We judge statements and behaviors based on their normative content, we want to sustain things as they are, and we understand culture as the existing ideas and behaviors of a group.

3   Life, humans and thinking

When we face a problem, we tend to concentrate on the value of the solution, and that decides usually, in a more or less automatic way, whether the solution will be seen as good or not. This is not strange. We want to solve the problem, and its solution is what really matters. However, in order to find a good solution, we need to think, especially if the problem is difficult or it is a kind of dilemma with no obvious right solution. Thinking is about judging and choosing between different, unclear, and often contradictory solutions. Process, i.e., thinking, and product, i.e., the solution, are definitely not the same, although they cannot exist separately. It even seems that life itself is an independent self-regulating thinking—or decision-making process to solve problems and guide actions in order to achieve an underlying goal, such as the spreading of genes or some other more profound and hidden goal (see, e.g., the "will" of Schopenhauer, 1966). If this is true, then the emergence of a separate living entity is dependent on a setting or on an environment where goals and processes to reach these goals are possible (Hedman, 2017). Without going further into this issue, we need to say that a goal seems to be necessary as a condition for the thinking process to start and run, but that this goal does not need to be something absolutely "right" or to be reached eventually (Camus, 1973).

We humans as living entities may think in a rational or a nonrational way or a mixture of these two. We may think in a right or in a wrong way. Right and wrong are here not about the product of thinking but of thinking itself; it is about the process of thinking. Of course, the product of thinking matters too. If we think in the right way, we have a better chance to find a good solution. In ethics, for example, choice matters most, in the sense that without choice there is no ethics. Thus, the way choice is made is of paramount importance (see, e.g., classical philosophy of

Platon, 1986, 1992a, 1992b; Kant, 2004, 2006; Arendt, 2003; Camus, 1973). Still, in real life, what feels most important is the result of this process: the achievement of the goal. And, accordingly, if the goal is reached, we usually do not care so much about the quality of the cognitive process that led to this satisfactory result. After all, reasoning may be something very humble just as a simple tool for our needs, feelings, and sentiments (Hume, 2002, 2004).

4   Society, culture and thinking

Something similar can we see in society. We talk a lot about democracy as a process to find solutions to political issues. Democracy is supposed to be better than any other political system, for example compared to an enlightened and tolerant oligarchy. But if a nondemocratic system could provide us with most of the things we want, would not this be good and sufficient for us?

Apparently, the most important issues here are control and power. We recognize this officially, and we declare our support to democracy since we want to have the control of our own lives, which would be missing in an oligarchic political system or in a dictatorship. In other words, we focus so much on democracy because what is delivered by a democratic system is something that is decided by us and not by people who may be good today but maybe not tomorrow. But what happens when "enlightened" lords are consistent in delivering a good life?

Democracy is not high living standards, social security, and protected environment. Democracy is a process, i.e., the way good policies about important issues are decided and achieved. The most common view is that democratic processes, institutions, and formal procedures define democracy. However, this is not enough. It is necessary that people involved in these procedures also have the ability to think and to have a dialog within themselves as a condition for a dialog with others.

We can easily understand that if people do not want to search solutions to their problems in a dialog together with others or if they cannot do it properly, democracy is not there (Popper, 1945). It is therefore important to try to understand the psychological conditions under which dialog is possible and base any effort to promote democracy on this (Piaget, 1932).

Democracy is in essence a dialog between people (Habermas, 1990). That means that people search for solutions to their problems by thinking together with others. But that presupposes that each person has a dialog with himself and that each person starts with the position that his own ideas and beliefs are not perfect and that they need to be examined and become better, even though they cannot be perfect, i.e., Socratic *aporia* (Platon, 1986, 1992a). This makes it possible to listen to others. Each participant in a democratic process, or in a dialog, feels always the need of other participants' standpoint and criticism because he is expecting them to help him to find a better idea than the one he holds for the moment. If one trusts his own beliefs and does not doubt or question his own knowledge, there is no reason to search for something better. Under these conditions, dialog and democracy become impossible.

Of course, the same is true on a societal level. Absolute truths in a society, for example taboos or political correctness, prevent critical thinking, dialog, and democracy just because there is nothing better to search for; we already have the answers. On the contrary, even a slight effort for dialog on something that is held as sacred can be understood as an offense or provocation that nobody wants to listen to, or worse it can be seen as a criminal offense leading to suppression and punishment. Thinking in society is the democratic process, and that means that this group-thinking process would not be needed if better laws and policies could be offered directly by AI, avoiding risks for conflicts when dialog between people may question established truths. AI producing satisfactory solutions makes even thinking in society, not only individual thinking, superfluous (see Figure 2.1).

It seems that the main political problem in our modern technology-dominated society is not the brutal use of force to suppress free thinking and dialog, or to abolish democratic institutions and procedures. Rather, the problem lies with running and upholding a culture of dialog that makes a democratic sense, in the way persons think, and in the way individual citizens and groups communicate and cooperate. If satisfactory solutions to problems are offered by political leaders, or as it may happen in the future by technology, then people will not need to run the personal or group dialog process, i.e., to think by oneself and together with others about these problems. The only thing people need to be bothered of is to elect leaders, or to design and use AI, capable of taking care of whatever goals individual and group-thinking processes are there for to achieve.

5   Difficulty of thinking

It is not easy to run the process of thinking, and it is even less easy to keep the process of thinking on the right track. This is because for individuals thinking implies strong effort, skill, responsibility, insecurity, and anxiety. For groups and societies, common thinking in form of dialog demands special procedures and processes, questioning of established truths and beliefs, belated decisions, and risk for conflicts. All these demands are very difficult for ordinary people to fulfill. It is very tempting then for a human to take the chance to escape from all this if satisfying solutions are offered by someone else, unless you are a superhuman (Nietzsche, 1983).

Detaching oneself from the goal or the solution in order to concentrate on and manage the thinking process itself is very difficult for both persons and societies. In real life, the satisfaction of a need, the achievement of a goal, or getting the right answer to an important problem is what really counts. Success is judged accordingly meaning that most of the attention is focused on the result and not on the process of thinking behind it. Thus, a high-quality solution offered directly by AI without involvement of the individual's own thinking would be relieving and very welcomed since taking responsibility oneself for the correctness of the thinking process or for the quality of the solution would not be needed.

The definition of the goal or what would be the right answer to a problem can be very difficult. Investigating this issue in a more systematic way, i.e., start thinking about this, will certainly cause higher insecurity. The same is true for defining the problem to be answered or the need to be satisfied. It is very difficult for a person or a group to answer the question of what they really want to achieve or what their real need is. In addition, this may certainly cause conflicts inside the person and among different persons and groups in a society. Again, if AI can make it possible to remove this stressful process away from the individual or the society by delivering a satisfying solution to an important problem it would be very welcomed by them.

We cannot think perfectly even if we are doing our best (see, e.g., Newell & Simon, 1972). Thinking is not easy. It demands a lot of effort and energy while it remains difficult to control and manage it right. Thinking hard for a longer period of time makes us feel mentally tired. This is so not only because of logical demands and cognitive difficulties but also because of neurometabolic processes in the brain. Thinking stimulates higher than normal accumulation of glutamate, a signal transmitter, in the lateral prefrontal cortex. Higher levels of glutamate, which is potentially harmful to the brain, make further cognitive effort and mental work difficult. Obviously, even the chemical function of the brain makes thinking difficult. When we are trying to keep thinking hard, we feel soon tired (Wiehler et al., 2022). So much better then, if AI can relieve us from this burden by delivering a satisfying answer to our problem.

Thinking implies investigation of all possible aspects of an issue that may be relevant for groups and society, but also for persons, since it often implies the discussion of sensitive subjects. Reasoning rationally and correctly during this investigation process is not compatible with limiting thought by leaving out certain beliefs and truths because they are sensitive or sacred. Questioning is an important principle for democratic societies and indeed this principle is solemnly and sincerely promoted ideologically and supported legally, for example by ascribing self-criticism a very high value and by introducing rules and laws allowing whistle-blowing. However, including generally accepted "truths" in the examination process of a democratic dialog, we are taking the risk of doubting them. For a person, that would mean losing motivation, feeling lost, or being offended. For a group or a society, the risk is that the common ground on which they are based may be lost or that conflicts between citizens or groups will occur. So, despite the nice rhetoric of openness and tolerance, in reality it is very difficult for us to start examining important principles. In fact, we can go rather far in order to avoid thinking properly ourselves, by eliminating sources of annoyance (Platon, 1992b), and by revering, relying on, and, most important, by following without daring to challenge our authorities who tell us what we should do and what we should not do (Milgram, 1963). For authorities themselves though, or for people who are in power and want to keep it, fear of being disturbed or offended is nothing they want to avoid. Old-time kings employed jesters, jokers, or fools just because they knew how important it was for them to have access to correct information and to have their reasoning examined by someone who not only was allowed doing this but who had this task as his duty. Ignoring or dismissing the jester was almost a certain way leading to catastrophe and tragedy (Shakespeare, 1969). But for most of us and most of the time, avoiding disturbance is highly appreciated, and an AI delivering satisfying solutions to societal problems, making well-functioning rules and laws, interpreting and applying the law correctly, and being fair in legal judgments and decisions would indeed be very appealing.

Satisfying needs is inherent in life, it lies deep inside us. It is therefore very difficult, probably impossible, to say no to solutions given to us by a technical process (in this case AI) that is vastly superior to the effectiveness of our own tool (which is thinking), and which at the same time is freeing us from insecurity, anxiety, and responsibility (see Figure 2.1). Given also that already and without the help of AI nonrational thinking is effective in achieving our goals most of the time, for example in the form of Aristotelian habits (Aristoteles, 1975) or heuristics (Kahneman et al., 1982; Sunstein, 2005), rational thinking is really not an absolute necessity. Actually, outside the metaphysical world of generalities, logic, and mathematics, i.e., in real life, there is no guarantee that the rational way of thinking leads to the right solution of a problem or to the optimal decision and action.

Still, thinking at some level is necessary, for example when one needs to know the relevant conditions of a certain problem situation or to take control of it. Even in the cases where nondeliberative thinking is successful in reaching goals, for example in the form of habits, a thinking effort, taking place earlier and described as *phronesis*, is a presupposition for effective control and knowledge of the conditions and the situation (Aristoteles, 1975). Thinking is also necessary in the leader-follower constellation. The leader has to think in order to be able to produce right answers which followers are expecting to receive. The less followers think, the more leaders need to do. Even in the world of mathematics, thinking plays an important role, and not only in calculating, since there is no fixed and unquestionable truth directly and automatically perceived, for example, once a solution to a mathematical problem has been reached (Nakada et al., 2023; Chapter 5 of this publication).

6   AI's goal

The interesting question is what we need AI for. AI can deliver products and services. All discussions about AI focus on its ability to solve problems we humans have and to satisfy our needs. Another topic very much discussed is AI's ability to make itself independent and go on setting its own goals and solve its own problems, against the interests of humans. All interest around AI is concentrated on achievement of goals, solutions of problems, satisfaction of needs. This is what we expect from AI to deliver and we design it accordingly: AI replaces and surpasses thinking by delivering higher-quality solutions directly.

But we may think of a different approach. Thinking is the biological tool for achieving goals, solving problems, and satisfying needs. However, by introducing AI in this process, thinking is affected. It might be possible to handle the issue of the impact of AI much better if we changed focus from the product to the process:

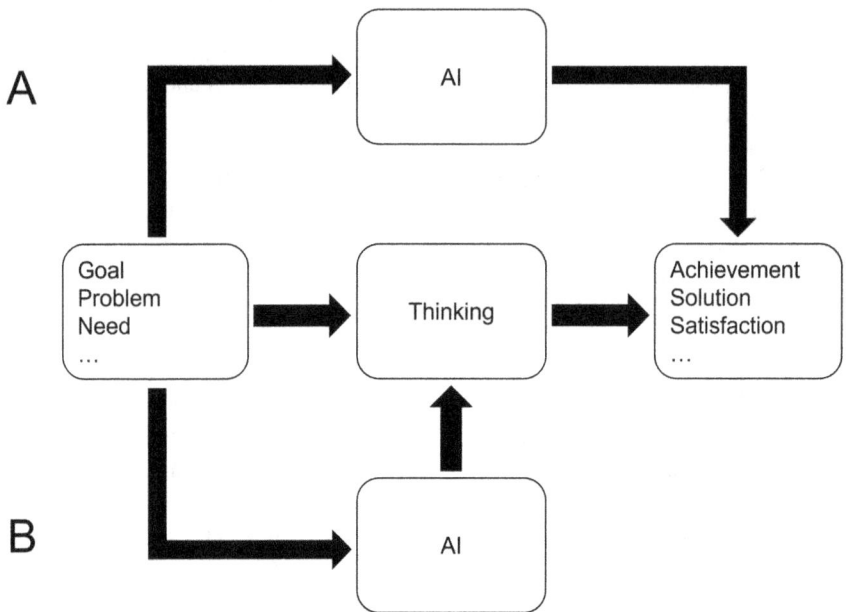

Figure 2.1 Two ways of AI usage: (A) delivering products and services and (B) supporting the processes of thinking.

that means an AI designed to help us use the "right" process of thinking to find answers, being it deliberative or heuristic, instead of using AI as the only tool to achieve goals and solutions delivering the answers to us directly before we get the chance to think (see Figure 2.1).

In order to be able to design such an AI, we need to know what we want. The answer to this question demands knowledge about what we are. Here, it is important to notice that it is impossible to make ethical judgments about AI's features, qualities, abilities or outcomes, effects, and products, by just focusing on them. Connecting explicitly and concretely to the values, strengths, weaknesses, and purposes of humans and other living organisms is necessary. So, we need to know what these values, skills, and purposes are. Like any other ethical or philosophical issue, we have to ask the questions about what we really want, what is the real goal, and what we really are. Are we recipients of services and products that we need according to our nature? Only that? Partly that? Are we recipients but through us, through our own thinking, and through our own choices? Or are we only thinking and choice-making creatures, without any other purpose, for example the satisfaction of needs?

7   Alternatives

If we think we are only recipients, and design AI in order to provide answers, solutions, and products (see alternative A, Figure 2.1), we will probably soon go to ruin like the old despots who could have all their wishes satisfied. Our thinking, making choices, and feeling anxiety will unavoidably languish and go away, in the first place. Later on, even the less rational and emotional reactions to handle our problems will follow suit.

It seems also that the fears expressed in the current discussion about the risks of AI may become true. By taking over our goals, AI would have a well-defined goal and that might lead rapidly to the emergence of an independent AI, not only because no one will be there to stop it, but mainly because there will be a very concrete goal from the very beginning for AI to work for the best it can, i.e., to provide answers, services, and goods to us.

If we design AI to help us think in the "right" way (see alternative B, Figure 2.1), that would fit our nature as it has been described above (e.g., thinking as a tool for sentiments; Hume, 2002, 2004) and also reveal the truth about our nature (Heidegger, 1977). Indeed, a human is a creature that needs a long time period of upbringing, education, and support in order to function properly as an adult. However, there is a risk that such an AI will disturb us. After all, we can reach our goals and satisfy our needs by not so rational thinking ways. An AI that always pushes us to think will never let us be in peace. It will soon perplex our mind to dissolution, meaning we will not be able to control our thinking and eventually cease to exist. If this is the case, that would be the perfection of the process of thinking itself unconnected to the satisfaction of real needs. Connecting thinking as a tool to the satisfaction of needs leaves room for compromises regarding the quality of thinking. A not-so-perfect thinking leading to a relatively satisfying solution or satisfaction can be accepted, but there is no meaning for this if the quality of the process is the only issue. In addition to that, AI would have a very clear goal to achieve and could very fast adopt this goal and make itself independent.

If we base the design of AI on the idea that we are both processors of handling problems, i.e., thinkers, and recipients of solutions produced by our thinking, it may be better. Here, there is room for compromises and accommodations so AI's support to thinking will not be unbearably disturbing leading to the avoidance of AI-provided support. Thus, we would not need to kill Socrates. On the contrary, the interconnection of thinking and answering as two sides of the same thing would allow the parallel existence of *phronesis* and habit (Aristoteles, 1975). The only issue would not need to be the perfection of *phronesis*. Instead, not-so-phronetic processes like habits or heuristics leading to satisfactory answers, solutions, and actions would have to be created and supported.

Then, it would be possible for us to exist in this world together with AI at least for some further time period in the future. This approach is also in accordance with the idea of thinking and knowledge (as a product of thinking) being interdependent, and of us thinking in order to solve our problems and to satisfy our needs. Moreover, the goal would not be well-defined: Delivery or choice? Both delivery and choice? AI or humans choose? AI or humans deliver? AI or humans are thinking? We can have a kind of confusion here and it could be that no clear goal is offered to AI so its possible way to independence is facilitated right from the start.

8   Conclusion

It is not possible for us, or for any living entity, not to strive for a purpose. We have our goals and needs, that is why thinking (or if we want to call it *logos*) emerges; in order to achieve these goals, even if in real life these cannot be satisfied fully and even if, in a metaphysical sense, they are illusions but necessary as a condition for the emergence and running of the thinking process. When we have reached the goal and have satisfied the need at an acceptable level, we do not need to continue thinking. We feel that we have the answer, so we do not need to search for it.

The task of AI we design and use is to do the same work as thinking does but much better: deliver what we need. Our own thinking becomes then redundant as an effect of a much more effective competitor taking over its task. If that is the case, then the most dangerous risk for being harmed, enslaved, or exterminated may not originate in AI itself but may come from our weakness to protect thinking.

It is though possible for AI to work in a different way: Support thinking and letting it alone deliver things and services we need. Thus, what we can suggest here is AI designed to help us think in the right way, or to help us keep thinking; disturb us and perplex us like a horsefly or a jester would do in order for us to keep the *dialegesthai* going, the dialog inside a person alone or together with others in a democratic process. Still, and in order to avoid the risk of annoying us too much, AI would also be designed to support us building habits, i.e., to repeat a successful way to reach a goal without so much thinking or to help us find a more quick and more simple way to reach our goals. Such an AI would not deliver anything to us. It would only help us run the process of delivering answers, goods, and services to ourselves by us using our own thinking in the right way.

Probably, and according to the above discussion, one of the most serious risks with AI, if not the most serious risk, is it substituting our thinking. However, AI properly designed and used can instead focus on thinking, support it as the sole tool for handling our problems, including our ethical problems, and by that make us sustain our nature.

Acknowledgement

This study was supported by the JSPS/STINT Bilateral Joint Research Project, "Information and Communication Technology for Sustainability and Ethics: Cross-national Studies between Japan and Sweden" (JPJSBP120185411) and Kakenhi (19K12528).

References

Arendt, H. (2003). *Responsibility and judgment*. Shocken.
Aristoteles (1975). *Ethika Nikomacheia* [Nicomachean ethics]. Papyros.
Bostrom, N. (2014). *Superintelligence: Paths, dangers, strategies*. Oxford University Press.
Burell, J. (2016). How the machine 'thinks': Understanding opacity in machine learning algorithms. *Big Data & Society, 3*, 1–12. https://doi.org/10.1177/2053951715622512
Camus, A. (1973). *O mythos toy Sisyfoy* [The myth of Sisyphus]. Boukoumanis.

Future of Life Institute. (2023). *Pause giant AI experiments: An open letter.* https://futureoflife.org/open-letter/pause-giant-ai-experiments/

Habermas, J. (1990). *Kommunikativt handlande: texter om språk, rationalitet och samhälle* [The theory of communicative action: Reason and the rationalization of society]. Daidalos.

Harrari, Y. N. (2015). *Homo Deus.* Natur & Kultur.

Hedman, A. (2017). *Consciousness from a broad perspective: A philosophical and interdisciplinary introduction.* Springer.

Heidegger, M. (1977). *The question concerning technology: And other essays.* Harper Perennial.

Hume, D. (2002). *Om förståndet* [Of the understanding]. Thales.

Hume, D. (2004). *Om moralen* [Of morals]. Thales.

Kahneman, D., Slovic, P. and Tversky, A. (1982). *Judgment under uncertainty: Heuristics and biases.* Cambridge University Press.

Kant, I. (2004). *Kritik av det rena förnuftet* [Critique of pure reason]. Thales.

Kant, I. (2006). *Grundläggning av sedernas metafysik* [Foundations of the metaphysics of morals]. Daidalos.

Kavathatzopoulos, I., Asai, R., Adams, A. A. and Murata, K. (2017). Snowden's revelations and the attitudes of students at Swedish universities. *Journal of Information, Communication and Ethics in Society, 15*(3), 247–264. https://doi.org/10.1108/JICES-02-2017-0009

Kurzweil, R. (2006). *The Singularity is near: When humans transcend biology.* Penguin Books.

Milgram, S. (1963). Behavioral study of obedience. *The Journal of Abnormal and Social Psychology, 67*, 371–378. https://doi.org/10.1037/h0040525

Nakada, M., Kavathatzopoulos, I. and Asai, R. (2021). Robots and AI artifacts in plural perspective(s) of Japan and the West: The cultural-ethical traditions behind people's views on robots and AI artifacts in the information era. *The Review of Socionetwork Strategies, 15*(1), 143–168. https://doi.org/10.1007/s12626-021-00067-8

Nakada, M., Kavathatzopoulos, I. and Asai, R. (2023). Truth and reality in the digital lifeworld: Departure from reductionism. In T. T. Lennerfors & K. Murata (Eds.), *Ethics and sustainability in digital cultures* (pp. xx–xx). Routledge.

Newell, A. and Simon, H. A. (1972). *Human problem solving.* Prentice Hall.

Nietzsche, F. (1983). *Etsi milise o Zaratustra* [Thus spoke Zarathustra]. Dodone.

O'Neil, C. (2016). *Weapons of math destruction: How big data increases inequality and threatens democracy.* Crown.

Persson, A. and Kavathatzopoulos, I. (2017). How to make decisions with algorithms: Ethical decision-making using algorithms within predictive analytics. In R. Volkman (Ed.), *Values in emerging science and technology, CEPE/ETHICOMP* (pp. 1–13). Università degli Studi di Torino.

Piaget, J. (1932). *The moral judgment of the child.* Routledge and Kegan Paul.

Platon. (1986). *Phaidros, Protagoras, Menon, Theaitetos* [Phaedrus, Protagoras, Meno, Theaetetus]. Zacharopoulos.

Platon. (1992a). *Politeia* [Republic]. Kaktos.

Platon. (1992b). *Apologia* [Apology]. Kaktos.

Popper, K. R. (1945). *The open society and its enemies.* Routledge and Kegan Paul.

Reese, B. (2018). *The fourth age: Smart robots, conscious computers, and the future of humanity.* Atria Books.

Schopenhauer, A. (1966). *The world as will and representation,* 1 & 2. Dover Publications.

Shakespeare, W. (1969). *Tragedier* [Tragedies]. Albert Bonniers Förlag.

Sunstein, C. R. (2005). Moral heuristics. *Behavioral and Brain Sciences*, *28*(4), 531–542. https://doi.org/10.1017/S0140525X05000099

Tegmark, M. (2017). *Life 3.0: Being human in the age of artificial intelligence*. Knopf Publishing Group.

Wiehler, A., Branzoli, F., Adanyeguh, I., Mochel, F. and Pessiglione, M. (2022). A neuro-metabolic account of why daylong cognitive work alters the control of economic decisions. *Current Biology*, *32*(16), 3564–3575. https://doi.org/10.1016/j.cub.2022.07.010

3 What is the problem to which AI chatbots are the solution? AI ethics through Don Ihde's embodiment, hermeneutic, alterity, and background relationships

Mikael Laaksoharju, Thomas Taro Lennerfors, Anders Persson and Lars Oestreicher

Prompt: Write an academic book chapter of up to 9,000 words including abstract and references using Don Ihde's four relationships to analyze the ethical implications of ChatGPT and its ilk. The style should be personal, preferably including the authors' experiences.

1 A nauseous exchange

On a Friday afternoon, after a week of hard work, almost at the psychological point when there was no energy left while work tasks were still far from done for the weekend, we prompted ChatGPT, an artificially intelligent conversation tool (or a large language model if you prefer a more precise description), the Neil Postman-inspired and somewhat passive aggressive question: "What is the problem to which ChatGPT is the solution?" The lackluster answer came within seconds:

> *ChatGPT is a language model designed to generate human-like responses to natural language inputs. Its purpose is to provide a conversational interface to users, enabling them to ask questions, seek advice, or engage in dialogue on a wide range of topics.*
>
> *In this sense, ChatGPT can be seen as a solution to the problem of creating an intelligent, natural-language interface that can understand and respond to a wide range of user inputs. Its development represents a significant advance in the field of natural language processing, and has the potential to revolutionize the way we interact with computers and other digital devices.*

In its usual style, the language model started by giving a short definition of the critical topic for conversation. Interestingly, it singled out a purpose: enabling users

DOI: 10.4324/9781003367451-4

"to ask questions, seek advice, or engage in dialogue". This was followed by a suggestion to what problem it was a solution, namely "*of creating an intelligent, natural-language interface that can understand and respond to a wide range of user inputs*". We were not happy with this answer. Skeptically and perhaps unconsciously trying to find fault with ChatGPT, we thought that there was an ethical issue in providing a service intended to compete with the almost 8 billion "intelligent, natural-language interfaces" that already populate our planet, but our negative mood got somewhat better when we saw the button that allows regenerating responses if the user is not satisfied with the one that was offered. After pushing it, we were supplied with the following:

> *ChatGPT is a solution to the problem of natural language communication between humans and machines. As a large language model trained by OpenAI, ChatGPT is capable of understanding natural language inputs from humans and generating appropriate responses in real-time. This enables ChatGPT to be used for a variety of applications, such as customer support, virtual assistants, and language translation, among others. With its ability to process large amounts of data and learn from it, ChatGPT is designed to improve its accuracy and understanding over time, making it a valuable solution for natural language processing tasks.*

There are clear similarities between the answers but the act of asking for a different answer prompted the model to slightly rebase the line of argumentation. Now the problem had shifted somewhat to a technical one: humans apparently want to communicate with machines using natural language. Including all its ambiguities, nuances, social signaling, and referencing, as well as the embedded power structures, we might wonder? Yet again disappointed, we clicked "Regenerate response" again and again but the closest we came to the model suggesting an actual problem (yes, in hindsight we could really question what we meant by an "actual problem") was the following:

> *[ChatGPT] can also be used for language translation, text summarization, and sentiment analysis. Additionally, ChatGPT can be used for generating creative writing prompts, summarizing long-form content, and assisting in the development of natural language processing applications.*

For disgruntled and tired teachers (i.e. us), this read: ChatGPT is a solution to the problem of having to do the actual intellectual work by yourself; it is a tool for generating text without purpose or authorship; it is the perfect tool for a society that has sacrificed the idea of *content* on the altar of *output*. Evil tongues could claim that it is the tool that makes the generalist, who knows a little bit about many things but nothing in depth, redundant. Below we will suggest that this could become the fate of experts as well.

We realized quite quickly that this was not the right mood in which to write a chapter about ChatGPT but also acknowledged that the first meeting with new technology often is of this affective and moody nature. These affects should not be suppressed and sublimated but should work as food for thought in the ethical process (Lennerfors, 2019). Thinking back on this afternoon, we need to concede that this technology will likely not go away, regardless of how many objections are voiced by social commentators. One example where it is already obvious that we need to relate to chatbots on a societal level is within education. At universities, there is already an intensive ongoing debate about whether the use of chatbots constitutes "cheating" during examinations. Regardless of which standpoint is taken in this matter, it is clear that it will not be possible to just carry on as if nothing has happened.

Like Neil Postman, we need to accept that we have come across a potentially world-changing piece of technology, one that could fundamentally alter how we work and communicate. In the same way as gunpowder weapons became inevitable after their invention, chatbots are now too powerful to dismiss. We need a better way of discovering the use of chatbots, or integrate the chatbots as part of our everyday practices, such as in teaching and student assessment. The question is not anymore what problem ChatGPT is a solution to, but how to live well with it and its ilk.

2   This changes everything

Almost overnight at the end of 2022, ChatGPT became a hot topic – a technology that seemed to "change everything." ChatGPT is an artificial intelligence (AI)-powered application, based on a large language model and can, thus, be subsumed under definitions of AI as a system that displays intelligent behavior and that may take actions to achieve specific goals with some degree of autonomy (Artificial Intelligence Act, 2021). At present, EU's definition of AI includes technology that is based on either of three approaches: machine learning approaches, logic- and knowledge-based approaches, and statistical approaches. There is also the distinction between artificial general intelligence (AGI) and the narrow type of AI, within delimited application areas, that has been realized thus far (Tegmark, 2017). The latter is used to address specific tasks and problems, such as calculating music and video recommendations, classifying images, or playing games. The former, AGI – which, like humans, can handle general problem-solving, reasoning and thinking – has received a lot of public attention but still remains a dream, or nightmare (Bostrom, 2014). Systems such as ChatGPT (GPT-3 and GPT-4) by the company OpenAI are by many described as approaching AGI capabilities (cf. Bubeck et al., 2023). The GPT series of systems are based on the creation of models from very large collections of text, where the resulting system is able to participate in dialogues with humans, write articles on prompted topics, and even write computer programs. The system is of course impressive, but the question remains whether it is intelligent in any sense or just the result of a massive statistical analysis of texts, resulting in an ability to "predict" texts from some given input.

ChatGPT builds on machine learning (ML) which has arguably been the most hyped type of AI in the last decade or so. ML is a set of techniques of using data, often a lot of data, to gradually improve an algorithm's output, to as accurately as possible categorize and predict outcomes, and sometimes even act based on those predictions. To a large extent, but not solely, ML is a mathematical and statistical approach that means: when given an input – essentially anything that can be represented as a numerical vector – return the probabilities that the input belongs to different categories. What is significant for ML is that the descriptions of the different categories are not programmed explicitly but are the result of feeding thousands, millions, or billions of examples to an algorithm that learns to recognize patterns that distinguish the categories.

AI has sparked our fascination, which is not the least present in media representations of it. Within the media, there are often diametrically opposed views of AI. For example, discussions about the "singularity" (Kurzweil, 2005, 2014; O'Lemmon, 2020), which is defined as the point in time when the AI has become more intelligent than humans, have mostly been from a fairly dystopian perspective, promoting the idea of the superhuman (artificial) intelligence, which will take over the ruling of the world as we know it, resulting in either AI as a benevolent super ruler or a malevolent self-centered despot. Both themes have a negative tone to them.

There are, of course, also positive instances of AI in media, although they are most of the time in a robotic disguise. Most people today, for example, have some reference to the AI-driven robot "WALL-E." In this context, it is almost impossible to not mention the very influential science-fiction writer Isaac Asimov, who already in 1942 wrote about how "The Three Laws of Robotics" could lead to absurd consequences, like a robot getting caught in an infinite loop (Asimov, 1942). Still, although they were the result of a series of science-fiction short stories and novels, they have more or less formed a standard ethical guideline for robots, which has also been applied to the general field of AI.

More nuanced views of the ethical implications of AI can be found in research. Similarly to Coeckelbergh (2020), one should eschew both the overly optimistic accounts of AI (AI will solve all humanity's problems) and the overly pessimistic accounts of AI (AI will overtake the world). Rather, it is important to reflect on the ethical effects of new technology, as the development is unfolding. As an instantiation of AI, AI chatbots such as ChatGPT have made AI concrete and able to be experienced by regular users.

As such, discussing the ethics of ChatGPT is in line with those who argue that we cannot speak generally about the ethical impact of technology, or even the ethical impact of artificial intelligence. Rather than such broader attempts, however interesting they might be, Peter-Paul Verbeek, situating himself within an empirical turn of the philosophy of technology (Verbeek, 2011), means that we should think about and analyze how specific technologies in their contexts affect our actions and perceptions. By this approach, he tries to escape both the instrumentalist view, namely that technology is a neutral means for humans to achieve their non-technological goals, and the substantivist view that technology is a controlling force on society. Verbeek contrasts his view with that of classical philosophers of technology such as Martin Heidegger (1977) and Karl Jaspers (2014), who saw all kinds of technology (e.g. clocks, airplanes, electrical infrastructure, medical technologies) as representing one and the same ideology. Both concluded that the essence of technology is fundamentally about human control and domination over nature and others (Heidegger), or that technology fuels a society of mass production and mass culture which dehumanizes individuals, turning them into cogs in the societal machinery (Jaspers). Verbeek instead recognizes a plethora of different technologies, not all neatly fitting into the schema of either control or dehumanization. The question of what the technologies do is instead answered when specific technologies are analyzed.

3   Ethics within human-technology-world relations

Focusing on a specific technology is precisely what we intend to do in our analysis of ChatGPT, or more generally, AI chatbots. Ethical reflection about ChatGPT is widespread in media, and papers are starting to come out within scholarly literature. For example, Laurence Brooks in Dwiwedi et al. (2023) wrote that there are some different ethical aspects relating to ChatGPT, namely how ChatGPT changes the relationship between people and technology, what it means for society when so many use the technology to gather information, while also many fear this technology, the power structures that are perpetuated by this new technology (e.g. the power of tech companies), security issues and the resilience to hacking, and finally fake news and information inflation. Brooks also quickly mentioned how Chat-GPT should answer in an ethical way when being asked questions, but also indicated research that argues that ChatGPT is a middle-class white male voice. Bernd Carsten Stahl, also in Dwiwedi et al. (2023), pointed out that there are different positive and negative consequences of ChatGPT. While it can certainly lead to fake essays and misconduct in academia, it can also level the playing field for those students who have difficulties with writing texts, perhaps due to their background. Stahl also argued that there is knowledge about evaluating new technology and this should be applied to ChatGPT, which will be done in this chapter.

In relation to previous works, we intend to situate the ethical issues of AI chatbots within a set of relationships between technology, humans, and the world. The reason for doing this is that chatbots are always co-constructed with humanity. As mentioned in the introduction to this book (Lennerfors & Murata, 2023), this idea of co-construction of technology and society, or technology and culture is widely shared. The mediation approach developed by Ihde (1990) and Verbeek (2005) views human and technology as intertwined and constantly shaping each other through their interactions.

It is in this conceptual terrain where we locate AI chatbots, as something that mediates relationships between the human, technology, and the world. What is important in our method is the focus on everyday experiences (cf. Ciborra, 2004). Rather than relegating the analysis from ourselves to a view from nowhere, we reflect upon others' and own experiences of the technology. This is also the reason why we include a number of vignettes of ChatGPT-related issues that we have experienced ourselves or have been told by others. We find Ciborra's notion of hospitality as interesting, because it expresses an attitude towards new technologies, not as friends nor enemies, but as strangers.

Following Ihde, we take four relationships between humans, technology, and the world as a point of departure, namely embodiment relationships, hermeneutic relationships, alterity relationships, and background relationships. We focus on how chatbot technology, in each of the four relationships, introduces new practices between humans and the world and renders some of these alternative practices more likely.

Ethics predominantly concerns the critical thinking about alternative courses of action with different ethical impact (positive and negative consequences, following or breaching duties, cultivating or breaking virtues, building or destroying relationships, upholding or downplaying justice, allowing or hampering freedom, etc.) and an evaluation of the desirability of these courses of action (Lennerfors et al., 2020). In this chapter, this means identifying changing practices following the introduction of chatbots, which could have an impact on desirable values in relation to our practices before the introduction of chatbots. Our analysis also includes the exposure of potential decisions that individuals or organizations need to make in face of the new technology. Our approach is pluralistic, acknowledging the openness of ethical judgment and therefore does not posit one value above all other values. Although our approach is centered around critical thinking, it does not mean that critical thinking is a given, but something that needs to be performed and practiced in relation to power, politics, and tradition.

In relation to general ethical assessments of new technology, Ihde's relationships allow us to structure our thinking, but also render new creative connections. The three first relationships in his framework are on the micro-level, i.e., the individual's connection with the technology, while the idea of background relationships incorporates more of the indirect effects of the chatbots, which might otherwise go unnoticed.

3.1   *Chatbots in embodiment relationships*

The first of the four relationships is the embodiment relationship. In this, humans use technology to act on the world, as a unit; as in Heidegger's ready-to-hand relationship in which technology, when it functions well, becomes an extension of one's own capabilities. Ubiquitous examples of this type of relationship include corrective eyeglasses, keyboards, and phones. When these technologies work as they are expected, we do not think about them; we think about what we achieve through them. We see the world in sharper detail rather than look at lenses; we type text rather than press buttons; we speak to someone, not to the phone. The formula, where the arrow indicates intentionality and the parentheses unity, is:

(human–technology) → world.

If we were to consider chatbots as constituting technology in this type of relationship, we would be looking at tools that help their users perform tasks and interact in the world. This type of chatbots could be described as digital extensions of their users, like "do what I mean" interfaces, that ideally can be used for acquiring information about the world and performing digital chores like formulating routine texts, scheduling meetings, organizing data, and other everyday tasks that do not spark joy. For these types of tasks, the technology is an enabler and we interact through it rather than with it. It does not require, nor necessarily benefit from, presenting itself as a human-like dialogue partner, as long as the tasks can be performed. For instance, if we were to design an AI assistant that could step in as a substitute for its user in a digital conversation with some other, the user's handling of this assistant does not require the same conversational abilities that it uses in its communication with the other. Precursory examples of this type of assistants (or user interfaces to AI chatbots) are the type of buttons, which can be found on various social media platforms, that suggest an appropriate answer to a chat message or posting; that let you choose between either different emoticons to indicate a reaction, or different short text snippets that relieve you from the burden of typing a sufficiently polite response. The logical next step is that the assistant fleshes out these emoticons and snippets into something more personal, based on your conversation style and chat history with the other. In this relationship, the user retains some sense of autonomy, although arguably directed and constrained by the affordances of the technology.

Consequently, a fundamental ethical issue in embodiment relationships concerns the extent of human agency. How active do we want a chatbot to be? Current chatbots are always in a stand-by mode waiting for a cue from the human individual. But if a chatbot has more comprehensive data as well as a record of the way the human being wants to live her life, is it then not a logical consequence that it will help the human being to accomplish her strategies?

3.2   *Chatbots in hermeneutic relationships*

In the second of the relationships, the hermeneutic relationship, humans read the world as technology represents it. The unity between the human and the technology from the first relationship is here replaced by a unity between technology and the world. This means that humans, to an increased extent, are directed to accept a world view as represented by technology. Thermometers come to represent temperature and are considered more trustworthy than our sensations, e.g., for assessing whether a rental apartment is warm enough to be comfortable, polymerase chain reaction (PCR) tests represent viral infections, the signal from a Geiger counter represents the level of ionizing radiation, and so forth. In some domains, technology is the only way for us to turn noumenal features into perceptible phenomena; in others, like in data visualizations, it can also reformulate phenomenal aspects of the world into representations. The formula is:

human $\rightarrow$ (technology–world).

Chatbots in this type of relationship take on the role of mediators. Both the input (the data that chatbots "learn" from) and output are curated by their designers, who, consequently, have power over the user's perceptions and responsibility to give an accurate representation of the world. This type of chatbot is already today used for learning about the state of the world through conversations. Simple versions can inform us about opening hours, and connect us with the appropriate human customer service agent while more advanced chatbots can act much more human-like and even mimic and detect the style, tone and valence in written text. They can be used as inspirational tools for essay writing, as assistants for summarizing long texts, and much more.

When asking ChatGPT and Bing AI in which of Ihde's relationships they are with their users, they both lean toward describing the relationship as hermeneutic. ChatGPT starts with replying the following:

> *As a language model, ChatGPT primarily fits into the hermeneutic relationship. It is a tool that humans use to mediate their experience of language and communication. ChatGPT is designed to understand and generate language in a way that is useful to humans, helping us to communicate more effectively and efficiently.*

At first, this seems like a well-formed and coherent statement. But what is the assumption behind the purported function to "mediate [humans'] experience of language and communication"? In what ways can ChatGPT help humans "communicate more effectively and efficiently"?

While we have discussed the AI chatbots' role to generate text for the user in order to schedule meetings and similar tasks, the AI chatbot becomes the window to the world in a hermeneutic relationship. Through this, it can potentially help us with reading the world in a better way than we would be able to without the technology.

One function it could take is to scrutinize emails in order to filter out the relevant information in an email, as well as diagnosing the intended social effects of the email, through the evaluation of pleasantries and social protocol.

An email from pre-GPT times may be illustrative for showing that this is an old "problem" that has previously been solved in less effective ways. In September 2014, one of the authors of this chapter received an email with the subject line "Can you make it to [Event name redacted] on 30th October Mikael?"

> *Hello Mikael*
> *I saw you weren't signed up yet to the next [Event] coming up on the 30th of October and wondered if you were able to make it along?*
> [Event information]
> *Spread the word!*
> *Regards,*
> [The event organizer, a remote acquaintance]

The casual tone made Mikael interpret this as a genuine personal reminder (from an "other") and signed up for the event, believing that it would be of interest for him. While at the event, which incidentally was packed with enthusiastic young tech people, he thanked the organizer for reminding him to sign up. The organizer looked puzzled, as if he did not understand what Mikael was talking about. After returning home, Mikael investigated the email more closely and realized that it was sent through an email automation service. Hence, the casual pleasantries around [Event information] were not really directed at him personally. More than anything else, he felt that his role in the event had been to contribute to the headcount. If he would have had access to an AI chatbot that diagnosed emails, the chatbot could possibly have signaled to him that this is an email generated by an automated service; that its function was to send a seemingly kind invitation since this is a more effective way of making Mikael show up to the event. With this hermeneutic reading of the email, Mikael could have made a somewhat more informed decision about whether he should have attended the event or not.

The logical next step is to use this technology for real-time assessment of text, including the statements that people utter. For teachers, this can become a convenient tool for assessing students' knowledge, but it does not stop there. It can also be used to check the veracity of sales pitches, politicians' claims and journalists' reports. Can we resist the possibility of knowing whether a politician is telling the truth or not? Whether the argumentation is consistent or non sequitur? We predict the real-time truth-o-meter, an instrument that rates the debaters' claims according to how "true" they are, to become a standard attribute of political debates.

Here, yet another problem appears concerning the models and their sources. The creation of the models requires huge amounts of texts, including factual reporting, subjective opinions and pure fiction. The intriguing question that emanates is whether the chatbots themselves can make the distinction between fact and fiction? This is the answer from ChatGPT:

> *GPT distinguishes between real and fictitious facts by relying on the patterns and context it has learned during its training. It uses the knowledge it has acquired from the training data to infer whether a statement is likely to be factual or fictional. However, the model's ability to differentiate between real and fictitious facts is not perfect and depends on the quality and comprehensiveness of the training data.*

Yet, the last sentence points to something we know next to nothing about as general users. Today, when there are many difficult topics that are subjects of heated debates (such as the existence of a climate crisis, the necessity of vaccinations or not, or even whether AI will become dangerous), a perceived "objectivity" could be a difficult issue when used in debates and discussions.

The ethical issues related to AI chatbots in hermeneutic relationships are to what extent the chatbot represents the world in a truthful way. What aspects of the world are picked up by the chatbot? Will we trust its assessment of the world and rely on it or will we have time and energy to second-guess the interpretations of the chatbot and engage in our own assessment of the world? Furthermore, in the longer run, perhaps a constant diagnosing of the world contributes to creating suspicion and to see all communication as much more instrumental than it actually is. Does having a chatbot as a hermeneutic aid lead to us constructing a world where the other is seen merely as a source of information and someone who potentially tricks us and needs to be constantly evaluated? Does it lead to the construction of a world in which pleasantries and social protocol should be subject to scrutiny? Does it lead to a world which places enormous demands on the sender of messages to tailor-make statements in order to really convince the AI chatbot that the sentiments are genuine? As most senders of emails know, even human-made manifestations of social protocol could be somewhat predictable and to some extent automatic. Could these increasing demands not, rather than increase efficiency, decrease efficiency by having two layers of chatbots mediating human relationships, one as an embodied tool (trying to create as powerful and genuine messages as possible to break through the skeptical hermeneutic chatbot of the receiver), and one as a hermeneutic meter (trying to help the human user as much as possible by really scrutinizing each message, trying to find if it has been written by a chatbot in an embodied relationship)?

3.3   *Chatbots in alterity relationships*

In alterity relationships, humans relate to technology as an "other" and the world recedes to serve only as the background for the interaction. In this type of relationship, technology is not a mediator but the focus of attention; the relationship is with the technology itself. Ihde brings up the special relationship that people can have with their "spirited" sports car as an example of affectionate relationships with technology, which falls into this category, like fondness toward a familiar and trusted piece of software or a cherished music instrument. Another example is religious icons, which are treated as sacred despite being made out of mundane materials. An important distinction from the previous relationships is that technology in an alterity relationship is not characterized by strict utility but is anthropomorphized into something resembling a human-human relationship; technology takes the form of a "quasi-other" (Ihde, 1990, pp. 97–108). The most troubling version of alterity relationships, according to Ihde, is when human characteristics, like intelligence, trustworthiness, etc., are ascribed to technology. While the technology is not independent and autonomous in a strict sense, the sentiment of the human in the relationship implies these qualities. The formula is:

human → technology (world).

The continuation of ChatGPT's reply from the previous section:

ChatGPT can also be seen as having elements of the alterity relationship, as it is an autonomous entity with its own agency and independent existence. However, its autonomy and agency are limited by the way it was programmed and the data it was trained on.

> *In other words, ChatGPT's "intelligence" is a product of human design and input, rather than emerging spontaneously or developing on its own. ChatGPT's behavior is largely predictable and its responses are determined by the patterns and information it was trained on.*
>
> *Thus, while ChatGPT may exhibit some characteristics of an alterity relationship, its behavior is still largely determined by humans and it is primarily used as a tool to mediate our experience of language, rather than being viewed as a separate entity with its own agency and goals.*

This part of the answer is wrong in the sense that it implies that alterity relationships are objective or potential features of technology. Ihde is clear that it is in the perception of the human, rather than a reciprocal relationship; the relationship does not require an "other" but will suffice with a "quasi-other". For users, it does not matter whether answers are predictable and "determined by the patterns and information" if they lack access to these patterns. The fact that ChatGPT answers with a judgment call that is backed by seemingly well-reasoned arguments suggests that the aim of the tool is in fact to form an alterity relationship with the user. Also, the excuse that is added to downplay its limitations, i.e., "I am a large language model trained by OpenAI and do not have the ability to …" is formulated to reinforce the image of personhood, as an "I", rather than impersonally as "this". Finally, text is not generated as an immediate chunk but instead as if it was typed by a human, further reinforcing the impression that it is in fact intended to be interacted with as in an alterity relationship.

Given the alterity relationship, ChatGPT might be seen as the perfect speaking partner. In Arifin and Lennerfors (2023), respondents state that voice assistants could be better if they would have a memory. ChatGPT has memory, at least during each session, and could thus evolve to become one of these proactive assistants that some respondents wished for. This would fulfill the wet dream of many – to have a speaking partner, nurturing digital solitude, or even solipsism: a speaking partner which you have full control over, leading to a range of potential ethical issues. The assistant has infinite patience, unlike human assistants who would most likely be angry after your nth prompt of the right behavior has failed. The assistant is always at your service, which once again makes it better than any human assistant, who is perhaps ill, or busy with something else, exactly when you need them. This "at your service" can spread to becoming the norm also for how we interact with other people. In the future, we envision possibilities where the chatbot can nudge the user to engage in certain behavior. For example, the human being can present herself as a liberal, friendly person but then engage in behavior that does not cohere with this self-presentation. Should the chatbot then intervene?

An interesting possible effect of advanced chatbots is that so-called experts need to become increasingly skilled in order to justify their roles when they are challenged by a conversation tool that fulfills their roles sufficiently well. A more positive angle on this possibility is that these chatbots can perform the tasks that are less engaging and creative, forming an embodiment relationship, which in fact is one of the sales pitches for AI. Within research, chatbots could, for instance, liberate researchers from performing time-consuming querying of databases to compile literature reviews so that they instead could spend more time on the creative parts of research.

Nevertheless, the need for expertise within many domains, also within research, could become challenged in the sense that the effort to acquire it may not be justifiable when the expert's role can be sufficiently fulfilled in a cheaper and faster way. For instance, human computers and telegraph operators are occupations that once required expertise but are now considered obsolete. How many occupations are facing a similar fate? Here, it should be noted that it is not completely absurd to claim that people should not waste their time performing redundant tasks for the sake of having an occupation, but it becomes a problem for a society when it loses its mechanism for distributing wealth among its population. As pointed out by The Future of Life Institute in their call for pausing AI systems training, societies are not yet ready for the economic and political disruptions that are introduced by powerful AI tools. Although this possible effect of chatbots could be seen as a hostile alterity relationship, in which the "quasi-other" is replacing you, it can also be described as a background relationship.

3.4   *Chatbots in background relationships*

In background relationships, technologies constitute, or are part of, the context for human experience. The technology is not being experienced itself but affects the lifeworld in some way. The way that air conditioners keep a room appropriately tempered means that we, without observing so, think and behave in different ways than we would if there was no air conditioning. Like in embodiment relationships, we notice the technology when it malfunctions, but we notice this in a different way because it was not an extension of ourselves. A breakdown of background technology changes the milieu so that we may become attentive to something being different, but we usually do not reflect over how the technology, while it was working, affected how we think and behave. As part of the milieu, technology can, nevertheless, have a pervasive effect on us. The formula is:

human (technology–world).

What then, would it mean if we established background relationships with chatbots? For this question, one of the responses generated by ChatGPT is interesting:

In this background relationship, ChatGPT would function as a seamless and reliable intermediary between human communication, without drawing attention to itself. Its responses would be so integrated and natural that we would not even recognize its role in our interactions, and it would become almost invisible, like the air we breathe or the background noise we tune out.

In this relationship, ChatGPT would essentially become a transparent interface for our communication, serving as a reliable and efficient tool for facilitating our language interactions. However, it would not be seen as an autonomous entity with its own agency, but rather as a tool or medium that we use to communicate with each other.

Overall, a background relationship with ChatGPT would be characterized by a sense of seamlessness and integration, where the technology becomes so embedded in our everyday communication that we no longer think about its role or even recognize its presence.

In a world in which we to an increasing extent communicate via technology, rather than face to face, this is a probable prophecy. Imagine your video calls being seamlessly translated, what you say being instantaneously backed up by illustrative facts and figures and your rambling being clarified with the help of assistive technology. If it sounds like it would be difficult to imagine this kind of assistance as a background relationship, you should think about the last time you noticed the autocorrect function in your mobile phone or word processor. Most likely, this happened when the technology malfunctioned rather than when it functioned as you expected. If autocorrect for ideas becomes a product, it will become an integrated part of the relationships in which we want to appear as the best version of ourselves. It will become like the business suit for your thoughts. For an individual, this will be experienced as an embodiment relationship but it will also have a pervasive texturing effect on how we communicate.

As an analogy: Neil Postman was famously critical of the way that television had taken over much of public discourse in the United States in the 1980s. When contrasted with printed media, he considered it to hamper rational discourse by necessarily dumbing down messages since the format does not afford abstract reasoning. In many respects, he follows the intellectual, arguably somewhat Luddite, heritage of Plato (who famously argued that the written word, much like what could be feared from chatbots, was bad for memory, as well as for wisdom, by enabling any fool to put forth eloquent arguments or misinterpret the meaning of arguments), with clear parallels with his contemporary Marshall McLuhan (McLuhan & Fiore, 2001). However, rather than claiming that media changes the way people think, he argued that media changes epistemology, i.e., the structure of discourse in a society, definitions of intelligence and wisdom, and the way in which truths are reached (Postman, 2005, p. 27). If chatbots are seen as a medium, their effects are not necessarily noticed in the direct relationships with them, but instead in the societal expectations arising from the availability of the technology, i.e., in the background.

There is yet another way in which chatbots can become part of the background, namely in a more indirect way. Rather than focusing on when chatbots are really mediating your activities in the background, this proposed dimension of background relationships concerns how our views on the world are altered by the mere fact that chatbots could be mediating our experiences.

At an event, a PhD student presented their research. Towards the end of their presentation, they made an impressive final statement about the relevance of their work, about the grand challenges of the future, and how we should all work towards solving them together. The sentence was long and while listening to it, parts of the audience started to giggle, look at each other, and smile. Something was quite remarkable with this young researcher. During the poster session, Thomas went to congratulate the student on their presentation, particularly the ending, and the student admitted that they had asked ChatGPT to write an ending for their talk. Then it suddenly made sense – perhaps the giggles and affect in the room was due to the fact that something was eerie, uncanny, uncomfortable, or perhaps too comfortable with this sentence. That it was more than human?

Or another vignette:

Thomas received an abstract for a special issue. It was well-written, but a bit grandiose since it discussed the future potential of a particular technology, and what role it could play in the future of humanity. It could easily have been written by a human author. But Thomas couldn't get ChatGPT out of his mind – who was really the author of the abstract?

This affect that is mentioned in these vignettes – was that something rare or something that will be more commonplace in the future? Here, we can see that ChatGPT becomes part of the background – of the social fabric. Perhaps what will materialize is a strange undertone of doubt and suspicion to the work done by others, and whether it has really been generated by a human other.

Furthermore, as some have pointed out, chatbots are the perfect tool to produce mediocrity (mentioned by Alf Rehn in a blog post). As chatbots rely on immense amounts of data and try to identify and reproduce common patterns, mediocrity is perhaps the expected consequence of increasing the penetration of the technology. While chatbots allow us to get a good starting point for our investigations and thought processes by telling us what people think in general about a topic, one should be wary that this must not be the end result, but merely the starting point. Mediocrity here does not only mean the lack of high quality. Also it means the perpetuation of the dominant forms of expressing oneself.

A final aspect of chatbots sneaking into the background is through the phenomenon of prompting becoming a ubiquitous mode of interaction.

At a webinar about ChatGPT, the value of prompting was highlighted. It was explained that ChatGPT, at first, perhaps provides a bland response to your prompt, but then you can improve it by learning how to prompt better.

Imagine if we start seeing the whole world as something to be prompted to get the responses you wish. Perhaps, the more widespread use of prompting in our dealings with chatbots will unveil a previously hidden dimension or our existence, how we perhaps will from now on think of others and the world as something to be prompted, rather than nudged. Prompting, in its essence, is about creating the stimulus that creates a certain response. In relation to ChatGPT, the nature of prompting seems straightforward and unproblematic. You write a prompt, you get an answer. You did not know why you got the answer, but you try to modify the prompt if you are not happy with the answer, similarly to what we did in the introduction to this chapter. Perhaps you get a better answer, but you still do not know the mechanisms behind getting a good answer.

By seeing the world as something to be prompted, you are, on the one hand, positioning yourself as that entity which has the world at its disposal like a standing reserve of promptable material, ready to give you the desired answer if you only find the right prompts. Yet again, turning prompting on itself, you also become that entity which is expected to be prompted by others. This could create a double negation, which in combination implies that you are the one able to prompt everyone and everything else, but you yourself need to be prompted for that purpose.

4   Concluding remarks

We have above presented a range of ethical issues classified into the four human-technology-world relationships by Ihde (summarized in Figure 3.1). Ihde's work is commonly used for classifying different technologies based on the kind of relationship we have with them. ATM machines, for example, are claimed to fit into the alterity category since they take the position of a bank teller, and refrigerators are moved into background relationships since we do not pay attention to them while they still exert influence on our behavior. However, Ihde himself was not strict in seeing the relationships as distinct so we have used the relationships as perspectives to explore one and the same technology. Through this, we believe that we have been able to highlight more aspects of AI-powered chatbots than if we had stuck to one of the relationships.

However, we have still seemingly compartmentalized the different implementations of AI-powered chatbots into four distinct categories. Indeed, we see a dynamic between the four categories. Not the least is the multiplications of the direct relationships through, with or to the AI chatbot leading to that the chatbots come to occupy an increasingly thick background.

Furthermore, it is likely that there are more positive consequences in the direct relationships we have with the AI chatbots, but the negative consequences are piling up in the background. As a tragedy of the commons, each individual is enticed (or even forced) to engage with the technology in order to keep up with social and economic expectations, but the emergent sum could be a reduction of happiness and humanity for all.

A final reflection is what happens if chatbots turn into a societal addiction. We have heard about the positive ethical implications of ChatGPT – that it will level the playing field and help those who have trouble formulating themselves to write. We have also seen negative aspects, that this leveling of the playing field indeed means that all will become mediocre. Perhaps ChatGPT will be seen as an equality-inducing method that all should have access to? At present, it is free of charge, but

EMBODIMENT RELATIONSHIPS	(HUMAN–TECHNOLOGY) → WORLD
Efficiency engine	Can increase efficiency for the person, which would allow them to focus on more creative tasks. But might agency be compromised?
Automated manners	Can automate social protocols with personalized greetings and other pleasantries.
The business suit	Can potentially help you be the one you want to be by adjusting its actions in a more fine-grained manner than you were able to.

HERMENEUTIC RELATIONSHIPS	HUMAN → (TECHNOLOGY–WORLD)
Transparency creator	The risk of misinformation and misinterpretation may decrease due to the hermeneutic support.
Truth-o-meter	It can constantly monitor and assess statements in the world, from teachers, ideologues, journalists, and from those near and dear.
Trustworthy curator	Can one trust the interpretations of the chatbot, and will one spend the time and effort to interpret the world directly, as a human?

ALTERITY RELATIONSHIPS	HUMAN → TECHNOLOGY (WORLD)
Tireless sounding board	A level of deceit involved, but one gets a possibility to engage in a deep dialogue and explore thoughts that might not be possible in relationships with humans.
Digital solipsism	This can lead to the development of novel thoughts but of course also contribute to the further compartmentalization and the creation of extreme ideas.

BACKGROUND RELATIONSHIPS	HUMAN (TECHNOLOGY–WORLD)
Generalized doubt	Can lead to the creation of a constant suspicion or doubt that texts, images, and thoughts have been created by a chatbot despite being created by a human. This increasing distrust might have damaging societal consequences.
Bot to bot	If many use chatbots as hermeneutic instruments, your own utterances might also be constantly evaluated, which means that you might need to prime your chatbot in an embodied mode to produce exactly the right interpretations of your utterances. This dynamic could create a layer of chatbots communicating with each other, leaving the human out of the loop.
Complicating relationships	The increasing importance of alterity relationships with chatbots might have a negative impact on the creation of human-human relationships.
A pattern of prompting	The world, including people, becomes something to be prompted.

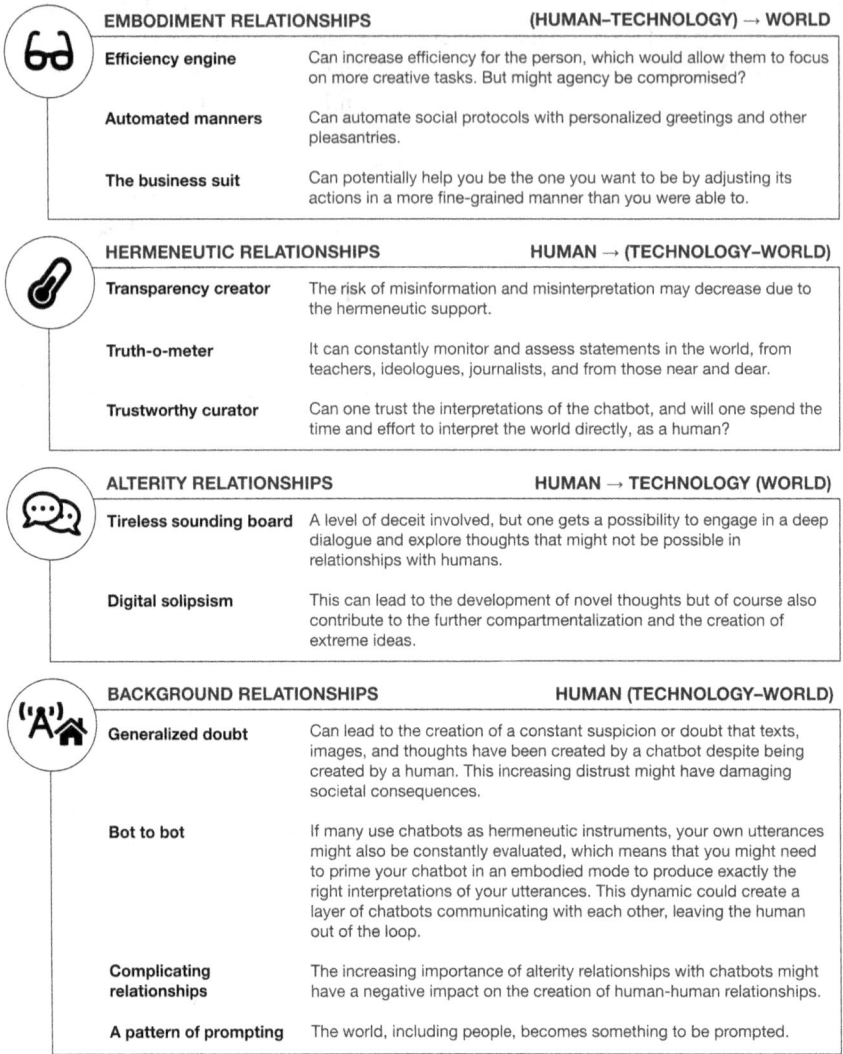

Figure 3.1 Examples of Ihde's four different relationships with chatbots and related ethically relevant features.

what will happen if it ends up behind a paywall? Will it be used only by those who can afford it, or will governments step in to provide chatbots for all?

This chapter has revolved around reflecting over answers produced by Chat-GPT in the light of Ihde's four different relationships. While it is unfair to expect philosophical depth from a language model that, based on statistical frequencies, simply guesses the next suitable word, it seems important to point out that its signaling of human capabilities leads to expectations from users, however unfounded these are. In the same way, as the model establishes patterns from exposure to data, also humans establish expectations based on experience. If the model is capable of producing linguistically coherent utterances, humans will, in anthropomorphic fashion, attribute also other capabilities to it. The oft-repeated caveat that Chat-GPT is just a large language model and not a knowledge engine is something that we intellectually recognize as true. Nevertheless, it has not prevented people from using the tool for various intellectual tasks. This dissociation between what we know and how we act is similar to our attitudes toward, e.g., substance abuse, the climate crisis, and Wikipedia: we know better but we act as if we do not. The dissociation between knowledge and behavior can be addressed as a social phenomenon (tragedy of the commons) or as a psychological one (conflicting perceptual cues). What can be concluded is that AI chatbots present themselves as useful, immediately available, and sufficiently capable tools to relieve a user from some chores, more specifically communication chores.

However, in a world where ChatGPT is used for greasing the communication between humans, expression of self is not anymore seen as a necessary part of relating with others. Rather, communication is seen as instrumental, tending toward merely exchange of outputs in relationships that are transactional rather than aiming for togetherness. The world in which this type of communication makes sense is a world in which the other is seen mainly as a source or sink of information and in which pleasantries are an inefficient vestigial part of a social protocol. Like in the tragedy of the commons, instead of doing away with them altogether, it makes sense for every individual to pretend like they have invested effort in the relationship-building part of their communication, regardless of how inefficient this is for the recipient.

Prompt: Write a letter to the editor to ask for a deadline extension, mentioning how much we appreciated meeting them in Japan during the last cherry blossom season.

Acknowledgment

This study was supported by the JSPS/STINT Bilateral Joint Research Project, "Information and Communication Technology for Sustainability and Ethics: Cross-national Studies between Japan and Sweden" (JPJSBP120185411) and Kakenhi (19K12528).

References

Arifin, A. A. & Lennerfors, T. T. (2023). A dumb spy? Ethical aspects of voice assistant technologies. In T. T. Lennerfors & K. Murata (Eds.), *Ethics and sustainability in digital cultures* (pp. 49–71). Routledge.

Artificial Intelligence Act. (2021). Proposal for a regulation of the European Parliament and of the council laying down harmonised rules on artificial intelligence (artificial intelligence act) and amending certain Union legislative acts. *EUR-Lex - 52021PC0206.*

Asimov, I. (1942). Runaround. *Astounding Science Fiction*, 29(1), 94–103.

Bostrom, N. (2014). *Superintelligence: Paths, dangers, strategies*, First ed. Oxford University Press.

Bubeck, S. et al. (2023). Sparks of artificial general intelligence: Early experiments with GPT-4. *arXiv preprint* arXiv:2303.12712.

Ciborra, C. (2004). Encountering information systems as a phenomenon. In *The social study of information and communication technology: Innovation, actors, and contexts* (pp. 17–37). Oxford University Press.

Coeckelbergh, M. (2020). *AI ethics*. MIT Press.

Dwivedi, Y. K. et al. (2023). So what if ChatGPT wrote it? Multidisciplinary perspectives on opportunities, challenges and implications of generative conversational AI for research, practice and policy. *International Journal of Information Management*, 71(102642). https://doi.org/10.1016/j.ijinfomgt.2023.102642

Heidegger, M. (1977). *The question concerning technology*. Translated by William Lovitt. Harper & Row.

Ihde, D. (1990). *Technology and the lifeworld: From garden to earth*. Indiana University Press.

Jaspers, K. (2014). *Man in the modern age*. Translated by Eden and Cedar Paul. Routledge.

Kurzweil, R. (2005). *The singularity is near: When humans transcend biology*. Viking.

Kurzweil, R. (2014). The singularity is near. In R. L. Sandler (Ed.), *Ethics and emerging technologies* (pp. 393–406). Palgrave Macmillan UK. https://doi.org/10.1057/9781137349088_26

Lennerfors, T. T. (2019). *Ethics in engineering*. Studentlitteratur.

Lennerfors, T. T., Laaksoharju, M., Davis, M., Birch, P. & Fors, P. (2020, October). A pragmatic approach for teaching ethics to engineers and computer scientists. In *2020 IEEE frontiers in education conference (FIE)* (pp. 1–9). IEEE.

Lennerfors, T. T. & Murata, K. (2023). Ethics and sustainability in digital cultures: A prolegomena. In T. T. Lennerfors & K. Murata (Eds.), *Ethics and sustainability in digital cultures* (pp. 1–16). Routledge.

McLuhan, M. & Fiore, Q. (2001). *The medium is the massage: An inventory of effects*. Gingko Press.

O'Lemmon, M. (2020). The technological singularity as the emergence of a collective consciousness: An anthropological perspective. *Bulletin of Science, Technology & Society*, 40, 15–27. https://doi.org/10.1177/0270467620981000

Postman, N. (2005). *Amusing ourselves to death: Public discourse in the age of show business*. Penguin.

Tegmark, M. (2017). *Life 3.0: Being human in the age of artificial intelligence*. Vintage.

Verbeek, P. P. (2005). *What things do: Philosophical reflections on technology, agency, and design*. Penn State Press.

Verbeek, P.P. (2011). *Moralizing technology: Understanding and designing the morality of things*. University of Chicago Press.

4 A dumb spy? Ethical aspects of voice assistant technologies

Anisa Aini Arifin and Thomas Taro Lennerfors

1   Introduction

In the sitcom *The Unbreakable Kimmy Schmidt*, a dialogue played out between Kimmy and Zach – the director of a tech company where Kimmy is working. Kimmy told her colleague Zach that she got advertisements for flights to London and horse insurance, since she is going to London and getting a horse. While Kimmy was perplexed by the fact that the advertiser knew exactly what she was looking for, Zach replied that the code that the tech company worked with was aimed at tracking and listening to users and see what their phone cameras are seeing. To Kimmy that sounded like spying, but Zach calmly replied that this is what all internet companies do – data mine.

This is not the only example of how ethical issues related to data are brought up in popular culture. Thus, at some level, these ethical issues related to data and artificial intelligence (AI) have entered into the public consciousness, and in this chapter, we are interested in deepening the understanding of how users relate to such technologies and the ethical issues therein.

The present chapter concerns voice assistant (VA) technology, installed on smartphones, smart speakers, or those on wearable devices. Although the technology is not transparent and most probably differs from provider to provider, in general, a VA is listening for a wake-up word, after which it starts to function. It is likely that this process is taking place without connections to databases through the Internet, but it is not clear. Using some other settings, the VA can also be activated by the push of a button instead of a wake-up word. When awake, the VA software records the request of the user, using speech recognition technology to convert it into text, analysing the text using natural language processing, all in connection to databases outside the device on which the VA is installed. How this collected data is used and stored by different service providers is not clear. Since the VA is 'listening' also for the wake-up word, the VA has been described as 'always listening'. However, in a sense, this 'always listening' feature is not necessarily limited to or dependent on the VA, as a device could potentially always 'listen' to its users whether the device has an active VA or not.

When this study was conducted, VAs were one of the fastest-growing AI applications in the market and a growth rate of about 25.4% was expected between 2019

DOI: 10.4324/9781003367451-5

and 2023 with an estimated 8 billion VAs installed in 2023 (Moar, 2019). The market leader Google Assistant was available on a billion devices by the end of January 2019 (Huffman, 2019), followed by Apple's Siri (more than 500 million devices), Microsoft's Cortana (approximately 400 million devices), and about 100 million installed Amazon Alexas. In line with this growth, there is a burgeoning scholarship on the reasons for adopting or rejecting this technology, as well as studies about how users engage with it, including ethical issues such as the possibility of privacy intrusion and the extent to which this technology is trusted by users (Arifin & Lennerfors, 2022; Arnold et al., 2019; Brill et al., 2019; Chopra & Chivakula, 2017; Foehr & Germelmann, 2020; Han & Yang, 2018; Liao et al., 2019; Lopatovska et al., 2019; Yang & Lee, 2018).

Although voice interactions using a phone to speak to others is not a new phenomenon, since the popularity of smartphones most of our interactions with the devices has been through the use of vision and touch. VA technologies re-introduce a way of relating to devices through voice interactions. Similarly to when smartphone cameras became ubiquitous leading to a range of privacy issues, it is likely that interactions using one's voice can also change the very nature of ethical issues. Further challenges can appear if and when one expands the widespread interaction with devices through smell, touch, and perhaps even taste. Through our focus on VA technologies, our chapter is incidentally aimed at exploring this shift to, or comeback of, the sound medium when interacting with connected devices. Don Ihde (2017) has described the absence of studies on the sound medium and has argued that there is a visual bias in science studies literature, which has also been discussed in adjacent fields.

As was argued in recent research on the ethical aspects of VAs, studies can be divided into two strands: quantitative (based on surveys) and qualitative (based on interviews and diaries) (Arifin & Lennerfors, 2022). In the former strand, the papers describe that VAs are used for information retrieval online (Arnold et al., 2019) and for enjoyment (Yang & Lee, 2018). Interpersonal attraction between the user and the VA contributes to user acceptance, while negative ethical issues such as privacy and security risks, as well as lack of trust to the service provider, lower the willingness to adopt the technology (Han & Yang, 2018; Liao et al., 2019). Thus, values such as utility and hedonism are important for the adoption of the technology, while privacy and security concerns are barriers to adoption.

In the latter strand, we find in-depth qualitative studies. Lopatovska et al. (2019) argued that interactions were often of a more leisurely or casual kind. It also became obvious that privacy risks were either misconceived or rationalized by users of smart speakers (Lau et al., 2018). Pradhan et al. (2019) explored how older adults anthropomorphize VAs, where the users move between seeing the VA as 'human-like' and 'object-like'. Kudina and Coeckelbergh (2021) studied the meaning-making processes of users of VAs, as a part of appropriation and techno-performances. These studies show the situatedness of ethical concerns related to VAs.

Previously, we conducted a study on media coverage of VA technologies, focusing on Indonesia but suggesting that since the news coverage almost exclusively concerned global matters, the study could plausibly be generalized (Arifin & Lennerfors, 2022). We concluded that there are several ethical issues discussed related to VA technology: gender issues, false marketing of the VA, ethical wrongdoing of the VA, ethically positive effects of the VA, misuse of VA, privacy and security. We further argued that represented users are sometimes the stereotypical adult user, but often people in distress and children. The VAs themselves are oftentimes represented as potential spies – always listening – but also as flawed, not fully conforming to our expectations and to some extent kawaii (cute) or stupid. This led us to conclude that the representation of VAs as cute or stupid could counterintuitively lead to increased acceptance to them 'as they seem less harmful and potentially less able to learn about the user' (Arifin & Lennerfors, 2022, p. 32), a sort of dumb spy.

The present study was designed to continue exploring users' perspectives on the ethical issues of VA technology. From the abovementioned literature, it is suggested that privacy and trust (both to the VA itself and the service provider such as Apple and Google) are commonly used concepts to discuss the ethical issues of VA technologies. We ask exploratively what ethical issues users perceive, as well as how they handle these issues. Furthermore, previous studies point out that VAs are often anthropomorphized, which points us in the direction of mediation theory which describes the relationship between human beings, technologies, and the world. Through that lens, we can explore users' relationships to VAs and whether it has any effects on the ethical aspects they perceive or do not perceive, as well as users' dreams about future VAs. This leads us to the following research questions:

- RQ1: How do users relate to VA technologies and what influence might that have to how they relate to the ethics of VA?
- RQ2: How do users of VA technology perceive ethical issues of such technology?
- RQ3: What strategies do users of VA employ to handle ethical issues?
- RQ4: How do users of VA perceive a desired relationship with VAs in the future?

Our country of choice for exploring these questions is Indonesia, a country which was also studied by us previously (Arifin & Lennerfors, 2022), arguing that this country is of interest not only because research related to VAs in it is scarce, but that such focus would be relevant given that Indonesia is the world's fourth most populated country, that it is experiencing significant economic growth, as well as an increasing Internet penetration of about 65%. In our previous paper, media coverage on VA technologies in Indonesia was studied, and it was identified that there was a discussion about a variety of ethical issues, in recent years mostly about the leaks of voice data stemming from VA technologies.

Despite the ethical issues of VA technologies, as well as the populous and developing Indonesian market, VA technology and the geographical context of Indonesia are both underrepresented in earlier research and this chapter, thus, seeks to advance the discussion on ethical aspects of VAs in this context. This national focus goes in line with a strand within information ethics, called intercultural information ethics, which has surged as a response to the a-contextual or even contextually insensitive or ethnocentric approaches of studies within information ethics (Ess, 2002). In this field, numerous empirical examples of how local cultures are resisting or translating and modifying information and communication technologies have been presented. However, essentialist assumptions underlying part of the field have been critiqued (Palm, 2016) and in response, culture is construed as non-essentialist and that we should rather study how 'culture' is used by actors in particular settings (Lennerfors & Murata, 2021). In this chapter, therefore, we are open to how respondents use culturally based arguments related to their view of the ethical issues of VAs, but we do not see their answers as determined by a shared culture, which is also the reason why we do not go into describing and analysing the cultural and religious context of Indonesia.

We aim to answer the research questions by interviewing a set of respondents. These respondents are in their mid-late 20s and have already used smartphones at this time in 2011 when VA technology was developed. All are Indonesian and many have extensive experience of living abroad as well as higher education. Therefore, the sample is seen as knowledgeable users of VA technology. Our reason for choosing this sample of respondents was to see them as a critical case (Flyvbjerg, 2006) in the sense that we suspected that even this group of users were not fully aware and dealt with ethical issues that might arise in relation to VA technology.

The remainder of the chapter proceeds as follows. Section 2 presents the theoretical concepts that are guiding our investigation. Section 3 presents the methodology, while Section 4 presents the results from the empirical study. Section 5 is a discussion which answers the research questions, and Section 6 contains concluding remarks.

2   Theoretical framework

2.1   Ethics

Although implicit in this chapter, the theoretical foundation of ethics that is used in this chapter is a pragmatic approach, which is drawing on the work by Lennerfors (2019). In his heuristic model, ethics is presented as a process characterized by awareness, responsibility, critical thinking, and action. Awareness is a precondition for ethics, and would in the case of VA technologies be to be knowledgeable about the potential ethical issues surrounding the technology. Technology is in this theoretical framework seen to affect the perceptions and actions of human beings, since humans are always seen as imbricated with technology. Responsibility would mean to internalize the identified ethical issue, viewing it from a first-person perspective, and seeing oneself as an agent able to handle the ethical issue. Critical thinking is based on awareness and responsibility and concerns an evaluation of what are the different options for action in relation to the ethical advantages and disadvantages, known or unknown, of the technology, and ethical action is expected to follow from a critical thinking process. This foundation is taking the user as the point of departure and is a way to structure his or her encounters with VA technology.

2.2   *Mediation theory*

Although the theoretical framework of Lennerfors (2019) is explicitly inspired by Peter Paul Verbeek and his ideas about how technology affects our actions and perceptions, it does not go into depth concerning the plurality of relationships that humans and technologies might have. To get a more nuanced understanding of this, since we are interested in how users relate to VA technologies, we draw on mediation theory (Ihde, 1990; Verbeek, 2015). Ihde (1990) described different relationships between humans, technologies, and the world (human-technology-world). The mediation approach lets us see humans and technologies, not as two opposing poles but instead as mutually shaping each other in the relations that come about between them (cf. Verbeek, 2015). This view changes the relationship we have to the world and possibly transforms our entire perception, experience, and understanding of it. Based on Don Ihde's (1990) work, there are four types of relations of human-technology-world (see also Laaksoharju et al. in this book).

In embodiment relationships, there is a unity of technology with human beings directed at the world. For example, we speak with other people through the phone, rather than to the phone. The formula would be: *(human-technology)* → *world.*

In *hermeneutic* relationships, technology forms a unity with the world, and humans read how technology represents the world. For instance, an MRI represents brain activity or the beeping of a metal detector represents the presence of metal. This could be represented as: *Human* → *(technology-world).*

In *alterity* relationships, humans interact with technology with the world in the background of the interaction. For example, human-robot interaction, getting money from an ATM, accessing built-in sat navs or operating a certain machine. Here, technology takes the form of a *quasi-other* (Verbeek, 2005), which means that they seemingly possess a kind of independence and apparent autonomy, even though technology is never a genuine other, like another human being. This relationship is characterized as *Human* → *technology (world).* However, recently, it has been argued that due to the advances in robotics, robots can appear not only as quasi-others but as what Kanemitsu (2019) calls another-other, in other words, beings with autonomy and interactivity closer to that of true other.

In *background* relationships, technologies are the context for human experience and action. The technology is a context of human existence not being experienced itself but can be felt in the sound of air conditioners and the warm air from heating installation. This relation would be represented as: *Human (technology/world).*

In this chapter, using mediation theory, we aim to discuss what kind of relationships users of VA technologies have to the VAs, and what relationships they fantasize about. Given the outcome of such an analysis, it is also possible to ruminate about if the kind of relationships that users have to VAs can influence the way in which they perceive ethical issues. In an example on Personal Digital Assistants (PDAs), Verbeek (2005) discusses how PDAs are mediating relationships in a variety of ways, such as embodiment relations (writing with the PDA) and alterity relations (relating to the PDA as a valuable object).

2.3   Privacy

In the framework on ethics described above, it was described how ethics relating to a technology is dependent on being aware of ethical issues. The major ethical issue that has been identified in previous literature in relation to VA technologies is privacy. Privacy, in a legal right, is synonymous with a 'right to be let alone' (Warren & Brandeis, 1890). Alan Westin has defined privacy as 'the claim of individuals, groups, or institutions to determine for themselves when, how, and to what extent information about them is communicated to others' (Westin, 1967). Although most theorists agree that privacy is fundamentally important to the human experience, there is no agreement on what that concept means or is encompassed by it. Then it is possible to justify an invasion of the right to privacy on another ethical basis since privacy has developed as a weaker right for not having such a clear definition (Charters, 2002).

Nowadays, in this AI era, people live with complex privacy implications. Also, the private companies that produce such technology collect and use data with interests that differ from the benefits of their users, which creates conflicts of interest. Users have to balance the trade-off between maintaining their privacy and the convenience afforded by the technology, which would in our vernacular be called critical thinking. Sometimes, people need to choose between comfort and giving up on their data being collected to get their gadget personalized by letting the technology 'learn' about the person.

Few studies have proved the privacy implications and perceptions of VA use (Lau et al., 2018). A study shows that talking to a phone-based VA, especially in public, can create discomfort (Mennicken & Huang, 2012). Moorthy and Vu (2013) found that smartphone-based VA's users were alert when sending private information such as their location (public vs. private space) and accessing options (keyboard vs. voice). Besides, Zeng et al. (2017) studied security and privacy concerns in smart homes, including smart speakers. They found that reasons for participant's lack of security and privacy concerns regarding smart homes were because of not feeling personally targeted, trusting potentially adversarial actors (like companies or governments), and believing their existing mitigation strategies to be sufficient (Zeng et al., 2017). Moreover, Lau et al. (2018) included the 'always listening' feature of VAs in their study as an additional privacy challenge.

2.4   Trust

Trust towards the VA technology and its service provider means that the user knowingly and willingly places herself in a situation of vulnerability, with the expectation that the other will not hurt the actor (Mayer et al., 1995; Rousseau et al., 1998) 'irrespective of the ability to monitor or control that other part' (Mayer et al., 1995). Additionally, by having trust in technology, McKnight (2011) stated that users face the risk of unfulfilled expectations and responsibilities. Li et al. (2008) even determine trust as a 'primary predictor' of technology use (Li et al., 2008, p. 39).

From a more philosophical point of view, Kiran and Verbeek (2010) describe trust as a central dimension in the relation between human beings and technologies.

They argued that most technological discourses place humans and technologies as two external entities that can give impact to each other but do not mutually constitute each other. This means relations of trust can vary from reliance (e.g., technology as the extension of human-help humans based on its aim) to suspicion (e.g., ethical precaution focused on the risk of technology). While in the notion of trust in the philosophy of technology, Kiran and Verbeek argued that using technologies does not imply an uncritical subjection to technology, rather more about trusting ourselves to technologies.

Having represented the theoretical literature that serves as an inspiration for the empirical study, a problem area starts to emerge. We are concerned with understanding how users are aware, take responsibility for, critically think about, and ethically act in relation to the VA technology. Privacy concerns of users are seen as a predominant risk of VA technologies, but such privacy concerns can potentially be mitigated by the user trusting the VA technology and the service provider. In many cases, users only have vague concerns related to privacy invasion, and thus trust providers to not infringe on the users' privacy. Also, the ethical concerns raised by the technology can be affected by the relationships that the user might have to the technology, whether it is one of trust or of suspicion. With this problem area in focus, we turn towards the empirical study.

3   Methodology

3.1   *Semi-structured interview as data collection method*

The method of semi-structured qualitative interviews with Indonesian users (see Appendix A) was chosen to generate rich empirical material (Bell et al., 2019) in order to explore their relation to VAs, ethical issues perceived, how they dealt with such ethical issues, and future visions of VA. The interviews were conducted online, due to interviewees living in different cities as well as because of the COVID-19 pandemic, in the spring of 2020. Within the interview, a short observation was also made about how respondents interact with their VAs. Twenty-two Indonesians agreed to participate. This group consists of an equal ratio of women and men, and most were 26–30 years old, and this group was chosen partly out of convenience and partly because they belong to a generation that already used smartphones in 2011, when VA technology was developed. Many have extensive experience of living abroad as well as higher education. Therefore, the sample is seen as knowledgeable users of VA technology. Our reason for choosing this sample of respondents was to see them as a critical case (Flyvbjerg, 2006) in the sense that we suspected that even this group of users were not fully aware, took responsibility for, critically thought about, and acted related to ethical issues that might arise in relation to VA technology. Seeing them as a critical case would mean that if even this group were not fully aware, it is likely that other users are even less aware.

All respondents were informed about the purpose of the study, how the data would be used, and that they could opt out of the study at any time. All interviews were recorded and answers to questions concerning ethics were transcribed. The interviews were conducted in Bahasa Indonesia – the official and national language of Indonesia. After about 12 transcriptions, the first author realized that she was starting to reach theoretical saturation, and then only transcribed and analysed those interviews which gave differing perspectives from the first 12. In the end 18 interviews were used for the study.

3.2   *Population and sampling*

Considering the limited time and resources to do this study, accessibility and availability of the interviewee took the most prominent part, which supported the choice of the convenience sampling method (Cooper & Schindler, 2014) using the first author's personal network. Based on Cowan et al. (2017), most of the VA users are infrequent users (people who use it occasionally), purposive sampling was conducted for this study. The first author selected those who aged 18 years or older, and both live in Indonesia and abroad, thinking that they may have different characteristics. To reduce self-selection bias, although the first author used convenience and purposive sampling when choosing the prospective candidate, some screening questions were used, namely: (i) whether they are using VA and (ii) how often they used it lately. The foremost requirement is that they supposedly have experience in using a VA. To make the data more appealing and reduce the possible theoretical saturation, which VA they used was also taken into consideration. Based on screening survey responses, the first author proceeded with those who use at least one VA and those who use and have experience using the VA for some time.

From those 18 respondents, more than 70% of them have experience living abroad or are still living abroad. The majority of the respondents, as many as 13 people hold a Master's degree, one person has a doctoral education, and the remaining four have bachelor's degrees. With the same population, seven interviewees are working in the Engineering or IT area and another seven are in Education, Health, or Science sectors. The remaining four are students at the postgraduate level. The vast majority are tech-savvy; three people have experience joining a tech-related workshop or training included in the experience group (see Appendix A). Otherwise, there are four people grouped in non-tech expertise or background. Additionally, Google Assistant and Siri were the most popular assistants among the respondents. There are also Alexa, Cortana, and Bixby mentioned once each. Additionally, all of the users the first author interviewed have their VAs installed on their smartphones, while some others also have a VA either in their smartwatches, tablets, TVs, or laptops. Only R08 has his Google Assistant on a smart speaker and no one lives in a smart home environment.

The interview was divided into three parts: first of all, some general questions were asked. Participants also got some basic questions about their experience or knowledge about VAs. Second, the interview was paused for some observation. Participants received some random 'cue card' – a question or command that they have to give to their VA. For example: '*Find out what the weather will be like in your city tomorrow'*. Subsequently, the interview continued with questions concerning (1) digital literacy of smartphone/laptop use concerning security & privacy, (2) general privacy and mobile data concerns, (3) VA use, (4) reasons for the adoption of the assistant (5) experiences and expectations, (6) privacy concerns, (7) perceptions and attitudes around 'always listening' and sharing of data, (8) privacy control, and (9) their foresight about this technology.

3.3   Data analysis

On average, the interviews were about an hour and 20 minutes long. Each recording was listened to, and answers were input into a table to keep track of the question group, questions, answers, who is giving the answers, and the answers. When writing up this chapter, interview questions concerning the specific research questions were transcribed. The data were related to the theories that we surveyed in Section 3.

4   Results

4.1   Background: Understandings of the technology

In general, all users understood how VA technology works to some extent. Some said that the VA technology records voices, recognizes their voice, converts the audio to text-based information, and that the companies have data training or data lakes for the collected information.

R04, R06, R11, and R16 mentioned how the technology will first record their voice when responding to a command or question. Specifically, R16 guessed that the machine will record detailed information from his instructions:

> it records the frequency, amplitude, vibration, and then it is translated to machine language, translated again into information, and later returned back, and they are stored...

Even though not all interviewees mentioned the voice recognition step, R01, R05, R07, R08, R09, R20, and R21 thought that the voice technology supposedly recognizes their voice. It will, then, imply that the users believe their VAs can only be called by them. In contrast, R14, R18, and R21 just realized that they can 'call' the VA without pressing a microphone button on their phone while having the interview.

4.2   Relationship between the user and the VA

In general, there are two distinct groups seen from the result of the interviews. A majority of 14 interviewees thought that their VA(s) is a machine or tool, while others said they are their friends.

The vast majority of interviewees (R01, R02, R03, R04, R06, R07, R11, R12, R14, R16, R18, R20, R21, R22) described their VA as a machine or tool to help them do something. The answers vary either from the VA's being a machine, assistant, or tool, which has limited ability yet is helpful when the users cannot access their phone. R21 defined his VAs as a driving assistant since he uses it only when driving and cannot access the phone by hand.

> It is an assistant when I am driving. Since I never use it at home. Ya well, I used it one or two times at home just for trying, asking Siri to tell a joke to my kids. But just for fun, I would say. It doesn't really give benefits whether you have Siri or Google Assistant. It's just like you have it for fun. So… it is more a machinery assistant ya…

Moreover, the others agreed to some extent that voice technology can help them look for something and help them when they are lazy or need a faster way to access a specific app or utility, productivity, and communication applications without tapping the phone. There are some applications that have already been integrated into the technology, 'I only ask it for something that I know the VA can do' mentioned by R07. This explains why some interviewees mentioned that the VA still has limited abilities. The immature technology at times also creates concerns for R01:

> Sometimes it can give me some trouble since it is not quite sensitive and it leads to more confusion.

From a different point of view, R16 joked that his VA is a slave, a machine one.

> … it is not human. Precisely the other way around, since it is a machine, I don't need to give it rights. I don't need to use my feelings when interacting because it is a machine.

In Indonesia, generally, people would define and see 'helper' and 'assistant' differently. Helper is a noun for hired people whose job (helping) takes care of household chores (cooking, washing, sweeping, etc.). Meanwhile, an assistant is a professional job that helps other respected jobs like artists, coaches, cooks, etc. '*It's an assistant. Not a helper since the level [of the helper] is lower than assistant*', explained R06 about her VA. In contrast, R07 and R18 say that VAs are helper machines with some limitations. R18 pointed out that:

> Physically it can't do anything. So, it's more like… mmm I assume it doesn't have feelings. I mean… yaa like normal, why should we put some empathy or behave and use some feelings toward a machine?

On the other hand, the second user group (R05, R08, R09, and R10) builds a closer relationship with their VAs. Empirical results indicate that they use respectful words like 'thank you', 'never mind', and 'please' when giving commands or asking questions to their VAs. Although there was some self-doubt, it reflects on R05s answer:

> I think of it as a friend, so... maybe... I often say please, thank you, like that... so... that... Do you think it makes sense...? hahaha... It always does what I say. So, I always say 'please blablabla', say 'thank you', 'never mind'...

She also compared when she treats people:

> ... personally, when I ask someone, I will say those respectful words. Even though it is a machine, I think I am more comfortable saying those words since he/she/it does something for me, and I'm grateful for it.

R08 and R09 also pointed out the ability of the voice technology to make jokes and how they enjoy it. Meanwhile, two odd pieces of evidence showed in R14 and R21 responses. In spite of being caught several times saying 'please' after giving command to the VAs and playing the VA for telling jokes with the kids, respectively, R14 and R21 are just seeing their VA as a machine.

Based on the interviews and the observational data, we suggest that the relations between the users in this study and their VA are often alterity relations. Such relations entail humans' (the users') interacting with the technology (VAs) with the world in the background of the interaction. The empirical data show that most of them perceived their VA as 'something' with the ability to answer questions and do as commanded. It does not matter how they think of Siri, Google Assistant, Alexa, or other VAs as an assistant, helper, machine, or friend. The role played by technologies in this set of relations can be characterized as that of a quasi-other. For the respondents, the VA appears in alterity relations as quasi-other, because while the technology we interact with seems to behave as an 'other', of course, it is never present as a real person. Although Kanemitsu (2019) argues that robots can now appear as another-others, our interpretation is that the respondents' relationships to VAs are not characterised by significant respect of the autonomy of the VA nor strong interactivity.

People tend to approach the technology that they use in anthropomorphic ways. Also, the technology in alterity relations possesses a kind of 'independence' and can induce increased 'interaction' between humans and technology. Simply, Ihde (1990) mentioned that humans have always been fascinated by a quasi-autonomy of technology. This fascination is very evident seen with how people perceived technology in the beginning. However, with this also comes a degree of trust involved in this relationship. Here, we see the possible connection between how these people have ethical perception and their relations to this technology.

Although it might seem straightforward to diagnose the relationships as belonging to the alterity category, it would be logical to assume that relationships to the VA should be embodiment relationships where one for example speaks through the phone to another person, or where one accesses some content through the VA.

Although part of the relationships that our respondents have with their VAs are of the embodiment kind, alterity aspects of relationships are prevalent. If it is the case that people's relationships to devices are previously often embodiment relationships and that our respondents can be interpreted as having alterity relationships, it is likely that the existence of VA technology has contributed to the shift from embodiment relationships to alterity relationships, which in itself can have ethical implications since devices are now harbouring a quasi-other, rather than being tools for accessing something beyond them.

4.3   *Perceptions and attitudes around 'always listening' and sharing of data*

The notion of 'always listening', in other words that the VA is always listening for potential commands or requests from the user, should be well-known among VA users, yet it is intriguing that only five people (R03, R05, R10, R20, R22) claimed to have heard about it. This notion plays the most prominent role in making this digital VA play its part in its user's life. To be able to assist effectively, it is advised to set the assistant to always be ready to answer questions or commands of the user. However, this also creates an opportunity for the invasion of user privacy. Since moving to the US, R05 has been exposed closely and more often by big tech campaigns. Her awareness of the VA advertisements shows it:

> I noticed what they were talking about, and the theme was always like that. Both Siri and Google (Assistant). It always identified itself as an assistant who would always listen to me. From their language. However, in my opinion, it's only one of their marketing strategies. Hahahaha... I mean, from their point of view, they branded these two characters as such.

Meanwhile, R04 doubted it: '*I never heard about it but I think it won't be efficient for the battery*'. They might hear or have chosen the option – with or without intention (some providers like Apple provide an option called 'always listening', while others do not), but they do not pay much attention to it. However, by being unaware of this situation and notion, there is a possibility that those people also are not aware of the way the technology could impact their lives.

Since R21 never 'calls' his VA by voice, he tried to clarify '*so by default active ya...? without pressing a button or anything? Then no, ya*'. Meanwhile, as a user who has enough knowledge, has options, and is not under too much pressure to use a VA, R20 showed an interesting insight. For instance, because of her understanding of the 'listening' process of VA, R20 became uneasy about activating the 'always listening' feature and turning it off. She actively seemed to engage in critical thinking:

> It will be active if we call the 'keywords' meaning that as long as it is active, it will listen continuously. That made me nervous; that is why I turned it [the always-listening feature] off.

Now when she needs her assistant, she has to long-press a specific button on her phone. It does not sound convenient, but she prefers to do it to protect her privacy.

Regardless of their awareness about the 'always listening' feature, the empirical data showed that more users were either comfortable, giving up, neutral, or did not care about that feature. Both R05 and R06 had similar views, respectively '*I am worried, I feel insecure, but ya what can I do?*' and '*... so yeah I heard about it but I was never really concerned about it hahaha*'. R10, however, felt comfortable with that ability although R04 was not really concerned about it:

> I think it will be just the same, it will listen to our conversation, even if we don't ask. Emm yaa I don't think there is a problem. What will they get [by doing that], I mean it won't make thieves come to our house, it is not used to threaten us, so there is no problem. It is only causing discomfort if every time we speak or discuss other parties listen to it.

On the other hand, R01, R07, R18, R20, R21, and R22 felt scared, afraid, insecure, paranoid, angry, or concerned when discussing the always-listening feature of VA. R01 and R18 felt like there was someone who stalked or bugged their conversation. While, in the R21 case, having an opportunity to be involved in a GDPR-related project made him more aware of data privacy concerns. His argument after learning more about always listening was anticipated,

> That's awful because then you don't have a chance to turn it off, right? I think if that's the case then they should... but that's already happened I think in a smart speaker. Because then you need to keep it active, otherwise, if you (still) need to press a button then it is not really a smart speaker, it's a dumb speaker then... but such a thing is already happening in this smart speaker. But ... I would hold, I need to check also, that it doesn't happen on my mobile phone. Because that should be awful. And I am not willing to consent to that.

Additionally, discussing VA technology was a bit challenging since, for most people, it is not a daily use application, neither productivity nor a communication app. Thus, when the interviewees got stuck with the question, the strategy was to associate the question with other AI technologies. Empirical data informed that no one had experienced feeling uncomfortable around their VA. Some might not really understand or care about the possible situation, however, R14 explained:

> When I bought the device, it means that whatever happened, even if it got hacked or whatsoever [...] and someone gets access to my data, it is a risk that I have chosen.

This may also be influenced by his educational background and technology skills.

Nevertheless, there were issues mentioned several times during the interviews. R02, R06, and R08 related their experience in using the VAs to the advertising phenomenon they encountered. They made some theories on how the phone through the microphone or the geo-tagging services can stalk them and the provider can send personalized ads to their phone. Also, mentioned several times, almost one-third of users felt uncomfortable around a digital camera on their phone and laptop.

4.4   *Measures taken for privacy protection from voice assistant*

Twelve users (R01, R02, R03, R04, R11, R12, R14, R16, R18, R20, R21, R22) suggested some ways they could imagine or had done to prevent the data privacy violation. R01, R14, R16, and R18 proposed to either uninstall the app, stop using the VA, or use the VA less frequently in order to not share any information with the VA.

While giving consent before activating VA was a really important measure for R21 to protect his data privacy, R01, R11, R14, and R18 proposed to either uninstall the app, stop using the VA, or use the VA less frequently. Also, the input for VA technologies is closely related to the phone's microphone which then suggests to R04, R12, and R18 to turn off that phone part when needed. Further, of those 12, there were only three users (R02, R20, R22) who mentioned switching off the 'always-listening' setting. R02 shared:

> Turn off the always-on (listening) feature, and turn on the airplane mode.

Some additional notes from R20:

> … you (still) can activate/call the VA by pressing a (certain) button. I also deactivate the long-press home button to activate the VA. I once set it, but now I forgot how to redo it. And it is simply because I am paranoid about this tech.

On the other hand, the last 30% of users were, shockingly, did not do any precautions to protect their privacy from the VA. R08 expressed:

> Hemm is there one…? I don't think there is anything like that because I never turn it off. They are always [in] standby [mode].

In fact, he just knew there was a switch to activate the always-on on his Google Home Mini when we did the interview. It was in line with what was stated by R05, R06, R07, R09, and R10 that they did not know how and have not done or tried to do something to prevent privacy violations from their VAs. R07 explained:

> I don't think there is any, ya what can we do? I give up and accept it as is since I trust… but I trust the 'VA' ya, not the company. But if you ask me about data privacy, in this era privacy is worthless, we cannot keep it safely.

Besides, to put some notes from the overall discussion, those who have some expertise in technology showed potential knowledge in reasoning how technology works. Decentralized data, anonymous storing processes, and locally stored data were mentioned by R10, R18, R20, and R22 because of privacy concerns, battery efficiency, and reliability of the VA answering question and command.

4.5   Future visions

All users have their own imaginative fantasies about their dream VAs. If there is no limit in the future, everyone agrees to have a smarter and more advanced VA. Specifically, R11 and R16 wished to have one as intelligent as Jarvis, a fictional character in the Marvel Cinematic Universe film franchise, '*One that can help, understand me like Jarvis*' said R11. Furthermore, being proactive, responsive, and understanding humane instruction and the context of discussions were also popular ideas among the users. R20 explained in detail, '*if today we discuss [a certain issue], then tomorrow we can continue discussing [the same issue]. So, it is not just a simple if-else like now*'. R14 specified his personalizable VA criteria, '*a VA that can be personalized, controlled, and defined its relationship status, age, skills, and personalities*'. R04 and R07 also voted for a VA that can be integrated into devices or wearable devices like earrings and rings, '*... so, it's not troublesome to "call" the assistant*', said R04.

Nowadays, the Bahasa Indonesia option is only available for Google Assistant users; thus, R01, R03, and R21 proposed a more advanced Bahasa Indonesia skill. Here, one could see a conflict between functionality and privacy, as large data sets generated by an 'always listening' mode could be used to improve the Bahasa Indonesia language skills of the VA, while it would at the same time potentially infringe the privacy of users.

Even though some dreamed of a VA that can support their productivity by having an ability to remind about flight or time schedule, R06, R08, and R12 cared enough to have their VA capability to take care of the user's health. '*It helps us in its capacity, mainly to take care of health*' described R08. It is also noticeable that several users were aware of the potential to be outsmarted by the voice technology that led to their prevention idea, '*do not give a response or react more than what we want, we have control*' by R01 or '*we are still the "brain" and the VA is the assistant which does things as we order/want*' mentioned R12. On the other hand, R21 was more philosophical:

> How the world will react, is it about money-capitalism or ethics? When we open up our data, we create a new business which is way more than 1–2 trillion or even more, even for the government. So, sometimes you need to give up your privacy. It may sound philosophical that people want to protect our society, humanity, but can we really do this in world war capitalism or economy-driven society? I don't think so. I don't see other choices.

At some point, all the interviewees were aware of what they need to sacrifice to have their reliable, dream VA – their data.

5   Discussion

We will discuss the research questions and relate our answers to them to previous research. The first research question was:

RQ1: How do users relate to VA technologies, and what influence might that have on how they relate to the ethics of VA?

The users perceived their assistants as something with the ability to answer questions and do as commanded. Some think of their VAs as friends, while others keep them as just machines; yet, this label has no direct influence on their relationship based on Verbeek's categories. We found that all users had an alterity and embodiment relationship to their VAs since it is indicated that they both describe their VA a means to an end but also as a quasi-other. We suggested that if the alterity relationship has become more prevalent because of the VA technology, this could be the beginning of a shift in how we relate to devices with VA technology installed.

The second research question was:

RQ2: How do users of VA technology perceive ethical issues in VA technology?

As mentioned, 'always listening' is most probably always associated with voice assistant technology. To make the technology 'smart', know when to respond, and respond fast and correctly, the assistant would need to be set to always be ready to answer questions or commands of the owner. Despite the popularity of the notion, it is intriguing that the majority of interviewees never heard about it. They are, then, triggered to believe that it is true that the VA is always listening to them and their surroundings. Some, first, denied it because it will cause inefficiency battery usage, can violate GDPR, or other reasons. Others tried to correlate the connection between that notion and their understanding, or rather guesses, as to how and when the VAs know when to respond or answer to questions. The ability to always listen and directly react when a certain keyword is mentioned means that it is highly likely the voice technology is readily listening to their surroundings. Yet, this also creates an opportunity for the invasion of user privacy.

The awareness of this potential violation, thereafter, brought two contradictory response groups. Some fewer people felt scared, afraid, insecure, paranoid, angry, or concerned when discussing the always-listening feature of VA. Some said that it feels like they are being stalked or bugged, while others decided to be more careful in using the VA and would check the permission for the application on their phones. A user mentioned that she became uneasy about activating the 'always listening' feature and turning it off. She teaches AI, and her knowledge of the 'listening' process of voice assistants gives her the awareness to think further on how to protect her privacy.

On the other hand, the empirical data showed that more users were either comfortable, giving up, neutral, or not caring about that always-listening feature. Some argued that they feel comfortable that the VA company, Google, keeps their data because they have stored many things while using the company's other applications. Others realized that it was uncomfortable and made them feel insecure, but what can they do about it? They still need to use the technology at some point since it is useful and there are no other options. Some also pointed out that it is probably true that the company collected their data, and someone out there listened and learned about their conversation. Yet, nothing has happened, and it has not triggered any criminal action yet. They felt that they were not important people, like any artists or political figures, that would be subjected to any criminal activity just because their conversations were being learned. This focus on the absence of criminal activity is interesting as it indicates that privacy concerns are not that central among many of our respondents as long as it does not lead to tangible criminal activity.

Furthermore, the next research question was:

RQ3: What strategies do users of VA use to handle ethical issues?

Only did one interviewee have taken precautions to protect her personal data from the VA during the study. Even though it is as expected based on their perception of ethical issues surrounding VA, there were some unique suggestions coming from those users.

Two-thirds of interviewees proposed to either uninstall the app, stop using the VA, or use the VA less frequently in order to not share any information with the VA. Some believed that if there were a way to turn off the microphone, it would be very effective to prevent the VA from listening to their conversations. Giving unambiguous consent to the voice technology to listen to their surroundings was another crucial suggestion during the research. This suggestion came from a user who lives abroad with enough exposure to the importance of privacy data protection and GDPR or regulation for data protection. He is living in a country where privacy is a top priority. Another unique idea came from a user who also lives in a European country.

Meanwhile, the other six did not know how and have not done or tried to do something to prevent privacy violations from their VAs. One user emphasized that he trusted his VA to keep his data and not the company who manages the VA technology. He believed that it is an era when data privacy is worthless.

Regardless of how they perceived the voice technology, and how they manage their privacy, overall, those who have some expertise in technology showed potential knowledge in reasoning how technology works. Some extra discussion also arose related to AI and privacy concerns such as decentralized data, anonymous storing processes, and locally stored data. Lastly, the fourth research question was:

RQ4: How do users of VA perceive the desired relationship with VAs in the future?

In general, the users want VAs that can be personalized, understand the context of a discussion, and be way more intelligent and more advanced than now, but still can be controlled and managed, like Jarvis. The ability to speak Bahasa Indonesia was also favoured among some users because until this report was written, only Google Assistant could speak the language among the VA. This may be related to the richness of the local language and the uncommon use of English within the general society in Indonesia. Hence, they were all well aware that they had to push aside their data privacy to make their dreams come true. One user elaborated on his philosophical thought that capitalism or an economy-driven society would be the main engine to give up our data privacy to produce technologies that will make our life easier.

6   Concluding remarks

Our study has indicated that the VA technology is perceived as a quasi-other that many users engage with in an alterity relationship. We also showed that many of our respondents were quite unaware of privacy risks of the technology, rather being influenced by the very interview to think more carefully about this issue. When it came to how to handle privacy risks, some users engaged in strategies of mitigating one's use or changing the settings, while others accepted the potential invasion of privacy. However, when it came to future visions, many wanted a more intelligent and proactive assistant, although this would potentially mean that they would give up their privacy.

On a more general level, we could see how users related to a still undeveloped technology, using it for throwing jokes, asking it to only do what it is able to do and nothing more, and that it often leads to confusions. This goes hand in hand with how the media portrays VA, and other AI (Arifin & Lennerfors, 2022). If one would follow their line of reasoning, could it possibly be that the harmless and undeveloped nature of VA technology is providing more user acceptance for also using the voice medium when interacting with their devices, feeding the service providers with more data and knowledge about oneself? That the VA is a dumb spy? Whether or not this speculation is relevant, it would be interesting for future researchers to follow the development of VA technology as it progresses beyond this undeveloped state towards and beyond the uncanny valley to become the more sophisticated VA technology being an assistance in all areas of life from work to health which is mentioned by our users. As an exercise in self-reflection, perhaps we as authors have also been fooled by VA technologies, since there is no need for a VA in order for a malignant actor to access the microphone on one's devices. And we have heard several stories both from Indonesian and other users about how creepy it is when you are discussing a particular product with a friend, and then when you turn on the phone, an ad for the product shows up, perhaps somewhat like what was showcased by the introductory dialogue between Kimmy and Zach.

As mentioned in the introduction, it is also likely that we have captured an empirical situation where voice or sound is in focus, while in the future the focus might be not on VAs, but on digital assistants that have all forms of sensory contacts with users, not only limited to touch and sound.

We have also noticed incidentally that the interview situation with users of AI technologies serves as an intervention in the sense that it creates awareness of the pros and cons of technology which might be beneficial for the users in how they use technology. In such a situation, it would be highly beneficial to be able to explain to users exactly how the technology works and is used, but although the first author partly had such knowledge, there are also several unknowns in the use of the technology, meaning that one can never have full certainty about the potential privacy invading effects of VA technology. In a sense, users are lingering between trust and conspiracy, and a challenge is to find a middle, reflexive ground. Users not wishing to either unconditionally accept nor reject such technology must therefore navigate an uncertain terrain, but we believe that way to do so is still to create awareness, promote the responsibility of users (rather than just accepting it without even considering options), promote critical thinking and action based on such thinking.

Acknowledgements

This study was supported by the JSPS/STINT Bilateral Joint Research Project, 'Information and Communication Technology for Sustainability and Ethics: Cross-national Studies between Japan and Sweden' (JPJSBP120185411) and Kakenhi(19K12528).

References

Arifin, A. A., & Lennerfors, T. T. (2022). Ethical aspects of voice assistants: A critical discourse analysis of Indonesian media texts. *Journal of Information, Communication and Ethics in Society, 20*(1), 18–36. https://doi.org/10.1108/JICES-12-2020-0118

Arnold, R., Tas, S., Hildebrandt, C., & Schneider, A. (2019). Any sirious concerns yet? An empirical analysis of voice assistants' impact on consumer behavior and assessment of emerging policy challenges. In *TPRC47: The 47th Research Conference on Communication, Information, and Internet Policy 2019*. https://doi.org/10.2139/ssrn.3426809.

Bell, E., Bryman, A., & Harley, B. (2019). *Business research methods.* 5th ed. Oxford: Oxford University Press.

Brandeis, L., & Warren, S. (1890). The right to privacy. *Harvard Law Review, 4*(5), 193–220.

Brill, T., Munoz, L., & Miller, R. (2019). Siri, Alexa, and other digital assistants: A study of customer satisfaction with artificial intelligence applications. *Journal of Marketing Management, 35*(15/16), 1401–1436.

Charters, D. (2002). Electronic monitoring and privacy issues in business-marketing: The ethics of the DoubleClick experience. *Journal of Business Ethics, 35*, 243–254.

Chopra, S., & Chivukula, S. (2017). My phone assistant should know I am an Indian, my phone assistant should know I am an Indian: Influencing factors for adoption of assistive agents, – MobileHCI '17. *Proceedings of the 19th International Conference on*

Human-Computer Interaction with Mobile Devices and Services, pp. 1–8, available at: https://doi.org/10.1145/3098279.3122137.

Cooper, D. R., & Schindler, P. S. (2014). *Business research methods*. 12th ed. New York: McGraw-Hill/Irwin (The McGraw-Hill/Irwin series in operations and decision sciences business statistics).

Cowan, B., Pantidi, N., Coyle, D., Morrissey, K., Clarke, P., Al-Shehri, S., Earley, D., & Bandeira, N. (2017). What can I help you with?: Infrequent users' experiences of intelligent personal assistants. In Proceedings of the 19th International Conference on Human-Computer Interaction with Mobile Devices and Services—MobileHCI '17, Vienna.

Ess, C. (2002). Computer-mediated colonization, the renaissance, and educational imperatives for an intercultural global village. *Ethics and Information Technology*, *4*(1), 11.

Flyvbjerg, B. (2006). Five misunderstandings about case-study research. *Qualitative Inquiry*, *12*(2), 219–245. https://doi.org/10.1177/1077800405284363.

Foehr, J., & Germelmann, C. (2020). Alexa, can I trust you? Exploring consumer paths to trust in smart Voice-Interaction technologies. *Journal of the Association for Consumer Research*, *5*(2), 181–205. https://doi.org/10.1086/707731.

Han, S., & Yang, H. (2018). Understanding adoption of intelligent personal assistants: A parasocial relationship perspective. *Industrial Management and Data Systems*, *118*(3), 618–636.

Huffman, S. (2019). *Here's how the google assistant became more helpful in 2018*. Retrieved 17 February 2020 from www.blog.google/products/assistant/heres-how-google-assistant-became-more-helpful-2018/.

Ihde, D. (1990). *Technology and the lifeworld: From garden to earth*. Bloomington: Indiana University Press.

Ihde, D. (2017). Sonifying science: Listening to cancer. *Nursing Philosophy*, *18*(1), e12152.

Kanemitsu, H. (2019). The robot as other: A post phenomenological perspective. *Philinq, VII* (1–2019), 51–62.

Kiran, A., & Verbeek, P. (2010). Trusting ourselves to technology. *Knowledge, Technology & Policy*, *23*(3–4), 409–427.

Kudina, O., & Coeckelbergh, M. (2021). "Alexa, define empowerment": Voice assistants at home, appropriation, and techno performances. *Journal of Information, Communication and Ethics in Society*, *19*(2), 299–312.

Lau, J., Zimmerman, B., & Schaub, F. (2018). Alexa, are you listening? In *Proceedings of the ACM on Human-Computer Interaction*, Vol. 2, (CSCW), pp. 1–31.

Lennerfors, T. T. (2019). *Ethics in engineering*. Lund: Studentlitteratur.

Lennerfors, T. T., & Murata, K. (2021). Culture as suture: On the use of 'culture' in cross-cultural studies in and beyond intercultural information ethics. *The Review of Socionetwork Strategies*, *15*(1), pp. 71–85. htpps://doi.org/10.1007/s12626-021-00080-x.

Li, X., Hess, T. J., & Valacich, J. S. (2008). Why do we trust new technology? A study of initial trust formation with organizational information systems, *The Journal of Strategic Information Systems*, *17*(1), 39–71.

Liao, Y., Vitak, J., Kumar, P., Zimmer, M., & Kritikos, K. (2019). Understanding the role of privacy and trust in intelligent personal assistant adoption. In Taylor, N. G., Christian-Lamb, C., Martin, M. H., & Nardi, B. (Eds.), *Information in Contemporary Society: 14th International Conference, iConference 2019, Washington, DC, Proceedings*. Berlin: Springer, pp. 102–113.

Lopatovska, I., Rink, K., Knight, I., Raines, K., Cosenza, K., Williams, H., Sorsche, P., Hirsch, D., Li, Q., & Martinez, A. (2019). Talk to me: Exploring user interactions with the amazon Alexa. *Journal of Librarianship and Information Science*, *51*(4), pp. 984–997.

Mayer, R. C., Davis, J. H., & Schoorman, F. D. (1995). An integrative model of organizational trust. *Academy of Management Review, 20*(3), 709–734.

McKnight, D. H., Carter, M., Thatcher, J. B., & Clay, P. F. (2011). Trust in a specific technology: An investigation of its components and measures. *ACM Transactions on Management Information Systems (TMIS), 2*(2), 1–25.

Mennicken, S., & Huang, E. M. (2012). Hacking the natural habitat: an in-the-wild study of smart homes, their development, and the people who live in them. In *Pervasive Computing: 10th International Conference, Pervasive 2012, Newcastle, June 18–22, 2012. Proceedings 10* (pp. 143–160). Berlin Heidelberg: Springer.

Moar, J. (2019). *The digital assistants of tomorrow*. Hampshire: Juniper Research Ltd.

Moorthy, A. E., & Vu, K. P. L. (2013). *Voice activated personal assistant: Privacy concerns in the public space*. M.S. California State University, Long Beach, United States – California. https://search.proquest.com/docview/1513579796/abstract/DFF130DF16554E2FPQ/1

Palm, E. (2016). What is the critical role of intercultural information ethics? In Collste, G. (Ed.), *Ethics and communication: Global perspectives, rowman and littlefield international*, pp. 181–195.

Pradhan, A., Findlater, L., & Lazar, A. (2019). *"Phantom friend" or "just a box with information"*. Proceedings of the ACM on Human-Computer Interaction, Vol. 3 No. CSCW, pp. 1–21.

Rousseau, D. M., Sitkin, S. B., Burt, R. S., & Camerer, C. (1998). Not so different after all: A cross-discipline view of trust. *The Academy of Management Review, 23*(3), 393–404.

Verbeek, P.-P. (2005). *What things do: Philosophical reflections on technology, agency, and design*. University Park, PA: Pennsylvania State University Press.

Verbeek, P.-P. (2015). *Beyond interaction: A short introduction to mediation theory*. Retrieved 22 May 2020 from http://interactions.acm.org/archive/view/may-june-2015/beyond-interaction.

Westin, A. F. (1967). *Privacy and freedom*. New York: Atheneum.

Yang, H., & Lee, H. (2018). Understanding user behavior of virtual personal assistant devices. *Information Systems and e-Business Management, 17*(1), pp. 65–87.

Zeng, E., Marc, S., & Roesner, F. (2017). End-user security and privacy concerns with smart homes. In *Proceedings of the Thirteenth Symposium on Usable Privacy and Security (SOUPS 2017)*. Santa Clara, CA.

Appendix A: List of respondents

ID	Age group	Gender	Highest level of education	Occupation	Career/ educational background in tech	Living alone/ together?	Which VA(s) that you use usually?
R01	26–30	Female	Master's degree	Education, Health, or Science	No	Together	Google Assistant
R02	26–30	Female	Master's degree	Education, Health, or Science	No	Alone	Siri
R03	31–35	Male	Master's degree	Engineering or IT Professional	Yes	Together	Siri, Google Assistant
R04	26–30	Male	Bachelor's degree	Engineering or IT Professional	Yes	Together	Siri, Google Assistant
R05	26–30	Female	Master's degree	Student (Postgraduate, Doctoral)	Little	Alone	Siri, Google Assistant
R06	26–30	Female	Master's degree	Student (Postgraduate, Doctoral)	No	Alone	Siri
R07	26–30	Male	Bachelor's degree	Engineering or IT Professional	Yes	Together	Google Assistant, Cortana
R08	26–30	Male	Master's degree	Student (Postgraduate, Doctoral)	Yes	Alone	Siri, Google Assistant
R09	26–30	Female	Master's degree	Education, Health, or Science	Yes	Alone	Google Assistant
R10	31–35	Female	Master's degree	Education, Health, or Science	Little	Alone	Google Assistant
R11	26–30	Male	Master's degree	Education, Health, or Science	Yes	Together	Google Assistant, Bixby
R12	26–30	Female	Master's degree	Education, Health, or Science	No	Together	Siri, Alexa
R14	26–30	Male	Master's degree	Engineering or IT Professional	Yes	Together	Google Assistant
R16	26–30	Male	Master's degree	Engineering or IT Professional	Little	Alone	Cortana, Google Assistant
R18	26–30	Male	Doctorate	Student (Postgraduate, Doctoral)	Yes	Together	Google Assistant
R20	31–35	Female	Master's degree	Education, Health, or Science	Yes	Together	Google Assistant
R21	36–40	Male	Bachelor's degree	Engineering or IT Professional	Yes	Together	Siri, Google Assistant
R22	26–30	Male	Bachelor's degree	Engineering or IT Professional	Yes	Alone	Siri

Questions

1 How would you be more comfortable to call or think of your VA? (e.g., Machine, helper, friend, assistant, spouse, partner, server, tool)
2 Perceptions and attitudes around 'always listening'
 a Could you explain to me how the VA would know when to respond to a command or question? And how do they proceed with the command/question?
 b Have you heard of the concept of the VA's 'always listening'?
 c How do you feel about the VA 'always listening'?
3 Have there been any instances where you felt uncomfortable around the device? Tell us about a recent time that this happened.
4 Are there any ways/measures you take to protect your privacy from your VA? Imagine
5 If there is no limit in the future, how do you want your VA?

5 Truth and reality in the digital lifeworld

Departure from reductionism

Makoto Nakada, Iordanis Kavathatzopoulos and Ryoko Asai

1 Introduction

Given the digitization of the contemporary world and the increasing reliance on digital technologies ultimately based on mathematics, it is of utmost importance to investigate this increasingly important mathematical foundation of reality. On the one hand, it can be argued that the increasing penetration of mathematical thinking is concurrent with a spread of truth, which pushes out human error, corruption, and injustice. This could imply that we will usher in an increasingly ethical and sustainable society, where processes are optimized, where decisions are evidence-based, and where power differentials and domination will be exposed and ended. On the other hand, digital technologies based on mathematical thinking can be seen as perpetuating already existing unethical practices, given that computer technology is perfectly malleable. Discussions about the different implementations of digital technologies and their ethical and sustainability impact are conducted by several authors in this book. However, in this piece, we take a different route.

Rather, we hold that the ethical and sustainability impact of digital technologies is ultimately dependent on the way in which we view the nature of mathematics. Therefore, we argue that it is crucial to discuss the meanings of truth and reality in mathematics, and their relationship to the technologies that we experience today; computers, robots, and artificial intelligence (AI). The ultimate purpose of this chapter is to problematize the so-called reductionism theory, which we believe is dominant in the research on the information society or human–robot interactions. Reductionism suggests that mathematics, machine learning, and robotics are the realms of *truth* and that our lifeworld is the realm of interpretation of meanings. We discuss this claim from the perspectives of critical information ethics and critical robot ethics, where truth is seen to be socially constructed. In the first half of the chapter (Section 2), we will try to unveil the hidden or invisible presuppositions included in mathematics, computing theory, machine learning, deep learning, robotics, and the research on AI. These presuppositions seem to enable the realms of mathematical methodologies, machine learning, and robot research to appear as realms related to proof of truth and conditions of emergence of reality.

We don't deny that mathematics, machine learning or robotics aims to realize *truth*. However, throughout our analysis, we highlight another aspect related to

DOI: 10.4324/9781003367451-6

the emergence of truth in mathematics. It seems that truth in mathematics is not separated from its procedures of calculation or treating functions. This means that the procedures to make complex functions simpler or the procedures to use an approximation calculation in the case of say machine learning are part of truth in mathematics. 'How does truth in mathematics emerge?' includes these procedures or the intention to employ them.

In short, we will argue that mathematics is a plural matter including plural qualities and presuppositions. It also includes the surprising fact that the relations among the mathematic functions are not fixed and can be rewritten into different and alternative forms. We believe that truth in mathematics can't be separated from the role of functions and that the relation between the function and the solution based on this specific function shows also the relation between the original problem and the answer. However, the transformed relation between the rewritten function and the approximate calculation makes these beliefs unstable and doubtful. This suggests that truth in mathematics is not so stable anymore.

The transformation of mathematics specifically includes the following points. These also show us that there is a difference between *pure mathematics* and mathematics as a tool used in machine calculations or deep learning.[1]

1 In machine learning and deep learning, complex functions are replaced by simple functions and approximations. The relationship between the function (starting point of problem setting) and the solution is not visible. This obscures the meaning of *truth* with which mathematics is concerned.
2 In machine learning and deep learning, numerical values and functions related to *error* are artificially set, and conversely, these artificial setting will set a hypothesized (virtual) structure of overall problems and questions that makes it possible to eliminate error. This setting makes it possible to create, for example, 'machines to judge cat images' and 'robots that can walk by themselves without being taught by humans.' This is said to be truth or *learning*, though 'machine judgment of cat images' is something that humans already know the correct answer about. It mechanically reproduces the relationship between the question and the solution that is already known. It is unclear whether this means the *appearance of truth.*
3 In machine learning and deep learning solutions, once complex functions are reduced to simple functions and formulas, then, again artificially, simplified functions and formulas are combined with a set of *activation functions* as a method for calculations using these simplified functions or approximated calculation. As a result, although it is an approximate calculation, a numerical solution can be obtained within a certain range.
4 The above 'simplified functions or formulas for approximate calculations and a set of activation functions' adjust the relationships between the numerical values in the form of a chain of partial differentials. However, in this method, the relationship between the calculation method and the solution depends on the *specificity* of the method used in the process of calculation (such as the selection of the activation function). In other words, the methodology for obtaining

the solution cannot be generalized. In that sense, truth is neither fixed nor stable (i.e., not a definite and reproducible relationship between questions and answers) anymore.

5 The *specific number* that is the starting point for solving (or involved in minimizing the error as a proxy for the solution) is actually the gradient vector, which is also a normal vector and has (or can have) a relationship with a tangent vector. Such latent relationships between numbers and variables will disappear in the process of approximate calculations.

6 Furthermore, explanations like the above cannot be understood in the closed world of machine learning and deep learning that do not use functions and concepts.

In order to think about these points, it is necessary to reconsider 'what the truth of mathematics is.' We will do this with referring to the arguments of French philosopher Gilles Deleuze and others. This needs to reconsider the preconceived notions showing that mathematics is a fixed truth and doesn't need human subjective interpretation. Furthermore, we have to revise the general frame of thinking about the meanings of truth.

In the latter half of this paper (Sections 3 and 4), based on such critical discussions on the meanings of truth, we will continue to do our effort to make our informatized environments more intelligible to us. We will do this by examining some of the related discussions and the findings of our empirical research. Here, we discuss the possibility of reconsidering the Japanese idea of looking at things and matters in this world in the state of *oneness* or the interaction between objectivity and subjectivity. Linking mathematics to Japanese thought is a difficult task. However, at least according to the research we conducted, the following point was made to be clear: the problems related to technology in the information environment, including the problem of human-robot interaction, would show a different state of appearance of truth depending upon the difference of people's awareness.

Awareness means that we need reflective ways of thinking. What seems to be a fixed reality or truth would turn out to be a kind of combination of human assumption, hope, Vorurteil (prejudice, bias), and so on. As German philosopher Martin Heidegger suggested in his book *Nietzsche* (Heidegger, 1997), the precise and rigorous description in modern Western epistemology is based on the presupposition or partial understanding that 'truth' should appear as a fixed form of proposition ('A is B') in front of our eyes in the state of 'what-is-present' (das Anwesen). Truth is related to knowledge. However, knowledge can take many forms such as technical knowledge (*techne poietike*), scientific knowledge (*episteme*), practical knowledge (*phronesis*), and knowledge of the first principles (*sophia*) (Capurro, 2019).

We also know that various authors in the West and in the East try to devise different forms of knowledge, i.e., different forms of preconditions for emergence of reality and truth. Japanese philosopher Kitaro Nishida said that reality or truth in the world appears as a kind of wholeness called *pure experience*. Pure experience is similar to *direct experience* (Nishida, 1950). Japanese psychiatrist Bin Kimura tried to invert the relationship between the knower and the known or the relationship between objectivity and subjectivity by using Husserl's (1941) terms of *noema* (something related to the known) and *noesis* (some aspects related to the knowing) (Kimura, 1988). According to Kimura, the sounds as matters actually take many forms as noise, message, and part of a melody. Though the sound in a melody is a physical matter, it emerges as a form of linked melody and rhythm. The latter can be experienced by us as a form of emotional experience with expressions such as 'this melody is good.' This expression can make the new relation between noema and noesis. In this case, the expression can create a new wholeness of noema and noesis (Kimura, 1988, pp. 44–47).

This suggests us that the 'principle of identity' of modern Western thinking, which was thought to be self-evident in the fixed form of subjectivity and objectivity, would turn out to be not so stable anymore. This can be rephrased as follows: if we follow Heidegger's explanation: *imagination* (Einbildungskrfat) can determine meanings before *intelligence*. This is the idea that Heidegger regards as decisive in thinking about the meaning of existence, and Kant once recognized it but ultimately rejected it (Heidegger, 2003, p. 192). Husserl (1975) also tried to note the potential reversal of logic and experience.

As we will see later, in Japanese cultural experiences, it seems that the relationship between subjectivity and objectivity is often found to be reversed as reflecting Japanese language structure or Japanese historical experience. That is to say, things are in the form of oneness or un-differentiated. Japanese linguist Motoki Tokieda explained that in Japanese *Ji* (a suffix or a dependent word like a post positional particle) wraps *Shi* (a part of speech representing a concept like a noun) (Tokieda, 2007, pp. 206–207). This means that in many cases the subjective use of *Ji* as something to express one's emotional experience determines *Shi* as something related to the concept.

Japanese philosopher Kitaro Nishida showed a similar idea. He pointed out that the *true* self-reflecting role of the self is not found in the descriptive and prescriptive (normative) role of language related to the position of the self as the subjective in a sentence (Nishida, 1987, p. 145). Rather, the self-reflective role of the subject is found to be in the work of the predicate.

Watsuji (1979) described the 'experience of "the coolness of the wind" by use of a folding fan on a hot summer day.' This case shows 'the wholeness or continuation of matters, time, person, experience and evaluation of this experience.' Here is no dominant position occupied by the subject nor the object. In this case, we exist there along with other entities and the forms of experiences fused with these entities. In other words, this is *Fudo* 風土 (climate and culture) as a form of climate in which we *ex-sistere* (ex-ist or to be there) as an agent of experience of departure from isolated existence (Watsuji, 1979, p. 12).

In this case, to 'ex-ist' means to 'be outside,' i.e., to go outside from the stable relationship between the nature and the culture or between the human-made and the natural. French philosopher Augustin Berque writes that in order to understand Japanese concepts of *Fudo*, we need to understand the term of 'trajectivité (trajectivity).' He writes that 'tra-ject-ivity' is going beyond two 'ject,' i.e., the ob-ject and the sub-ject (Berque, 1992, p. 191).

It can be said that the idea of *Fudo* by Watsuji is close to the idea of Arne Naess's deep ecology (Naess, 1973). Their ideas are apart from human-centrism. Naess's thinking lacks the 'human-centered' tendencies, in contrast to the fact that the current *sustainability* has a strong aspect of 'reversing the de-human-centered' ideas within the original environmental problems (see Wakabayashi, 2004; Hiraki, 2017).

2   Mathematics as a plural matter of 'truth' and human intention

2.1   *The fixed truth to disappear in mathematics and the end of determinism*

As a preparatory work for the discussions here, let us first summarize the aim of the discussions on mathematics in the following passages. We need to view mathematics as a complex thing rather than a static and fixed truth.

Mathematics requires an exchange of viewpoints. The arguments of Deleuze (2007) and Osawa (1992) are to confirm the direction of the anonymous (third-party's) viewpoint in the text of mathematics from one's own viewpoint. Actual mathematics and physics theories, not an abstract story, require such a shift of viewpoint.

According to Wataru Hiromatsu, a Japanese philosopher, in the theory of relativity, when an observer sees from the moving system to which the observer belongs and from another moving system, the appearance of an object (e.g., the arrival time of the light emitted from the center of a train running at a constant speed on the front and back walls of the cabin) is different (Hiromatsu, 1977, pp. 204–224). This means that the theory of relativity requires the existence of multiple observers.

Even in the case of the so-called 'Lorentz transformation equations,' it is necessary to compare these equations with those for the Galilean transformations. A discussion of the 'change of time' in the theory of relativity also emerges from the comparison of these formulas (for this point, see Hiromatsu, 1977, p. 143).

Masachi Osawa, a Japanese sociologist, wrote that the mathematical proposition 'A is B' is usually accepted as something absolute to show the pure *truth* which exists without human existence. However, he said that the proposition 'A is B' in fact includes implicit prescription such as 'you should read this proposition as a fixed rule.' 'A is B' appears to belong only to a specific third party (Osawa, 1992, p. 112), but in reality, it can also belong to 'we' as a concrete existence too. As a concrete existence, we can see that 'A is B' includes and excludes various possibilities. In other words, 'A is B' can also be 'A is not B' (Osawa, 1992, p. 104).

This is similar to what Heidegger (1997) said. In order for 'A is B' to become a rigorous proposition, a fixed category on which the rigorous description can be possible must first be created. However, the proposition 'A is B' itself cannot do this. This is made possible by the creative nature of reason or by the imagination (Einbildungskraft) (Heidegger, 1997, pp. 142–143).

Osawa told that the idea that the logic exists before human observation is a kind of an unawareness (Osawa, 1992, p. 122). He suggested that to become aware of this plurality mentioned above in mathematics needs human *reflective* or *recursive* thinking. This is not included in pure mathematics itself as a form of descriptive formula.

Deleuze pointed out the importance of this kind of human *reflective* or *recursive* thinking in the case of understanding the meanings of mathematics. For example, the ratio used for differential calculation needs this kind of duality to understand it. The ratio represents the *change rate* of a curve and also the point of contact of *truth* and imagination. The ratio requires 'imagination' and also represents 'emergence of theory' to 'replace the rate of change of the curve with the slope of the tangent line.' Metaphors and mathematical formulas work together and are irreducible to one or the other. This is analogous to the relationship between 'reason' and 'schema' (or the power of judgment) suggested by Kant (Deleuze, 2007, p. 135).

The case described below clearly shows us that mathematics has a plural situation. The 'rigorous' formulas can be written in order to rewrite a difficult function with easier functions.

The *truth* contained within mathematics itself does not simply exist there. In fact, Euler's formula, $e^{i\theta} = \cos\theta + i\sin\theta$, has no meaning as long as it remains in the form of a formula. Sometimes it is used in the form of an exponential function containing complex numbers, and sometimes it is used as a trigonometric formula. In such practice, this formula makes sense. However, this complexity is usually forgotten. In mathematics, theory and practice exist in an ambiguous way without being distinguished. In short, here is a misconception that mathematics represents a static and fixed truth.

The changes in mathematics that accompany the advancements of deep learning and machine computing are making things more invisible. These can only deal with discrete variables (digitalized 1 or 0), so they cannot deal with the idea of continuous quantities or limits. Therefore, the derivative cannot be calculated. Instead, they use approximation calculation. In this sense, the fixed relation between the fixed mathematical formula and the answer doesn't exist anymore. Mathematics itself often uses such a rewriting technique called the Fourier transform, which replace complex functions with simple functions. However, the original mathematical formulas themselves don't exist anymore in computing machines.

Seen from this point of view, the idea that mathematics represents the stable truth is no longer viable in our information society. The reductionist thinking to reduce human experience to poor copies from science and mathematics is almost invalid, too.

2.2   *Contrast arrangement of sameness and difference in mathematics*

As we mentioned above, this chapter aims to analyze the so-called reduction-theoretic model of the information society. People usually seem to believe that the fields of mathematics, machine learning, and robotics are the realms of *truth* and that our lifeworld is the realm of interpretation of meanings. However, the conditions for emergence of truth in mathematics are not so clearly defined as people imagine.

Deleuze (2007) pointed out an important matter to the discussions here. He told: *truth* in mathematics includes its potentiality of transformation as the ratio in mathematics shows. The ratio keeps the numerical homogeneity and the contrastability (difference) of the quantity. It shows an appearance of both quality and quantity or extensive quantity (quantity to be measured in materials) and intensive quantity (quantity to make the measurable quantity in materials) (Deleuze, 2007, p. 206). This means that some sort of state which can't be interpreted as quantity neither quality enables the work of the ratio.

We observe similar states in mathematics. In fact, the transformation among scalar quantity, vector outer product, and vector inner product is not a mere change in terms of physical quantities and geometric properties. Similarly, vertical, horizontal, and height in the space are not considered to be as a purely homogeneous quantity, but another property (intensive quantity) works to enable vertical, horizontal, and height to emerge. This means that the vertical, horizontal, and height maintain both homogeneity and difference within them (Deleuze, 2007, p. 163). Without this duality of sameness and difference, no coordinates exist. The vectors can be rewritten as a set of inner and outer products which creates a new set of tangent vector and normal vector (gradient vector), and this gradient vector is then used in deep learning. This rewriting is both mathematical logic and human intention. In this sense, truth in mathematics emerges through this kind of arrangement of mathematical logic and human intention.

Deleuze's understanding of mathematics has been criticized in some cases: his understanding is at the mathematics level of the 19th century; the understanding of the limit of differentiation is confused; the singular point of differentiation is misunderstood in his idea (e.g., see Kondo, 2020).

However, on the other hand, we can imagine that Deleuze tried to reverse the impossible and the possible. The singular point brings into mathematics 'the point of impossibility of differentiation,' yet it is also 'the point that makes differentiation possible.' This is the difference itself. This duality of impossibility and possibility is a logic in a deep sense.

2.3   *Disappearance of continuity of formulas in pseudo-calculation in deep learning*

As we know, solving the differential equation accepts the hypothesis that the rate of change of a curve can be approximated by the slope of the tangent. Although this is a hypothesis, the fulfillment of the hypothesis leads to the fulfillment of other hypotheses. Here is a constant continuity of mathematical formulas and the hypothesis.

However, machine calculations achieve similar calculation results by *forgetting* the continuity of these hypotheses. This causes an interesting and at the same time serious problem. If '$y = 3x + 2$' is differentiated by x, '3' appears as a result. However, '2' as a constant in the original formula remains somewhere after this calculation. In fact, integral calculations make hidden constant(s) reappear. Differential calculation and integral calculation have such a reciprocity.

Osawa (1992, p. 262) put this in an interesting way. He told that by forgetting this 'real 2' as a constant, we can enter the abstract and universal functional relationship. However, in the case of machine learning or deep learning, the original calculation formula does not exist from the beginning. Variables and functions that indicate *error* are artificially created, and based on this, an original calculation formula is virtually constructed. In this sense, '2' as a hidden constant doesn't exist anymore, either.

The mathematical calculation mechanism called gradient descent method using gradient vectors is a method which is used in machine learning and deep learning. This method seeks to find the virtually constructed chain of partial differentiation through the use of artificially constructed *error function*. This method doesn't use the *real* partial differentiation but use the *virtually constructed differentiation* instead. This is because no real functions exist in the world of machine computing. This means that the reciprocity of differential calculation and integral calculation doesn't exist anymore. Similarly, the relation between *original* function and the *rewritten* function doesn't exist anymore, either. This suggests us that we can't follow the process of emergence of truth in mathematics which would appear as the relation between the original function as a question and the rewritten formula as an answer. This makes us face a kind of reversal of reality and virtuality.

As Ohzeki (2021), a specialist of information science pointed out, machine learning is a kind of approximate calculation, and a machine learning system works with the precondition that although we know the relationship between input and output, we do not know what kind of functional relationship makes it possible to connect them. In this case, an approximate calculation is used as a substitute for the function. In other words, this is not learning but an approximate calculation. In a sense, this shows us that mathematics is not the source for truth. Truth exists somewhere beyond mathematical description. Mathematics can only select one of the potential answers. We can imagine that the truth exists as a plural form with plural potentiality to emerge. Truth seems to emerge even through an artificial setting of pseudo-calculation in deep learning. In this sense, the arguments of Deleuze and Osawa are correct. They refer to plurality of reality.

Electromagnetically charged objects create a kind of field called electrostatic potential around them. This potential is a scalar potential, but the field related to the current is a vector potential.[2] This can be processed mathematically by changing the function related to the gradient. However, in approximate function processing, the process of interrelationship between such functions would disappear.

2.4   Sameness and difference in quantum theory and machine learning

What we can know through the discussions above is that the emergence of *truth* in mathematics and theory of computing depends on various conditions including human intention and imagination. In this sense, it seems that we have already reached the exit to leave the so-called techno-determinism. However, in fact, beyond that matter, it seems that the foundations of natural science, technology, and mathematics, which have been regarded as sources of truth, are being shaken.

As Deleuze's discussion of difference and the singular point suggests, the space considered in mathematics is not just a homogeneous space. It contains *differentiable points* and *non-differentiable points*. However, in the world of machine calculation, where formulas themselves cannot be established as continuous ones, such 'points of non-differentiability' disappear. In other words, from the perspective of machine computing, there is no longer a unified mathematical truth.

Furthermore, if it is impossible to rewrite the formula, the quantum theory would disappear. Let us consider a concrete example by taking the Schrödinger equation as a case. The formula used here, e.g., Euler's formula, makes a different aspect appear through the rewriting of it. Specifically, the direction of rewriting mathematical formulas is as follows:

> The exponential function can be rewritten with a trigonometric function. The trigonometric function appears in the form of sine and cosine when differentiated. When the exponential function is differentiated, the constants contained in the exponential can be independently extracted outside the exponential.

The conditions for rewriting such a series of mathematical expressions are used in the Schrödinger equation. In a sense, we experience a situation in which some kind of *intention* involves the practice of formula rewriting, which allows us to witness the emergence of things. In this sense, the electric field energy appears through co-work of mathematical logic and human intention.

In the Schrödinger equation, two characteristics of electron, wave nature and particle nature, are taken out through the transformation of the equation. It can be done, for example, by using Euler's formula. This well-known formula is a formula related to the mutual conversion of trigonometric functions and complex exponential functions as suggested above. The Schrödinger equation makes use of this rewriting.[3] However, the approximate calculation in machine learning can't deal with such rewriting of mathematical formulas to work as functions. Therefore, in a sense, we witness here the crisis of the quantum theory.

2.5   *The robot seen from mathematics with human intention*

On the basis of the discussions so far, we might say that most of the images concerning the robot based on a set of ideas of techno-determinism must be reconsidered. The image of robots such as *independency* and *autonomy* is not meaningful in many cases. Such images need the explanation concerning the conditions with which *autonomic* aspects would appear as in the case of the concept of *learning* of machine learning. This suggests us that we should regard the robot as something to have the potential interaction with the other entities which lie outside mathematical logic. As Heidegger (1960) suggested, most of the entities would emerge through a reciprocating system of appearance and hiding. This means that the entities would emerge not as mere something to follow the principle of cause and effect or the strictly prescribed written system showing the fixed relation between the subject and the object (see also: Heidegger, 2001).

Conversely, in some cases, some aspects of technological matters can be rewritten through the viewpoint on 'in what ways do those matters relate with other matters in our everyday life?' This means that even in the case of the technological function such as the angle of a robot arm, the meanings of the function depend upon its potential role in a broader context outside the fixed role of intelligence and mathematics. In fact, it seems that in the case of the angle of a robot arm, this is considered to be related to the ratio Deleuze (2007) argued. He stated that mathematics involves the work of ratio which is in the state of the intermediate situations as the relation between 'intensive quantity' and 'extensive quantity.' Moreover, the ratio makes *adjustment* possible. The angle of a robot arm can be adjusted through the adjustment of the ratio. Then, *intelligence* or *knowledge* is found to be related not just only to the fixed relation of cause and effect but also to this adjustment. It is clear that this kind of adjustment is also related to the validity of our act of *selection* or the work of *phronesis* as the fluent function of intelligence. This means that the plural functions of the ratio enable the plural states of a robot arm to emerge in plural contexts.

In fact, according to Aoi (2015), in the case of a walking human, the three movements of the hip, knee, and ankle joints have intra-limb coordination that changes together while maintaining a single linear relationship during walking. That is, there is an emergence of autonomous coordinating action in the context before control in the central nervous system. Control of robot arms and joints also needs to be considered from this perspective.

In a sense, this kind of adjustment can be expressed in mathematical formulas. For example, we can express the role of the sole of a robot foot by mechanical description of the plane of the sole, including the tangent and normal vectors. However, this doesn't mean that this kind of expression can help us to solve the problem of 'how can we make the robot walk naturally.' This means that knowledge can't make this kind of adjustment possible, but rather, conversely, 'how to manage to make the robot walk naturally' determines the content of knowledge or the contexts (or *frame* as in the case of frame problem, the conundrum for AI) for intelligence to work. The experience determines the emergence of knowledge or intelligence. This is what Brooks (1986) showed through his model of a robot without intelligence and is also what van Gelder (1995) explained about the weakness of 'the classical computational model.'

3 Japanese, Western, and Eastern views on robots in the cultural and social contexts

3.1 *The robot appears in our awareness*

The discussions so far seem to enable us to be free from the constraints of technological deterministic views. At the very least, we were able to confirm that mathematics and theory of computation are not separated from human intentions and interpretations.

In the next step in this chapter, we need to see the meanings of the cultural and social contexts in which some aspect of the meanings of robots and other technological products is expected to emerge. The authors have already discussed this matter elsewhere (Nakada et al., 2021), finding out that people's views on robots and technological products are correlated with people's views on life in the lifeworld. In what follows, we will further advance our discussions.

It seems that the views on life themselves are not fixed. The encounter with robots, AI, and AI artifacts in general itself gives people a motivation to renew their awareness. One of the findings in this chapter is that even the technological aspect of function of robot is not in the state of stability. As we have considered before, a robot arm is in the state of *adjustment*. In this sense, any robot is not in the stage of finished product. The adjustment of the arm is performed via the *angle*, and this angle is related to our awareness, too.

3.2   *The positions of Japanese views on life and robots in the information environments*

The world is in the intermediate state of appearance and hiding (Heidegger, 1960). Mathematics and the robot are also related with this kind of intermediate states. In this sense, Japanese views on the world, the meanings of lifeworld, and human interaction with technical products, which have sometimes been criticized as a 'strange dualism' of existence and logos or as animism, might be seen on a different horizon.

For example, Kojin Karatani, Japanese philosopher and critic, stated that modern Japanese views on the modern societies reflect the divided views kept in Japanese historical, existential, and cultural situations. People continue to live in the world of historical and cultural existence. Within the scope of these views, people share a traditional and existential view called '*mono no aware* もののあはれ,' i.e., a sensitivity to emergence of the momentary harmony of beauty and the consciousness of its fragility. On the other hand, there has been being a realm of institutionalized and rationalized views. Within this range of the views, people believe that the propositions as logos or science would determine the meanings of life (Karatani, 2001, pp. 160–161). In short, there is a distance between logos and existence in life in Japan in the modernized era.

Karatani told that Kitaro Nishida's position is in a kind of orientation to imagined Japanese Romanticism. This Romanticism is related to search for cultural identity in the modernized environments. Nishida's thought on *Mu* (nothingness) is interpreted to be something referring to Japanese duality, i.e., duality in the sense that orientation to existence would lead to the finding of Romantic homeland. This homeland is located outside the institutionalized or systemized description and prescription; in other words, the modern Western logos and science (Karatani, 2001, p. 161). This is an important explanation on Japanese dualism in a way. On the other hand, however, we might say that this separation would turn out to be a motivation for newer awareness. In a sense, everything in this world needs a newer awareness. If the matters are in the intermediate state of appearance and hiddenness (as Heidegger suggests), they need newer human awareness each time.

Karatani's ideas focus not only on Japan' modernization but also on the Western modernization. He believes that the structure of modern Western logic is not absolute (Karatani, 1978). Western identification is dependent on a linear description system in such a way 'A is B.' However, 'is' related to a linear prescription. This means that this 'is' doesn't allow the existence of different predicates. On the other hand, 'A is not B' is possible at a latent level. This is similar to Osawa's (1992) idea.

3.3   An angle to see the scenes in life in Japan

If 'A is B' is related to 'A is not B,' an experience with new awareness emerges. This is a logic, newer expression, and newer human relationship as well.

According Masakatsu Fujita, a student of Nishida's philosophy, there is a scene in one of the Japanese films in the war time (WWII) which made Augustin Berque, French scholar on Japanese cultures as climate, surprised. That scene depicts an expression of affection by a nurse to a doctor in a situation where the danger is imminent. When asked why she wouldn't run away soon, she said just 'Suki desu' with showing her back to the doctor. 'Suki desu' can be translated into French as 'Je vous aime' (I love you). However, in the original expression, there are no 'I' and 'you.' Fujita pointed out that here is an oneness, or temporary/spatial/semantic continuity, which includes the potential presence of 'I' (nurse) or 'I select to stay with you here.' The audiences can experience this oneness as their own experience and their own angle to see the scene (Fujita, 1998, pp. 117–123). We might say that here is a typical situation to *pure experience* in Nishida's terms (Nishida, 1950). Things can be seen in oneness. In a sense, this oneness or not-articulated situation can be seen in technological products and mathematics too.

This kind of oneness seems to be not limited in Japan. According to Introna (1998), one of the roles of language, the role as combination of expression and autopoiesis, is related to this kind of oneness. Through this kind of oneness, the plural aspects in languages as prescription and *poiesis* would work well.

4   The angle with width seen in empirical findings

4.1   Japanese views on life with the newer awareness in the informatized environments

The important point here is that the ideas discussed above are not just mere speculation. The authors have performed empirical research on Japanese views on the meanings of life in the past years. We have confirmed that the meanings of techno-logical products and also the meanings of life are in the process of newer awareness. This is similar in other cultures and societies as well, in some important respects.

In a way, the informatized environments seem to be full of voices coming from matters themselves just as we have seen in the case of *Fudo* mentioned above. In this sense, *truth* in the information environment seems to be rooted in a variety of sources, not just mathematics and scientific knowledge.

To measure people's attitudes toward technology, AI, robot, and important meanings in life, a survey with 400 respondents (males and females in age of 25–44 living in Fukushima, Miyagi, and Iwate Prefectures) was conducted in Japan in 2020. This survey (the 2020 survey) was designed as quota sampling, and ratios of gender and age were quoted from the official statistical report of the Japanese government about Internet users in Japan. The outcomes of the survey show that technology doesn't prescribe people's minds unilaterally, rather people's orientation to a better and ethical way of life prescribes the potential role of technology in the lifeworld. In phenomenology, a hypothesis is presented that our lifeworld comes first in the sense that the experiences in the lifeworld would determine the meanings of the mathematical and natural science worlds. What we have seen in the research based on the 2020 survey is the fundamental validity of these claims.

Table 5.1 shows the correlation between 'the views on robots and the related matters'[4] and 'the views on life.'[5] The results of the 2020 survey show that: in this world, people were exploring the meaning of their lifeworld in various ways; this orientation to search for 'existential meanings' defined the direction of understanding and interpretation related to the meanings of encounters with robots, self-driving cars, and diagnosis by machines. What is interesting here is that the quest for the meaning of life was linked to a certain kind of orientation to, or empathy with, values such as *mono no aware*, the awareness of meanings of this world as transience and depth. What we have found is that *mono no aware* and similar orientation to meanings of life were found to be strongly or fairly strongly linked to a positive interest in social issues as a whole, including evaluation of robots.

These results are amazing, and the findings were repeatedly confirmed in our previous research (Nakada, 2010, 2011, 2012, 2021). Concerning the point about awareness in life, Capurro (1997), in the tradition of Heidegger and hermeneutics, stated: modernity looked for some stable structure of the knowing subject, some primordial *a priori* categories, according to which we would irrevocably see beings and be ourselves within this kind of stable perspective; in the ancient Greek, the logos is fundamentally thought as a stable one, but this does not mean that this idea should be grasped through stable medium or ways to approach to the truth; in fact, Plato believes that a living *logos* is important; this is related to the work of a *dialogue* which can *co-respond* to the pure definitions or ideas. Capurro (1997) said that our life as an aspect of ethos is also full of this kind of intermediate situations of stability and un-stability.

4.2   *Orientation to plural phases of reality in the West*

Another survey with a purpose similar to the 2020 survey in Japan was conducted in Sweden in 2019. Responses to the survey (the 2019 survey) from 109 males and females who were/had been students at Uppsala University were collected by the authors. Because the number of respondents was limited, we examined the data carefully. Consequently, we found that even in the West, people's views on life extended beyond the reductionist framework. Table 5.2 shows the correlations between Swedish respondents' views on robots and other social problems[6] and the

Table 5.1 Correlations between Japanese people's views on robots and their views on important matters in life (the 2020 survey)

		The views on robots and the related matters						
		Care robots	Education by robots	Requiem service for broken robots	To give a name to a robot	Judgment for death by AI cars	Astroboy's final episode	Importance of smartphone or SNS
The views on life	Lonely death	.373**	.266*	.287**	.344**	.330**	.409**	-.015
	To help people in disasters	.411**	.291**	.326**	.414**	.338**	.412**	-.026
	Places for my memories	.289**	.253**	.258**	.328**	.229**	.318**	.002
	Mono no aware	.346**	.345**	.343**	.417**	.296**	.514**	-.074

** = $p < 0.01$, * = $p < 0.05$, without ** or * = ns = non (statistically) significant.

views on life.[7] *Mono no aware* and similar attitudes toward life were related to the views on robots. People's ways of thinking and feeling about the meanings of life in the informatized environments were supported by their sensitivity to meanings located in a deeper level in the lifeworld.

Table 5.3 shows that there were strong links among Swedish people's sympathies with various matters in life.[8] On the other hand, there were other kinds of link among the meanings or the evaluations in life for Swedish people. 'Judgment on life and death by robots' was found to have a strongly negative correlation with the view 'Automobile driven by robots with AI will enable a better situation in our society to emerge, because it increases safety compared to human driving.' The correlations coefficient was $-.490**$ ($p < 0.01$). This suggests that the horizon on which people's lives unfold includes various values and ethics. This is an acceptable result. As Deleuze (2007) pointed out, this world is developing with differences. It is interesting that while there are rational values, there is also a recursive

Table 5.2 Correlations between Swedish people's views on robots and other social problems and views on important matters in life (the 2019 survey)

		The views on robots and other social problems				
		Care by robots	*To prevent maltreatment for robots*	*Judgment on life and death by robots*	*Loss for choice of education and medical service*	*Interest in environmental problems*
The views on life	*Mono no aware*	.127	.221*	.165	.092	.342**
	Flowers for victims	.109	.201*	.065	.209*	−.094
	Sympathy for Astroboy	.168	.301**	.157	−.060	−.044
	Requiem service for robots	−.216*	.341**	−.120	.192	−.077

** = p < 0.01, * = p < 0.05, none of ** or * = statistically non-significant.

Table 5.3 Correlations among Swedish people's views on and sympathies with important matters in life (the 2019 survey)

The views on important matters in life	*Flowers for victims*	*Meanings of sacrifice*	*Sharing sympathy with co-workers*	*Sympathy for craftsmanship*	*Sympathy for Astroboy*	*Requiem service for robots*
Mono no aware	.309**	.244**	.285**	.506**	.301**	−.049

** = p < 0.01, * = p < 0.05, none of ** or * = statistically non-significant.

perspective that reconsiders the meaning of rationality. Also, as Foucault (1997) suggested, it seems that this world is made up of the exchange of 'rational things' and 'de-rational things' or things that can be seen as 'de-rational' from a narrow reason perspective.

4.3   *People's views on robot in China and their views on life*

We have found a similar tendency in China as well. One of the interesting findings in the survey conducted in 2015, to which the respondents were 300 men and women aged 25–44 living in Beijing, Shanghai, and Guangzhou, is that there were (fairly) strong correlations between 'people's view on "life as a wholeness" including "orientation to nature" or "denial of material happiness as the first and the dominant value"' and 'their views on robot.' This shows us that people in both the East and the West live side by side in a certain layer of life.

5   Concluding remarks

The first half of this chapter and the second half have dealt with the fundamentally similar issue, although it might not have seemed so. The former is related to the objective areas of mathematics and machine calculation, and the latter is related to the subjective area or the *inter-subjectivity*. In many cases, we see logical thinking, mathematics formulas, relations between a cause and an effect in our daily life as something lying outside our awareness or our existence.

However, when we try to pick up reflectively 'how do we see these matters from what kind of angles?,' the matters might start to change their appearance. The first page of phenomenology started when Husserl became aware of the fact that when we see the figure of parallelogram, there are two kinds of ways of seeing (or seeing based on a sort of thematization), i.e., one is of the act of seeing of the parallelogram as an object and the other is of seeing as 'what we see now is one of the appearances of this figure with the potentially plural appearances.'[9] In fact, a parallelogram may be a rectangle or a square when viewed from an oblique angle. This dualism is the starting point of phenomenology (Tani, 2002, p. 63). In this case, thematization, i.e., 'orientation concerning how to see matters in what kind of schema' determines two different aspects of the same matter. These two aspects are equal in relation to truth.

In this sense, imagination and reflexive thinking do not lag behind logic. Here is also the 'primordial or original truth' before the work of logic and experience which is discussed in phenomenology (Tani, 2002, pp. 212–213).

In the case of mathematics, similar things happen. It is we ourselves, who reflectively see the transformation of Euler's formula into exponential and trigonometric functions, but this reflection is only possible on condition that exponential and trigonometric functions are intermediary states. If mathematics deals with 'pre-temporal' and 'fixed truths' that concern only 'truths that are manifesting here and now,' such changes in formulas are impossible. At least, the 'idea' of combining trigonometric functions and exponential functions containing complex numbers by equation does not come from pure mathematics alone. If that is possible in pure mathematics, then, as Deleuze puts it, the 'truth' of mathematics contains 'difference' from the beginning (Deleuze, 2007, p. 23).

Concerning this matter, Merleau-Ponty refers to the intersection of 'the existing and real aspect of life' and 'matters that do not exist in reality but bring possibilities to life' at the level of 'intentionality' (Merleau-Ponty, 1967, p. 192). What we see now as something real is surrounded by the possibility of what we can do about this matter next time. And vice versa, 'what we can do next' changes 'what we see now.' In other words, the world does not exist as a complete, solidified one. For example, a person can point out on another person's arm the stimulation point received on one's own arm. What people are seeing and experiencing right now is not just something that actually exists there but has the meaning of a 'parameter' or 'model.' Here is a duality way of matters or recursive way of existence of matters (Merleau-Ponty, 1967, p. 223).

Mathematics is the act of creating mathematics itself. As Gödel, Cantor, Bertrand Russell, and even Alan Turing himself argue, mathematics also creates the impossibility and non-computability of mathematics, that is, self-denial of mathematics. Some people have also pointed out that the Turing machine, the prototype of the computing machine, actually executes 'calculation of mathematical things by performing a set of restricted and selected rules prepared to start a procedure' and 'executing the procedure of proving mathematics not by knowledge but by calculating itself.' This might be interpreted to mean that machine itself creates mathematics (Gödel, 2006).[10] If this is the case, it is possible to consider that differential calculation by a computer is not just an 'approximate calculation' but the 'realization of differentiation by a set of procedures to create differentiation conditions through calculation.' This means that the machine can create differentiation. However, what deep learning does is not just differential calculations. It is 'a device that mechanically reproduces human recognition of cat images.' Without this act of human recognition, differential calculation in deep learning would not work.

We emphasized the idea of 'oneness' in the descriptions in this chapter, but this 'oneness' itself is not a closed one. According to Berque, the naturalness of Japan, the unity of human existence and nature, essentially includes 'difference.' The pottery by Furuta Oribe pursues beauty, and this beauty includes something irrational to nature too, in the sense that his work emphasizes 'nature as what is produced' rather than 'nature as what is' (Berque, 1992, pp. 258–260). In other words, Japan's 'oneness' includes 'difference.'

In fact, our 2020 survey shows that 'oneness' itself is not a closed schema. We found that there are statistically significant correlations among the following items: 'the attitude shown by the agreement to the view, "I want to give priority to buying eco-friendly products even if the price is a little high,"' 'sympathy for *mono no aware* or social sacrifice,' 'orientation to creation of a society in which we live together,' and 'orientation to social change due to use of robots.' In other words, interest in sustainability, empathy, and orientation toward social change exists in the form of 'oneness' which in fact includes both 'oneness' and differences. The relation between '*mono no aware*' and 'interest in robots' found in our research seems to include some sort of plurality of 'oneness' and difference too.

Berque raises another interesting issue in this regard. The Japanese 'view of nature' and orientation toward 'oneness' has an aspect that was constructed as a kind of national policy in the process of modernization. This seemingly traditional view of nature in the modern era is not rooted in a specific place, and gives rise to an artificial view of nature, a view of the nation that is connected to an abstract 'oneness,' and a production-oriented view. It leads to the severe destruction of nature during the period of high economic growth (Berque, 1992, p. 305).

That is certainly true, but in a sense, it seems that the contradictions and 'differences' contained in this situation challenge the preconceived notion that 'science and mathematics = universal and fixed truths' too. This also leads to the 'original reality' discussed in phenomenology. This reality is also connected to the 'recursive recognition' that phenomenology emphasizes. Our 2020 survey shows the reality of 'oneness' in this sense.

Acknowledgment

This study was supported by the JSPS/STINT Bilateral Joint Research Project, 'Information and Communication Technology for Sustainability and Ethics: Cross-national Studies between Japan and Sweden' (JPJSBP120185411) and Kakenhi (19K12528).

Notes

1 Concerning these points, we have consulted and read a lot of mathematics textbooks as well as the contents of various sites related to mathematics, machine learning, etc., including many online texts of mathematics used in universities. There are so many of them that we can't list them all here.
2 This point is summarized basically relying on the explanation made by Maeno (2022).
3 For this rewriting of mathematical formulas, our understanding is based on the explanation by specialists on chemistry (see Ohtaki, 2016). We also examined the content of Namiki (1992) and others on physics and mathematics.
4 The contents of the views on the robot and others are as follows.

- Care robots: To leave handicapped or elderly persons in the care of robots worsens isolation of them from societies even though this idea seems to be appropriate at first glance.
- Education by robots: It's good to use robots for children's education at school to improve learning effect level.
- Requiem service for broken robots: Just as there is a ritual called a needle memorial service, robots and computers that are no longer in use may be offered similar memorial services.
- To give a name to a robot: It is understandable that if you give a name to a robot and treat it gently and carefully, the heart of the person who handles it will be kind too.
- Judgment for death by AI cars: Autonomous driving robots with AI seem to be convenient, but there is a problem with easy use, considering that machines are left to make decisions about life and death.

- Astroboy's final episode: I am moved when I know Astroboy's final episode as self-sacrifice for saving the earth.
- Smartphone or SNS as important tools for my life: What is important for me is 'to know the movement of the society through smartphones, PCs and SNS (Facebook, Twitter, etc.).'

5 The contents of the views on life are as follows.

- Lonely death: When I hear the stories of lonely death in newspapers, I feel 'this is a very important and serious matter for me,' even if they happen outside my community.
- To help people in disasters: If there are old people nearby when a disaster occurs in my community, I would like to help them evacuate.
- Places for my memories: In my community there are many places to make me remember precious memories.
- Mono no aware: I can sometimes feel that the fireworks or the glow of a firefly in the summer are beautiful because they are transient or short-lived.

6 The contents of the views on meaningful life in this table are as follows.

- Mono no aware: The Japanese think that the summer fireworks and the fireflies are beautiful because they are ephemeral and short-lived and/but because such beauty remains long in memory with its beautiful echoes. This is just like that case that the heart is fascinated by the lingering sound of beautiful music. We can understand such ideas to a certain extent.
- Flowers for victims: When I see a flower bundle offered to the memory of victims in the spot of the traffic accident or the scene of the incident, I sometimes reconsider the transience of the world and want to think about life more deeply.
- Sympathy for Astroboy: I am moved when I know that the final episode of Astroboy, Japanese comics hero, is about his self-sacrifice for saving the earth.
- Requiem service for robots: We sometimes have a kind of feeling to say 'thank you' to our broken robots and computers that we have used for a long time and we can't use anymore just as in the case of Japanese ceremony called 'a needle dedication' to give thanks to a needle that is no longer used.

7 The contents of the views on robots and on the other important problems in the society are as follows.

- Care by robots: To leave handicapped or elderly persons in the care of robots sometimes worsens isolation of them from societies even though this idea seems to be appropriate at first glance.
- To prevent maltreatment for robots: To provide robots with capability of expression of their emotions would be good in order to prevent (avoid) cruelty or maltreatment of robot.
- Judgment on life and death by robots: Although automobile driving robot by AI seems to be convenient at first glance, there might be a problem if that means to leave important judgment on life or death to the machine.
- Loss for choice of education and medical service: In Sweden, education is free from the elementary school to the university and medical expenses are also inexpensive, but in a sense it means to weaken the range of choice from some kind of medical service and education service.
- Interest in environmental problems: To have a strong interest in global environmental problems is important.

8 The contents of the views on important matters in life are as follows.

- Meanings of sacrifice: When I listen to the story of a person who helped others by sacrificing themselves during disasters, I also want to spend my life in a valuable way as suggested by this kind of episodes.

- Sharing sympathy with co-workers: Even though work at the workplace is tough, we can't simply leave the workplace where we have our colleagues struggling together with hard work and I think it is a natural feeling for most of people feeling sympathy to others.
- Sympathy for craftsmanship: When we pick up products carefully made such as watches, toys, dishes etc., we can feel a bit of soul of the maker within them. We are interested in such a kind of traditional Eastern idea that a human-made product reflects the maker.

9 Concerning this matter, see, for example, Husserl (1974, p.180). On this page, he discussed the problem of box perception.
10 This point is explained by Susumu Hayashi and Mariko Yasugi, translators and commentators of Gödel (2006), on p. 264. See also Penrose (1999, p.180).
11 We have examined the contents of printed text books and lecture notes (online version) on mathematics and physics. These are so numerous that we cannot name them here.

References[11]

Aoi, S. (2015). ヒトの適応的な歩行を生み出す低次元構造と感覚－運動協調：運動学・筋シナジーと位相リセットの機能と応用 [Low-dimensional structure and sensory-motor coordination that produce adaptive gait in humans]. 計測と制御 [*Journal of the Society of Instrument and Control Engineers*], *54*(4), 278–283. https://doi.org/10.11499/sicejl.54.278

Berque, A. (1992). 風土の日本 [*Japan in Fudo*]. Chikumashobo. (Berque, A. (1986). *Le sauvage et l'artifie*; *Les Japonais devant la nature* [*The savage and the artifice*; *The Japanese in front of nature*]. Gallimard.)

Brooks, R. A. (1986). A Robust layered control system for a mobile robot. *IEEE Journal of Robotics and Automation*, *2*(1), 14–23. https://doi.org/10.1109/JRA.1986.1087032

Capurro, R. (1997, March 19–21). *Stable knowledge*. Presented at the Workshop: Knowledge for the Future—Wissen für die Zukunft, Brandenburgische Technische Universität Cottbus, Zentrum für Technik und Gesellschaft.

Capurro, R. (2019). Ethical issues of humanoid-human interaction. In A. Goswami, & P. Vadakkepat (Eds.), *Humanoid robotics*: *A reference*. Springer, 2421 2435. Original article available at http://www.capurro.de/humanoids.html

Deleuze, G. (2007). 差異と反復 下 [*Difference and repetition 2*]. Kawade Shobo. (Deleuze, G. (1972). *Différence et Répétition* [*Difference and repetition*]. Presses universitaires de France.)

Foucault, M. (1997). 精神疾患とパーソナリティ [*Mental Illness and Psychology*]. Chikuma□Shobo. (Foucault, M. (1954). *Maladie mentale et personnalité* [*Mental illness and psychology*]. Presses Universitaires de France.)

Fujita, M. (1998). 現代思想としての西田幾多郎 [*Nishida Kitaro in the modern thoughts*]. Kodansha.

Gödel, K. (2006). 不完全性定理 [*Gödel's incompleteness theorems*]. Iwanami Shoten. (with commentary by Hayashi, S. and Yasugi, M.) (Gödel, K. (1931). Über formal unentscheidbare Sätze der Principia Mathematica und verwandter Systeme I [On formally undecidable propositions of principia mathematica and related systems]. *Monatshefte für Mathematik und Physik*, *38*, 173–198.)

Heidegger, M. (1960). *Der Ursprung des Kunstwerkes* [*The origin of the work of art*]. Reclam.

Heidegger, M. (1997). ニーチェ2 [*Nietzsche 2*]. Heibonsha. (Heidegger, M. (1997). *Nietzsche II (1939–1946)*. (Gesamtausgabe VIII). Verlag Vittorio Klostermann.)

Heidegger, M. (2001). *Sein und Zeit* [*Being and time*]. Max Niemeyer Verlag.

Heidegger, M. (2003). カントと形而上学の問題 [*Kant and the problem of metaphysics*]. Iwanami Shoten. (Heidegger, M. (1991). *Kant und das Problem der Metaphysics* (Gesamtausgabe Band 3) [*Kant and the problem of metaphysics*]. Vittorio Klostermann.)

Hiraki, T. (2017). アルネ・ネスのディープ・エコロジー再考 ― 存在論からケア倫理の方へ ― [Rethinking of Arne Naess's deep ecology]. *Bulletin of the Faculty of Humanities and Social Sciences, Iwate University, 101*, 71–93.

Hiromatsu, W. (1977). 科学の危機と認識論 [*Crisis of science and epistemology*]. Kinokuniya Shoten.

Husserl, E. (1941). 純粋現象学及現象学的哲学考察(上)[*Ideen 1*]. Iwanami Shoten. (Husserl, E. (1950). *Ideen I: Ideen zu einer reinen Phänomenologie und phänomenologischen Philosophie, Erstes Buch: Allgemeine Einführung in die reine Phänomenologie, Halbband: Text der 1.-3, hrsg. von K. Schuman, Husserliana, Bd. III. [*Ideen I: Ideas for a pure phenomenology and phenomenological philosophy, first book*]. M. Nijhoff.)

Husserl, E. (1974). 論理学研究 3 [*Logical investigations 3*]. Misuzu Shobo. (Husserl, E. (1928). *Logische Untersuchungen. Zweiter Band; Unterzuhungen zur Phänomenologie und Theorie der Erkentnis. I. Teil. Vierte Auflage* [*Logical investigations, second vol.; Investigations into the phenomenology and theory of knowledge*]. Max Niemeyer.)

Husserl, E. (1975). 経験と判断 [*Experience and judgment*]. Kawade Shobo Shinsha. (Husserl, E. (1972). *Erfahrung und Urteil: Untersuchungen zur Genealogie der Logik* [*Experience and judgment: Studies in the genealogy of logic*]. Felix Meiner.)

Introna, L. D. (1998). Language and social autopoiesis. *Cybernetics and Human Knowing, 5*(3), 3–17.

Karatani, K. (1978). マルクスその可能性の中心 [*Central point of the potentiality of Marx*]. Kodansha.

Karatani, K. (2001). ＜戦前＞の思考 [*Thoughts before war*]. Kodansha.

Kimura, B. (1988). あいだ [*Between-ness*]. Koubundou.

Kondo, K. (2020). ドゥルーズが『差異と反復』で言及していた数学はどのようなものであったのか、そしてそこにドゥルーズは何をみていたのか [Deleuze's understanding of mathematics in Difference and Repetition]. *Cultural Science Reports of Kagoshima University, 87*, 1–26.

Maeno, M. (2022). 物理学概論 [*A lecture note on physics at the Faculty of Science*]. Ryukyu University. http://www.phys.u-ryukyu.ac.jp/~maeno/cgi-bin/pukiwiki/index.php?FrontPage

Merleau-Ponty, M. (1967). 知覚の現象学 I [*Phenomenology of perception 1*]. Misuzu Shobo. (Merleau-Ponty, M. (1945). *Phénoménologie de la perception* [*Phenomenology of perception*]. Presses Universitaires de France.)

Naess, A. (1973). The shallow and the deep, long-range ecology movement: A summary. *Inquiry, 16*, 95–100. https://doi.org/10.1080/00201747308601682

Nakada, M. (2010). Different discussions on roboethics and information ethics based on different cultural contexts (*BA*). In F. Sudweeks, H. Hrachovec, & C. Ess (Eds.), *CATaC'10 proceedings: Cultural attitudes towards technology and communication 2010*. Murdoch University, 300–315.

Nakada, M. (2011). Ethical and critical views on studies on robots and roboethics. In M. P. Decker, & M. Gutmann (Eds.), *Robo- and informationethics: Some fundamentals*. Lit Verlarg, 157–186.

Nakada, M. (2012). Robots and privacy in Japanese, Thai and Chinese cultures: Discussions on robots and privacy as topics of intercultural information ethics in 'Far East.' In F. Sudweeks, H. Hrachovec, & C. Ess (Eds.), *CATaC'12 proceedings: Cultural attitudes towards technology and communication*. Murdoch University, 200–215.

Nakada, M. (2021). The Orientation to oneness of technology and meanings of life by people in Japanese technological environments. In J. Pelegrín Borondo, M. Arias Oliva, K. Murata, & A. M. Lara Palma (Eds.), *Moving technology ethics at the forefront of society, organizations and governments*. Universidad de La Rioja, 463–474.

Nakada, M., Kavathatzopoulos, I., & Asai, R. (2021). Robots and AI artifacts in plural perspective(s) of Japan and the West: The cultural-ethical traditions behind people's views on robots and AI artifacts in the information era. *The Review of Socionetwork Strategies, 15*(1), 143–168. https://doi.org/10.1007/s12626-021-00067-8

Namiki, M. (1992). 量子力学入門—現代科学のミステリー [*Introduction to quantum mechanics*]. Iwanami Shoten.

Nishida, K. (1950). 善の研究 [*Studies on goodness*] (the original version was published in 1911). Iwanami Shoten.

Nishida, K. (1987). 哲学論集1 [*Collection of philosophical essays 1*]. Iwanami Shoten.

Ohtaki, M. (2016). 基礎化学結合論 [*A lecture note on basic chemistry at Interdisciplinary Graduate School of Engineering Sciences*]. Kyushu University. http://www.asem.kyushu-u.ac.jp/~ohtaki/wp-content/uploads/2015/05/H28chemicalbond03.pdf

Ohzeki, M. (2021). 機械学習における物理が果たす役割——量子機械学習とその現状 [The role of physics in machine learning: Quantum machine learning and its current situation]. 日本物理学会誌 [*Journal of the Physical Society of Japan*], *76*(4), 194–201.

Osawa, M. (1992). 行為の代数学—スペンサー=ブラウンから社会システム論へ [*Algebra of action*]. Seidosha.

Penrose, R. (1999). 心は量子で語れるか [*Is the mind explained by quantum theory?*]. Kodansya. (Penrose, R. (1997). *The large, the small and the human world*. Cambridge University Press.)

Tani, T. (2002). これが現象学だ [*This is phenomenology*]. Kodansha.

Tokieda, M. (2007). 国語学原論（上）[Basic course for the study of Japanese language (1)]. Iwanami Shoten.

van Gelder, T. (1995). What might cognition be, if not computation? *The Journal of Philosophy, 92*(7), 345–381.

Wakabayashi, A. (2004). アルネ・ネスの環境哲学—ディープ・エコロジーとエコソフィ [Arne Naess's environmental philosophy]. *The Journal of Chiba University of Commerce, 42*(3), 335–352.

Watsuji, T. (1979). 風土 [*Climate*]. Iwanami Shoten.

6 Telework for a sustainable society

Lessons from the remote work boom during the COVID-19 epidemic in Japan

Hiroshi Koga, Akio Sato and Sachiko Yanagihara

1 Introduction

This chapter discusses the possibilities of transforming the way of working into ethical and sustainable one using information and communication technologies (ICTs). The discussion will focus on telework—remote work or work from home (WFH), in particular—in Japan that gained attention during the COVID-19 epidemic in the country as a way to secure business continuity. As is generally known, the Japanese work culture is characterized by long working hours, long commutes, and consequently, a small amount of time for family and private life, making it difficult for individual workers to achieve a work and life balance. When the Japanese corporate culture was praised in the 1980s, it was encouraged to conform individual workers to such work culture. However, it is now recognized as an issue that must be overcome in order to achieve higher individual, as well as corporate, productivity. Remote work is seen as a potential solution to alleviate many of the problems Japanese organizations face caused by their conventional business practices. This chapter attempts to explore the sustainability implications of remote work, in particular. To this end, the authors conducted a diary survey on remote work practices during the COVID-19 epidemic in Japan.

Before giving an overview of the survey, we would like to review the background of the demand for change in the conventional ways of working. Needless to say, many researchers and business leaders have pointed out that a new way of working is required in the new era. We focus on two representative researchers: Charles Handy and Linda Gratton.

Charles Handy, an Irish philosopher, once noted that the era of full-time jobs is over (Handy, 1989). He referred to the new era as characterized by the shamrock organization, which consists of three types of work arrangements: (1) full-time employees consisting of core personnel and professionals, (2) temporary contract employees, and (3) part-time employees. He further highlighted the phenomenon of reduced working hours (which he predicted would decrease to half of what they were at the time) as another characteristic of the new era. He argued that these two characteristics would lead to a diversification of work styles in the new era: core personnel working thick and short, part-time workers working thin and

DOI: 10.4324/9781003367451-7

long, and those returning to work after a period of interruption for childcare or nursing care.

On the other hand, Lynda Gratton, a British organizational theorist, argued that new ways of working will be essential, focusing on the increase in healthy life expectancy as a feature of the new era (Gratton & Scott, 2016). She noted that we are now in a time when people live to 100 years old and that they will need to work until they are 75–85 years old. However, she also pointed out that people should not end their long lives working. Therefore, she argued, it has become important to spend a long and prosperous life by enjoying multiple stages of work. Furthermore, she emphasized that the key factor for this new way of life is intangible assets, which specifically refer to the following:

1 productivity, the ability to create something in society, which requires constant learning;
2 vitality, which involves physical and mental health and a person-centered life;
3 transformation, which refers to the ability to create a different self.

Behind Handy's and Gratton's arguments, there is a hypothesis that if companies realize work styles that contribute to work-life balance, they will be able to secure a diverse workforce, and employers will be able to lead richer lives. Simply put, a logic other than economic efficiency is indispensable for leading a prosperous business life. Such a logic could be said to be oriented toward "sustainability" and "ethics."

Telework (in English-speaking countries, the terms "mobile work" and "work-from-home" are usually used, but in this chapter, as will be explained later, "telework" will be used) has attracted attention as a way of working in line with such a new logic. That is, telework is expected to adapt to a business environment characterized by volatility, uncertainty, complexity, and ambiguity (VUCA), as a result of the flexibility of the working style. Not only that, it has been expected to be an enabler of sustainable and ethical development of business organizations.

Despite these high expectations for telework, however, its introduction had been slow in Japan. Disincentives such as "lack of suitable jobs for telework," "concerns about security and other environmental factors," and "difficulty in managing invisible subordinates" were generally highlighted as reasons for this slow introduction (Japan Telework Association, 2007).

However, the COVID-19 pandemic that shook the world in 2020 led to the rapid spread of telework in Japan. Before the pandemic, the telework penetration rate in the country was only 17.6%. By April 2020, however, 55.9% of companies had adopted telework (Tokyo Shoko Research, 2022). However, once the reduced risk of infection from COVID-19 became widely recognized, 27.2% of companies stopped telecommuting, and a telecommuting implementation rate went down to less than 30% (Tokyo Shoko Research, 2022). These trends indicate that teleworking during the COVID-19 pandemic was mainly used as a means of business continuity in the crisis situation. Notably, the benefits of conventional telework, such as improved work-life balance, were overlooked. In other words, the emphasis on the aspect of measures for preventing the spread of infection has probably weakened the original significance of telework. Of course, however, that significance has not been lost.

We do not deny the concept of telework as a means to realize such a business continuity plan. However, we are concerned that the practice of telework as a measure to prevent the spread of infection (developed in response to the government's request to refrain from leaving the house unnecessarily) is biased in favor of performing duties online and, if anything, overlooks the discussion of what kind of work style is desirable. Of course, in the face of a pandemic, it was important to consider how to prevent it, there was no time to discuss the "ideal way of working," and companies were forced to implement measures to prevent the spread of the disease. Nevertheless, if telework is to be understood as a new form of work practice, questioning the way we work should be just as important as practicing it.

Therefore, in this chapter, we would like to emphasize the importance of telework as a new way of working. To this end, the remainder of the chapter is organized as follows. In the next section, an overview of the definition and current status of telework in Japan are described, and the situation of telework under the COVID-19 pandemic and the problems identified therein are examined. In Section 3, a tentative discussion of the potential of telework for achieving a sustainable society will be conducted.

2   Current status of telework in Japan

The most widely used definition of telework in Japan is that of Japan's Ministry of Internal Affairs and Communications (MIC). They defined telework as "a flexible work style that uses ICT and enables effective use of time and place" (MIC, 2019a). In this very simple definition, there are two key words: flexible work styles and ICT.

2.1   Significance of telework

The most significant feature of telework is that ICT has made it possible to work in locations other than the original office. As a result, it has become possible to realize a way of working that is free from the constraints of time and space.

The term "telework" covers a wide variety of activities. Therefore, we categorized telework into six types according to employment status and work locations. Employment status is classified into two types: (1) employees and (2) self-employed workers. Work locations are classified into three categories: (1) home, (2) a fixed location outside one's own home, and (3) unspecified locations outside one's own home. Combined, these can be typified as shown in Table 6.1.

However, the focus of the discussions on telework during the COVID-19 pandemic was on WFH or telecommuting. Needless to say, the reason for this focus (rather than on working at home as a self-employment worker, which tends to be discussed as a way of working during the start-up phase) was to ensure business continuity during the pandemic and to prevent the spread of infection. In the following subsections, we will discuss the impact of COVID-19 on teleworking, focusing on employees.

Table 6.1 Types of telework

		The location of telework		
		At home	Outside home	
			Fixed location (Outside the original office)	Diverse locations (e.g., where to move to)
The type of employment relationship	Employees	Working from home Telecommuting	Satellite office Shared office	Mobile work
	Self-employment workers	Working at home	Shared office	Self-employed Mobile work

Source: Drawn by the authors.

2.1.1 Benefits of telework

Simply put, telework is characterized by the elimination of commuting and greater flexibility in work hours. This implies liberation from two features of industrial society. Namely, work is performed based on the time set by the clock, and workers have to spend time on the activity of commuting, which was created as a result of the separation of residence and workplace. In particular, in modern society, people who have sought housing in the suburbs are now forced to commute for long hours. In addition, traffic congestion in urban centers has become so intense that commuting is a very painful and arduous activity. For this reason, telework, which allows people to avoid commuting, sounded attractive to many people before the COVID-19 pandemic. It was especially appealing to elderly people and people who were physically challenged, as well as to those with nursing care or childcare issues, as it enabled them to work continuously.

Freedom from commuting brings diverse benefits. In addition, greater freedom in work hours tends to promote so-called autonomy. Increased autonomy allegedly leads to higher productivity and effectiveness. Therefore, various advantages of telework have been acknowledged, as shown in Table 6.2.

2.1.2 COVID-19 and telework

Because of these various advantages, the Japanese government has positioned telework as an effective means of overcoming the stagnant economic situation and has actively promoted this practice since 2000. Nevertheless, the penetration rate of telework has remained at a low level of approximately 10% in December 2019. However, as mentioned above, the COVID-19 pandemic led to its rapid spread. The driving force behind this was the Japanese government's request for the implementation of telework.[1] As a result, the prevalence of teleworking quickly increased. As the risk of infection gradually decreased, however, there was a movement to return to traditional ways of working due to "communication problems" and other

Table 6.2 Advantages of telework

Benefits for companies	Benefits for employees
• Secure and develop human resources • Support business process innovation • Reduce business operation costs • Ensure business continuity (BCP) in case of emergency • Improve business competitiveness by strengthening collaboration within and outside the company • Secure and develop human resources, reducing job turnover, and providing support for continued employment • Enhance corporate brand and images	• Improvement of work-life balance • Productivity improvement • Autonomous and self-managed work style • Strengthening of cooperation with workplaces • Increase in overall job satisfaction and work ethic

Source: Drawn by the authors based on MHLW (2016).

Table 6.3 Types of telework experienced by respondents (multiple answers allowed)

	Before the pandemic (%)	As of April-May 2020 (%)	As of October 2021 (%)
Telecommuting	27.0	45.4	35.2
Mobile work	5.3	6.1	5.5
Satellite office work	5.1	4.7	5.7
Other	0.2	0.6	0.2
Never engaged	71.2	52.6	63.0

reasons.[2] These trends were found in our previous work as shown in Table 6.3 (Koga et al., 2022).

This "swing back" in the introduction of telework in Japan should not be understood as an orientation toward the restoration of daily life due to the reduced risk of infection or as an aversion to mandatory implementation. Rather, it can be seen as the true factor that inhibits the practice becoming apparent in a situation where teleworkers are forced to execute a voluntary curfew of going out. Specifically, Goto and Hamano (2020) pointed to a sense of anxiety and loneliness arising from the loss of opportunities for face-to-face conversations with colleagues and a "*HANKO* (personal seal) culture" that requires employees to come to work to stamp documents.[3] In addition, there was a (often sterile pseudo-mannerism) debate in the street as to whether the faces on the screen should be in order of position, or whether people should leave the meeting after those in higher positions leave, if meetings could be conducted only online.

These raise the following questions:

1   Was telework during the COVID-19 pandemic "telework as an emergency measure" and "nothing like" conventional telework?

2 Did the mirage of excessive expectations for telework fizzle out, and did its true nature emerge, through the COVID-19 pandemic?
3 Did the ideal form of telework emerge, through the experience of the self-restraint period (regardless of success or failure)?

The first question emanates from a place of positioning the conventional image of telework as an ideal and criticizing the inadequacy of reality. The second question comes from a cynical standpoint that the reality of telework is not as good as expected. The third question comes from the social constructionist position that a new image of telework has emerged through practice. Which question or position is more realistic? In the next subsection, we will consider the results of a survey conducted during the COVID-19 pandemic as preparatory work for examining these questions.

2.2   *Current status of telework based on survey results*

We obtained responses via an online survey vendor from monitors registered with the same vendor. The final number of respondents was 508. The survey period was November 8–14 in 2021, during the "time when the number of newly infected cases bottomed out," between the "5th wave" and "6th wave" of the COVID-19 disaster in Japan.

The respondents were male (262 of 508; 51.6%) and female (246 of 508; 48.4%), and in terms of employment status, 400 (78.7%) were full-time employees and 108 (21.3%) were part-time employees (including 81 dispatched or contracted workers and 27 part-time workers). The most common type of employment was clerical (180 of 508; 35.4%), followed by technical (95 of 508; 18.7%). Excluding side jobs, on average, 42.4 hours were worked each week, which is approximately the same as the national average for full-time workers (42.5 hours). In addition to their main job, 64 respondents (12.6%) were engaged in a second job.

Almost without exception, existing telework surveys (cf. MIC, 2019a, 2019b; MLIT, 2021) ask respondents to indicate their "telework hours per month." However, responses to such questions inevitably contain errors due to assumptions made by respondents. In this survey, we attempted to minimize such errors by limiting the question to "yesterday's activities and location." Thus, we were able to accurately determine the duration of activities by using a time-of-day survey method that asked respondents to respond in 15-minute increments regarding their activities during the survey period in the places where they performed the activities.

2.2.1   *Fact/data that previous studies missed*

In this study, those who engaged in telework for at least 15 minutes per week, either at their main or secondary jobs, were considered teleworkers, accounting for 53.3% (271 of 508) of the total number of respondents. This percentage is notably higher than in the previous surveys.[4]

There are two implications of the high teleworking rate in our survey results. The first is that short-time teleworkers have been overlooked in existing surveys. If the survey asked respondents how many hours per week they teleworked on average, those who worked less than one hour per week would be unlikely to answer the question. We asked respondents to record their activities in 15-minute increments in a diary format. As a result, those who were engaged in telework at their main job worked 62 hours per week for the most frequent teleworker, and 15 minutes per week for the least frequent teleworker. Telework hours (5,142.8 hours) accounted for 48.5% of the respondents' total hours worked at their main jobs (10,604.8 hours). Among them, 9.2% of all teleworkers worked less than one hour per week.

Secondly, there were uncompensated teleworkers. In Japan, teleworking has frequently tended to increase "unpaid overtime" (Koga & Sato, 2022). The reasons that have given for this include "the difficulty of time management, which tends to result in the erosion of daily living hours," and "the fact that while working hours have been reduced, the amount of work assigned has not decreased, leading to a situation in which work cannot be performed without unpaid overtime." Our survey results show that 250 of the respondents were engaged in telework at their main job. Of these, 235 were compensated by telework in their main job. Conversely, 25.5% (69 of 271) were engaged in unpaid telework. The average duration of unpaid telework was 3 hours for the corresponding group. Such unpaid telework hours are considered to have never been explicitly indicated in the data, although their existence had been pointed out.

The adoption of the diary survey research strategy contributed to our ability to shed light on such a "phenomenon that has been overlooked in conventional surveys" and visualize it as data.

2.2.2   *Features of telework during the COVID-19 pandemic*

An outstanding feature of teleworking during the COVID-19 pandemic was the preponderance of WFH types; prior to the COVID-19 pandemic, the "mobile work type" was generally the most common form of telework (MIC, 2019a, p. 13; MIC, 2019b, p. 50).

The period when our survey was conducted was between the fifth and sixth waves of the COVID-19 disaster, a time when WFH, which was recommended as an infection control measure and rapidly spread, is said to have decreased as the number of new infections declined. However, survey results indicate that even when infection was at its bottom, the majority of teleworkers were still engaged in WFH. Since the pandemic, a general awareness of "telework means WFH" had spread, and the results of our survey confirmed this awareness. Specifically, WFH hours (total WFH hours for respondents during the survey period was 4,528.3 hours) accounted for 88.1% of the total telework hours (5,142.8 hours) during the survey period for those who teleworked at their main job (250).

The second characteristic of teleworking during the COVID-19 pandemic was evident in the number of days in the office per week. The most common response was "1 day" (69 of 250; 27.6%), and the next most common one was "5 days" (60 of 250; 24.0%). A similar trend was seen when limited to regular teleworkers (1 day: 58 of 218, 26.6%; 5 days: 50 of 218, 22.9%). The figure of the "five-days-a-week teleworkers" is interesting when compared with the results of the Telework Population Survey in Japan by the Ministry of Land, Infrastructure, Transport, and Tourism (MLIT) (MLIT, 2021). According to the results of it, (1) comparing the survey results for the four years from 2016 to 2019, the percentage of people who telework (WHF, satellite work, or mobile work) at least one day a week was increasing (Increased from 60% to 78% of all teleworkers). (2) From 2020, the percentage of those who practiced telework 2–4 days a week increased, and in the 2021 survey, the percentage of those who practiced more than once a week increased to about 78%. (3) Comparing the survey results for the four years from 2016 to 2019, the average number of telework days for employment-type workers who telework one day or more a year had remained around two days a week. (4) According to the survey results from 2020 onwards, that number had increased to 2.4 days a week.

On the other hand, the results of our diary survey showed an average of three days a week. According to the survey results of MLIT (2021), the percentage of teleworkers working five days a week was 15.1% in the 2020 survey and 17.2% in the 2021 survey. It may be wild to attribute the difference in the average number of days to the "5 days per week" response rate. It may also be criticized as being due to the representativeness of the sample and errors. Nevertheless, the diary survey would suggest that the average number of days engaged in teleworking had increased due to the experience of the COVID-19 pandemic. This is another feature of teleworking during the pandemic.

However, those who engaged in telework five or more days a week were not necessarily "full-time teleworkers," who engaged in telework for all of their working hours (it refers to those who never showed up for work during the week); some of the respondents who teleworked five or more days a week were "at work" during the survey period. This shows that they were not only fulfilling their duties by teleworking but also were engaged in teleworking on the days they were at work. On the other hand, there were also those who achieved a full-time teleworkers status, working during the survey period and not going to work at all. The number of full-time teleworkers was 46, corresponding to 9.1% of all respondents (508), 17.0% of teleworkers (270). From these, it can be understood that teleworking during the pandemic was the first step toward establishing a full-fledged telework environment. In turn, it can also be seen that even with telework during the pandemic as a measure to prevent the spread of infection, it was difficult for workers to achieve full telework, and they were effectively forced to go into the office.[5]

The survey results revealed a third characteristic of teleworkers during the COVID-19 pandemic; when working at home, they engaged in work that did not use ICT. The average number of hours worked by full-time teleworkers was 42.6 hours per week, which was hardly different from other teleworkers and non-teleworkers. However, during working hours, full-time teleworkers completed an average of 40.2 hours of telework, indicating that they were engaged in work other than telework during their working hours. This temporal difference is likely because respondents did not regard non-ICT work outside of their main office, such as at home, as part of teleworking. For example, that the respondents spent time reading paper documents and other materials while working at home and did not count such non-ICT work as their teleworking may have caused the difference in those average hours. However, as long as respondents' work was conducted outside the office (at home, in a satellite office, on the road, etc.), it should be included in the "telework" category, regardless of whether ICT was used or not. This is connected to the definition, and people's perception, of telework, which can be considered after investigating the reality of non-ICT work outside the office.

2.3   Characteristics of telework during the COVID-19 pandemic

In our survey, we asked people engaged in telework at their primary job to choose one of several alternatives regarding the merits of telework. We asked about the disadvantages as well.

2.3.1   Benefits of telework during the COVID-19 pandemic

Overall, the most frequently cited benefit was "I have more free time (for hobbies, self-development, etc.)" (64 of 250; 25.6%). By gender, however, female respondents ranked it first (37 of 122; 30.3%), while male respondents ranked second (27 of 128; 21.2%) following "increased sleep" (29 of 128; 22.7%) by a narrow margin.

"Increased sleep" was most frequently chosen by full teleworkers (10 of 46; 21.7%). Nevertheless, by number of days worked, it was selected first by those who worked 2 days a week (7 of 30; 23.3%), second by those who worked 5 days a week (12 of 51; 23.5%), and even by one person among the three who worked 7 days a week. This suggests how teleworkers were experiencing an increase in sleep time.

Similar to the previous survey results (cf. MIC, 2019a, 2019b; MLIT, 2019), a higher percentage of female respondents (15 of 122; 12.3%) than male (9 of 128; 7.0%) selected "more time for housework." Women were also more likely than men to select "ability to cope with unexpected situations (children's illness, transportation delays, earthquakes, etc.)" (female: 15 of 122; 12.3%; male: 6 of 128; 4.7%).

Conversely, the item with the higher selection rate for male than women was "parenting time." While 4.7% of men (6 of 128) selected this item, only 0.8% of female (1 of 122) did. This may mean that dual-earning female spend as much time as possible on childcare, regardless of whether they were engaged in telework.

Furthermore, when we look at respondents' marital status (we asked respondents whether they were "married" or "unmarried"), clear differences were found in three items.[6] Overall, 50.8% of respondents (127 of 250) are married. Then, "increased time for household chores" was selected more frequently by married respondents (married: 15.7% of female (8 of 51) and 11.8% of male (9 of 76); unmarried: 9.9% of female (7 of 71) and 0.0% of male (0 of 52)).

On the other hand, "hobbies/enlightenment" had a higher selection rate among unmarried persons, that is, 30.8% for male (16 of 52) and 38.8% for female (27 of 71). The selection rate for "dealing with sudden emergencies" was notably higher among married female (17.6%; 9 of 51).

Furthermore, a fairly low percentage of full teleworkers perceived that whether one is at the main office or not had a significant positive impact on "work efficiency." While 13.2% of respondents who went to work on some days during the survey period (27 of 204) chose this option, only 2.2% of full teleworkers (1 of 46), who had zero attendance, chose it. However, the survey results alone do not tell us whether there was no room for full teleworkers to improve their work, or whether there were some tasks that could more efficiently be conducted on site than remotely. To clarify these, further research is necessary.

The survey asked about their activity history in 15-minute increments, which allowed us to determine the specifics of their free time. Table 6.4 shows the time spent by teleworkers (271) and non-teleworkers (237) engaged in various activities during a week. Not surprisingly, there is a large difference between the two groups in terms of the hours worked on their main job using ITC. In addition, teleworkers spent more time engaged in household chores and slept more than non-teleworkers.

Although not shown in Table 6.4, "more time for self-development" was often pointed out as a benefit of telework in Japanese government surveys before COVID-19. However, the survey results show that the number of hours per week spent on self-development was, on average, less than 15 minutes (conversely, teleworkers spend most of their free time on "entertainment"). This discrepancy in the survey results (e.g., MLIT, 2021) may be due to the fact that the survey was conducted by the government, as well as to the respondents' own psychological factors (e.g., pride in being selected as a teleworker by the company). The contribution of our survey is that we clarify the amount of time spent on the activities listed as advantages.

Looking at the number of days spent commuting to work, those with the highest response rate for increased sleep were those who had 5 days of commuting to work (12 of 51; 23.5%), followed by those who worked 2 days a week (7 of 30; 23.3%), and even by one respondent among the three who worked 7 days a week (there were three people in total who went to work every day). This suggests how teleworkers felt that they were getting more sleep than when had not engaged in

Table 6.4 Hours engaged in various activities (hours per week)

	Teleworkers (271)	*Non-teleworkers (237)*
Main work hours (with ICT use)	37.2	24.1
Main work hours (without ICT use)	5.1	18.4
Secondary work hours	1.1	0.4
Sleeping time	53.7	52.9
Housework time	9.3	6.9
Caregiving time	0.1	0.1
Recreation time	23.8	28.4

telework. Nevertheless, since it is difficult to ascertain sleep hours before telework implementation, it is extremely difficult to know to what extent sleep hours had actually increased.

On the other hand, it seemed clear from the results of the analysis of activity time data that the time spent commuting was being spent on hobbies and entertainment. If this is the case, it is possible that the respondents feel that they are able to refresh their minds and bodies by engaging in hobbies and entertainment and that this improves the quality of their sleep, resulting in an increase in the amount of time they spend sleeping. Of course, such an interpretation is a matter of speculation. Nevertheless, it seems necessary to emphasize that sleep time is a central concern for teleworkers.

In total, less than 10% of respondents (24 of 250) answered "none in particular" for the benefits of telework. However, among those who work five days a week, this answer was the most acknowledged benefit: approximately a quarter (13 of 51; 25.5%) of them chose it. This may be because they were almost non-teleworkers and therefore tended to perceive no advantage. However, looking at this from the opposite perspective, it can be understood that more than 70% of them considered that there was some sort of advantage of telework, albeit a limited one (the most perceived advantage was the increase in sleep time as mentioned above).

2.3.2   *Disadvantages of telework during the COVID-19 pandemic*

The perceived disadvantages of telework were also revealed by the survey results. First, the item with the highest response rate overall was "nothing in particular" (106 of 250; 42.4%), regardless of gender difference or the number of days worked. This indicates that many respondents had a positive view of the telework environment. Of course, previous surveys (cf. MIC, 2019a, 2019b; MLIT, 2021) had also shown that many experienced telecommuters expressed a positive view of WFH or telecommuting. However, as noted above, telecommuting prior to the COVID-19 pandemic was generally done at the request of the individual, so it is not surprising that the evaluation of telecommuting was positive. However, most of the respondents to our survey became (or were forced to become) teleworkers regardless of their wishes due to the unintended circumstances of the pandemic. For this reason, we expected that there would be more than a few respondents who could not dispel the impression that the WFH was "forced" and "formal" and that there would be many who would evaluate it negatively. However, the results of this survey proved that the evaluation of telework was relatively positive, even in the case of a semi-mandatory telework practice during the COVID-19 pandemic.

The item with the second highest response rate was the "difficulty in communicating with coworkers and others who are at work." Some companies are concerned about this point, and are canceling telecommuting and other forms of telework that have been practiced as a measure to prevent the spread of COVID-19 infection, and are reverting to the traditional work arrangements in the main office (cf. Koga et al., 2022). However, the survey results show that just a small number of respondents (35 of 250; 14.0%) perceived this disadvantage.

The third most perceived disadvantage was "Due to telework, private time has become shorter," with a response rate of 12.8% (32 of 250). However, there was a large gender difference, with 8.6% for male (11 of 128) and 17.2% for female (21 of 122). Moreover, the percentage of unmarried female who agreed with this opinion (15 of 71; 21.1%) was nearly twice as high as that of married women (6 of 51; 11.8%). Conversely, only male (5 of 128) responded that "My private life has taken away my time at work" (We discuss later).

The next largest difference between genders was found in the opinion "working long hours." In particular, the male response rate was 8.6% (11 of 128), while the female response rate was 4.9% (6 of 122). However, since nearly half of the respondents chose "nothing in particular," the numbers of respondents who agreed with this opinion were only 11 for male and 6 for female. Another item selected only by male (three married and two unmarried) was "My private life has taken away my time at work." As mentioned above, note that the results of this survey indicated that the blurring of the boundary between work and life was mostly due to "Due to telework, private time has become shorter" (32 of 250; 12.8%), with the opposite being less common (5 of 250; 2.0%).

Another item that shows differences by marital status was that of "loneliness/feeling of alienation." Most of the respondents who selected this item were unmarried. In other word, the overall response rate was 4.4% (11 of 250), but the response rate for unmarried people was 8.1% (10 of 123) and that for married people was 0.8% (1 of 127). There have been a number of surveys on the loneliness and alienation caused by teleworking (e.g., Persol Research and Consulting, 2021). Nevertheless, to the best of our knowledge, we are the first to identify differences by marital status.

Comparing the perceived biggest disadvantages among respondents by number of days in the office, the respondents who go to their office two days a week (7 of 30; 23.3%) selected "I have difficulty communicating with workers who come to work." Full-time teleworkers selected this item at a lower rate (6 of 46; 13.0%) than the overall average (35 of 250; 14.0%). Elucidating the reasons for this paradoxical phenomenon (i.e., less dissatisfaction with communication issues among full teleworkers than among partial teleworkers) is a topic for future research.

In the next section, we will discuss the challenges inherent to telework based on these survey results.

3 The potential of telework for realizing a sustainable society

As described above, we have identified the reality of telework during the COVID-19 epidemic in Japan. For teleworkers, the practice of telework was perceived to have advantages such as "more free time" and "more sleep time." On the other hand, the disadvantages of teleworking perceived by respondents who engaged in telework were few; we would venture to say that the "sense of loneliness and alienation" was perceived by unmarried people and "difficulty in communicating with commuters" was perceived by partial teleworkers.

These characteristics can basically be understood as a phenomenon that forced many companies to implement telework during the COVID-19 pandemic, when the government asked them to refrain from doing so.

In this section, we discuss the state of teleworking after COVID-19. In particular, from the perspective of realizing a sustainable society, which has been the focus of much attention in recent years, the practice of telework, a flexible work style, may be beneficial as a solution to social problems such as Japan's declining birthrate and aging population.

Based on the results of the survey, we would like to develop a vision of what teleworking should be. In this section, we discuss the following three perspectives in particular.

1 Expansion of the existing concept of telework
2 Expansion of the target: from individuals to groups
3 Expansion of objectives: from task execution methods to process management

3.1   *Extension of the telework concept: Telework without ICT*

First, it is important to confirm the constraint that the characteristics of teleworking obtained from the results of this survey are basically a phenomenon that forced many companies to implement teleworking during the COVID-19 epidemic, when they were asked by the government to refrain from doing so. Nevertheless, we find that the reality of this phenomenon is very different from the form of telework promoted by the Japanese government prior to COVID-19.

The bog-standard concept of telework as a sophisticated work system means working outside the designated offices using ICT. However, the survey results reveal the following characteristics.

1 Before COVID-19, mobile work was common, but during COVID-19, WFH was the mainstream.
2 The number of commuting days was not necessarily zero, and many teleworkers go to work (although some are engaged in full-time WFH).
3 There were full-time WFH workers who engaged in work that was not necessarily required to use ICT.

The last point may seem obvious at first glance. However, Japanese telework studies have not strictly included the time spent in non-ICT work in WFH as telework. Because this time is not included in telework, Japanese researchers have considered that WFH consists of telework and non-telework. From an academic point of view, such a strict division may be important, but from a practical point of view, it may not be useful. Our study is probably the first to show the reality of non-telework WFH.

These three characteristics suggest that the key to the future diffusion of telework is to extend the concept of telework from an academic perspective and adapt it to the actual situation, and to emphasize the flexibility of using ICT and working at different places and times, rather than only working outside the office.

3.2   *From individual to group*

Before the spread of COVID-19 impacted on society in various ways, telework had often been viewed, if anything, as a "silver bullet" that would solve management's most difficult problems all at once (Markus & Robbey, 1997; Koga et al., 2022). Correspondingly, conventional surveys on people's attitudes toward telework often included responses such as increased productivity.

However, our survey results indicate that many respondents pointed to the personal benefit of reduced commuting time (and the resultant increase in personal free time) as an effect of telework practice. Perhaps, some bias was at work in previous surveys because the responses were from people who were allowed to telework. In contrast, our survey is thought to reflect the honest opinions of those who were forced to telework under the influence of the COVID-19 epidemic. In particular, teleworking during the epidemic had a strong aspect of voluntary curfew as a preventive measure. This meant that people did not have time to realize the traditionally expected benefits of teleworking.

The focus of telework as an infection control measure has been on curtailing employee travel. In other words, the purpose of telework has been the short-term restraint of travel rather than the transformative factors of the past. This has weakened the intention to realize new ways of working, such as the much-publicized digital transformation through the practice of telework.

Thus, teleworking during the COVID-19 epidemic was more of a temporary measure to prevent infection than a way of working for a sustainable society. This may be evidenced by the fact that many companies have stopped practicing telework since the beginning of 2022 (see also Table 6.1 above).

What, then, is telework for a sustainable society? To conclude quickly, we focus on the following two elements as the keys: (1) a shift in mindset from an individual-oriented to a collaborative, future-oriented approach, and (2) the introduction of a business process management (BPM) perspective (cf. Weske, 2007).

The first element involves a change in the perception of telework. At any rate, telework during the COVID-19 pandemic seemed to focus on finding and practicing tasks that could be performed from home. This attitude can be said to position telework as a means for individuals to perform their duties. However, work has both future-oriented and collaborative elements. In other words, work includes the nuance of collaborating with others to bring things closer to the way they should be (Sugimura, 1997). Our survey results strongly reflected only one aspect of telework that of providing a means for individuals to perform their duties. However, to promote a sustainable society that includes diversity, a form of telework that is oriented toward collaboration with others may be needed.

Specifically, rather than dividing tasks into those that can be performed by individuals, tasks should be positioned as a series of business processes from the perspective of the entire workflow. In other words, understanding the significance of the tasks to be executed by the individual in the overall workflow is expected to alleviate feelings of isolation and anxiety. As described above, telework can be positioned as an opportunity to find the significance of one's tasks in the overall workflow, rather than as a means of separating tasks that can be performed at home. This is the first element of telework reform to achieve a sustainable society.

3.3   *From task execution methods to process management*

The second element of teleworking for a sustainable society is BPM. Simply put, this is a business management technique that is used to understand the current status of daily business processes and make changes and improvements to continuously move closer to the processes as they should be (Weske, 2007).

Through the practice of telework in the narrow sense of a way of working that utilizes ICT, records of diverse events and activities are accumulated. By utilizing such data, the actual status of business processes can be accurately grasped. In addition, through process tracking and analysis, it is possible to envision what the process should look like (process mining). As a result, the Plan – Do – Check – Act (PDCA) cycle of business process transformation can function. Such an approach is none other than BPM.

The research on business process reengineering, which gained attention in the 1990s (cf. Hammer & Champy, 1993; Davenport, 1993), focused on core system and back-office reforms. BPM, on the other hand, focuses on improving front-office operations, sales activities, and system restructuring (cf. Weske, 2007).

Of course, the idea of a data-driven PDCA cycle is not new. Rather, it may be criticized as nothing more than a discussion of production management from a generation ago. However, in these days of buzzwords such as "big data" and "data science," it can be said that many companies share a common awareness of the problem of gaining and maintaining a sustainable competitive advantage through the use of data such as customer purchase histories and web browsing and behavior histories. Therefore, PDCA is both old and new. The essence of BPM is to improve business processes by focusing on the company's own action history rather than that of its customers.

Needless to say, once again, it is not enough to simply understand customer needs to obtain and maintain a company's competitive advantage. It is essential to have the organizational capabilities to respond to such needs. Conventional theories of organizational capability have tended to focus on core competence and business systems. In other words, discussions of organizational capability have focused on the "moment of truth" of customer value creation in manufacturing and service. This is perhaps evidenced by the repeated discussion of productivity improvements in office work (cf. Gallardo & Whitacre, 2018; Kazekami, 2020). However, from a different perspective, if a business organization can achieve productivity improvement in office work, they should be able to gain a significant competitive advantage over their competitors, and this is why BPM is expected to drive the organizational capability research in the 21st century.

From the above, it is clear that a new item needs to be added to the list of the benefits of telework, as we will explain. In other words, telework is an attempt to provide basic data for accurately grasping business processes through its practice and from there to explore and realize the ideal form. By visualizing business processes and accumulating them as data, the ideal form can be clarified, just as data are analyzed to extract customer needs. Telework can be understood as a substrate that provides clues.

However, traditional telework research has tended to overlook this perspective. When productivity improvement through telework is discussed, the focus has been on improving the efficiency of individual work, and there seems to have been little discussion from the perspective of overall workflow and business processes. In fact, we did not explicitly include questions from a BPM perspective in our survey results. Nevertheless, the small number of respondents who chose increased productivity as one of the benefits of teleworking gave us the strong impression that individual productivity improvement may have reached its limits. Therefore, we advocate for broadening our understanding and adopting the BPM perspective.

4   Conclusions and discussion

In this chapter, we have summarized the expectations for telework prior to the COVID-19 epidemic in Japan, clarified the reality of telework during the COVID-19 pandemic, and identified its peculiarities. Our findings reveal that teleworking during the COVID-19 pandemic was more nuanced as an emergency measure to continue work and that the emphasis was on enhancing personal life through telework.

One of the most important issues in achieving a sustainable society is the enhancement of workers' own personal lives. Nevertheless, just as cannot being separated like the two sides of a coin, it is difficult to enrich individual lives without sustaining companies individuals are working for. For the sustainable development of companies, it is necessary to change the current awareness of and attitude toward telework. This is the conclusion of this chapter.

Specifically, we believe that a change in awareness is required in the following three areas. First, the concept of telework itself should be expanded to focus not on the use of ICT but on the diversity of workplaces; second, the work performed by teleworkers should be positioned not by individuals but by the organization as a whole; and third, from the standpoint of process reform through the documentation of practices, the perspective of the use of telework activity history data should be adopted. It is important to emphasize again that telework is not a "silver bullet." We should understand that the touchstone of telework will be whether or not it can promote a change in the awareness of "work" through its practice, and we should consider that a reform of work styles can be implemented only through the practice of telework. Therefore, in the future, it is essential to raise awareness of telework, which will exist alongside COVID-19.

Acknowledgments

The studies in this book were supported by the JSPS (Japan Society for the Promotion of Science)/STINT (Swedish Foundation for International Cooperation in Research and Higher Education) Bilateral Joint Research Program "Information and Communication Technology for Sustainability and Ethics: Cross-national Studies between Japan and Sweden" (JPJSBP120185411 and JA2017-6999). And this work was supported by JSPS Grants-in-Aid for Scientific Research JPS21K01917 (Principal: Sato Akio), JPS20K01899 (Principal: Koga Hiroshi), JP21K01650 (Principal: Yanagihara Sachiko), and Kansai University Research Organization for Socionetwork Strategies Joint Usage and Research Center Project (The research was supported by the Joint Usage/Collaborative Research Center Project of the Research Organization for Socionetwork Strategies, Kansai University (Representative: Koga Hiroshi).

Notes

1 The government did not block urban areas but asked citizens to refrain from going out unnecessarily (request to refrain from going out).
2 A statement purportedly made by Tesla CEO Elon Musk has been making the rounds on SNS, and the results of various surveys have also reported a decline in the teleworking rate. In addition to the abovementioned survey by Tokyo Shoko Research (2022), similar trends can also be seen. For example, according to the Cabinet Office data (2020), the telework penetration rate was only 10.3% in December 2019 before the declaration of the state of emergency, but after the declaration was issued, the rate rose to 27.7%. Additionally, according to Persol Research and Consulting (2021), the telework implementation rate was 13.2% in March 2020 but increased to 27.9% in April of the same year.
3 Goto and Hamano (2020) summarized the results of 70 published surveys on the actual situation of telework during the corona disaster.
4 According to a survey by the Tokyo Metropolitan Government, the figure for the same period was 39.5%. On the other hand, according to the Ministry of Land, Infrastructure, Transport and Tourism (MLIT)'s 2020 survey of the teleworking population, the percentage of employed teleworkers was 22.5% (See https://www.metro.tokyo.lg.jp/tosei/hodohappyo/press/2022/12/14/02.html and https://www.mlit.go.jp/toshi/daisei/content/001469009.pdf).
5 The results of our survey showed that 84 respondents were not obligated to attend work at all. Of those, however, only 37 actually did not show up for work.
6 It is predictable that married respondents would choose "parenting time" and "family time."

References

Cabinet Office. (2020). *第2回 新型コロナウイルス感染症の影響下における生活意識・行動の変化に関する調査* [Second survey on changes in attitudes and behaviors in lifestyles under the influence of the new type of coronavirus infection]. https://www5.cao.go.jp/keizai2/wellbeing/covid/pdf/result2_covid.pdf

Davenport, T. H. (1993). *Process innovation: Reengineering work through information technology*. Harvard Business Press.

Gallardo, R. & Whitacre, B. (2018). 21st century economic development: The telework and its impact on local income. *Regional Science Policy & Practice, 10*(2), 103–123. https://doi.org/10.1111/rsp3.12117

Goto, M. & Hamano, W. (2020). 新型コロナウイルス感染症流行下でのテレワークの実態に関する調査動向 [Trends in surveys about home teleworking during the COVID-19 pandemic in Japan]. *INSS Journal: Journal of the Institute of Nuclear Safety System*, *27*, 252–274.

Gratton, L. & Scott, A. (2016). *100-year Life: Living and working in an age of longevity.* Bloomsbury.

Hammer, M. & Champy, J. (1993). *Reengineering the corporation.* Harper Business.

Handy, C. (1989). *The age of unreason.* Business Books.

Japan Telework Association. (2007). テレワーク白書*2007* [Telework white paper 2007]. Impress R&D.

Kazekami, S. (2020). Mechanisms to improve labor productivity by performing telework. *Telecommunications Policy*, *44*(2), Article 101868. https://doi.org/10.1016/j.telpol.2019.101868

Koga, H. & Sato, A. (2022). 雇用型テレワークの日常的実践に関する実態調査 [Actual conditions of daily practice of teleworking employee]. *Journal of Informatics (Faculty of Informatics, Kansai University)*, *54*, 65–80. http://doi.org/10.32286/00026082

Koga, H., Sato, A., Yanagihara, S., Takagi, S., Nakai, H. & Kano, I. (2022). COVID-19パンデミック下におけるテレワーク行動の実態と課題 [Status survey and issues of telework behavior during the COVID-19 pandemic]. *Journal of Informatics (Faculty of Informatics, Kansai University)*, *56*, 65–80.

Markus, M. L., & Benjamin, R. I. (1997). The magic bullet theory in IT-enabled transformation. *Sloan Management Review*, 38(2), 55–68.

Ministry of Health, Labour and Welfare (MHLW). (2016). テレワークで始める働き方改革：テレワークの導入・運用ガイドブック [Teleworking and the start of workplace Reform: A guidebook for the introduction and operation of telework]. https://roumu.com/pdf/nlb0787.pdf

Ministry of Land, Infrastructure, Transport and Tourism (MLIT). (2021). 令和3年度 テレワーク人口実態調査－調査結果－ [Telework population survey report in 2020]. https://www.mlit.go.jp/toshi/daisei/content/001471979.pdf

Ministry of Internal Affairs and Communications (MIC). (2019a). 令和元年 通信利用動向調査報告書（企業編） [Survcy on telecommunications usage trends in 2019: Enterprises]. https://www.soumu.go.jp/johotsusintokei/statistics/pdf/HR201900_002.pdf

Ministry of Internal Affairs and Communications (MIC). (2019b). 令和元年 通信利用動向調査報告書（世帯編） [Survey on telecommunications usage trends in 2019: Households]. https://www.soumu.go.jp/johotsusintokei/statistics/pdf/HR201900_001.pdf

Markus, M. L., & Benjamin, R. I. (1997). The Magic Bullet Theory in IT-Enabled Transformation. Sloan management review, 38(2), 55-68.

Persol Research and Consulting. (2021). 第四回·新型コロナウイルス対策によるテレワークへの影響に関する緊急調査 [Fourth urgent survey on the impacts of the new coronavirus countermeasures on telework]. https://rc.persol-group.co.jp/thinktank/assets/telework-survey4-1.pdf

Sugimura, Y. (1997). 『「良い仕事」の思想―新しい仕事倫理のために』中公新書 [The Idea of Good work: For New Work Ethics] ChuoKohron Sha.

Tokyo Shoko Research. (2022). 第22回「新型コロナウイルスに関するアンケート」調査 [The 22nd questionnaire survey on new type of coronavirus]. https://lp.tsr-net.co.jp/rs/483-BVX-552/images/20220622_TSRsurvey_CoronaVirus.pdf

Weske, M. (2007). *Business process management architectures.* Springer.

7 The ethics of body modification

Transhumanism in Japan

Kiyoshi Murata, Yohko Orito, Andrew A. Adams,
Mario Arias-Oliva and Yasunori Fukuta

1   Introduction

This chapter deals with the ethical issues relevant to body modification with cyborg technologies such as wearables and implantables and other aspects of transhumanism, taking Japanese culture into consideration. Wearables are defined as electronic devices that are incorporated into clothes and/or accessories or attached to a human body (exoskeletons, smart contact lenses, etc.) for non-medical purposes and which interact with the user to enhance human ability beyond the norm. Implantables are defined as electronic devices that can surgically be implanted in a human body (brain chips, implantable radio-frequency identification (RFID) tags, etc.) for non-medical purposes and which interact with the user to increase his/her innate human capacities such as mental agility, memory and physical strength, or give him/her new ones such as a capability to control machines remotely (Benabid et al., 2019) or provide additional senses (e.g. Neil Harbisson (Guler et al., 2016, p. 146) or Dann Berg (Guler et al., 2016, p. 150)) (Murata et al., 2017a).

We are now at the threshold of the potential widespread use of these cyborg technologies, as described in the next section. This means that we are at a good point to proactively address the ethical and social issues that the development and usage of the technologies could cause. To clarify such issues, interviews with experts in the research fields relevant to body modification ethics such as life science, health science, robotics, information security and social security were conducted, and the results of these interviews were analysed, taking transhumanist ideals and Japanese culture regarding body modification into account. Interviews were conducted in Japanese. Quotes from interviewees are given in English, checked by both native English (Adams) and native Japanese speakers (Murata, Orito and Fukuta), as well as by the interviewees, to ensure an accurate representation of interviewees' opinions. From the analysis of the interview results, a research agenda for further investigating the ethics of body modification in Japan is derived. In addition, British and Spanish perspectives on body modification are discussed to relativise and clarify the Japanese sociocultural circumstances surrounding it.

DOI: 10.4324/9781003367451-8

2   Transhumanism through human cyborgisation

2.1   *Development and usage of cyborg technologies*

The use of wearables like smart watches, smart glasses and fitness trackers is currently spreading among innovators and early adopters (Rogers, 1962) at least. According to the Yano Research Institute (2016), the global unit shipments of wearables in 2015 were 71 million, and their forecasts for 2016 and 2020 were 117 million and 323 million, respectively. In Japan, the volume of shipments in 2015 was 2.1 million, and it was forecast to grow to 3.6 million in 2016 and to 11.6 million in 2020 (Ministry of Internal Affairs and Communications, 2016). These figures show that the market for wearables was expected to steadily grow, though the spread to the majority of the population was expected to take a little extra time. However, the Ministry of Internal Affairs and Communications (2020) claimed that the percentage of households owning wearable devices in Japan was only 5.0% in 2020.[1] In the meantime, new types of wearables have been developed: for example, an American start-up developed "technological tattoos" made from small pieces of hardware that could be stuck to the skin and could collect and store data such as heart rate and body temperature and then send that information to a smartphone application (Firger, 2015). A prototype of a multitasking wearable which can continuously measure multiple biomarkers of its users has been developed (Labios, 2022).

On the other hand, it is believed that implantables have been used so far only by innovators, although the usage of various kinds of implantables is slowly spreading. For example, an RFID chip implanted under the skin of a person can be used as a key to his/her house and/or office, or for other identification purposes (Adam & Wilkes, 2016; Eveleth, 2016; Nora, 2015). A Dutchman uses subdermal near-field communication chips to access his bitcoin wallet (Clark, 2014). In Japan, the authors have not found any actual case of implantable usage so far. However, the trend of electronic device development towards implantables seems unstoppable. Internal compasses, implantable micro-computers and internal headphones have already been developed (Monks, 2014). Sony has been awarded a patent for a smart contact lens that can be controlled by the user's deliberate blinks, recording video on request (Starr, 2016). The US military has been spending millions on the development of an implantable that opens the door for cyborg soldiers—an advanced implant that would allow a human brain to communicate directly with computers (Browne, 2016). In December 2020, it was reported that the French military ethics committee approved in principle the French armed force's development of augmented or enhanced soldiers using implantables to improve physical, cognitive, perceptive and psychological capacities, to prevent pain, stress and fatigue, and to improve mental resilience if they were taken prisoners (BBC News, 2020; Guy, 2020).

2.2   *What is a cyborg?*

The term cyborg—cybernetic organism or cybernetically controlled organism—was coined by Clynes and Kline (1960) to represent the exogenously extended organizational complex functioning unconsciously as an integrated homeostatic system. The cyborg, they said, deliberately incorporates exogenous components extending the self-regulatory control function of the organism in order to adapt it to new environments such as outer space. However, it seems that with regard to cyborgs, the emphasis has increasingly been on their enhanced physical and cognitive capabilities rather than on their ability to maintain self-regulating functions as an organism in an extraordinary ecological environment.

There is so far no generally accepted definition of a cyborg, though several definitions have been proposed, such as a human-machine hybrid (Haddow et al., 2015) and a human being with an electronic device implanted in or permanently attached to their body for the purpose of enhancing their individual senses or abilities beyond the occasional use of tools (Park, 2014). Greiner (2014) pointed out that whether technology is external or internal to the body does not matter, but the interfaces and relations human beings create between biological receptors and technological sensors, biological and technological information processing, and biological and technological modes of interaction with the environment do matter from the viewpoint of a cyborg. On the other hand, Clark (2003) insisted humans are natural-born cyborgs, based on the fact that an unusual degree of cortical plasticity brings about a sense of unity between humans and the tools, machines and technology they use and thus non-biological elements are effortlessly incorporated in the formation of their mind and personality, making the presence of such elements transparent and unconscious.

It is hard to draw a clear line between a human being and a cyborg, because human beings as Homo Faber have developed a human environment of *conjunction technologique* (technology-mediated environment; Imamichi, 2009) where technological artefacts are intricately intertwined with each other and with natural objects, and people use various kinds of technological devices as if such devices were parts of their bodies. Moreover, owing to the tremendous advances in and widespread use of information and communication technology (ICT)-based services, people seem to have already become a cyborg intellectually; a significant portion of their cognitive function, memory, calculation capability and reasoning ability are outsourced to (in many cases ostensibly cost-free) online services.

Assuming, for example, that a cyborg is a human being whose physical and/or intellectual ability is enhanced by wearable or implantable devices or equipment developed based on scientific technology, people with eyeglasses or contact lenses can be considered as cyborgs. Thus, those who engage in research on cyborg ethics need to appropriately define cyborgs and cyborg technologies that will be investigated and examined in their research, clearly articulating these related concepts. In this study, the authors focus on the ethics of body modification using wearables and/or implantables for the enhancement of capability outside human norms.

Restoration of vision using glasses or contact lenses is usually not ethically controversial. However, improvement of vision by ICT devices far beyond the innate human ability might be problematic. If a nanochip originally developed for the treatment of dementia is used to produce an artificial super genius, or if an advanced artificial leg helps an athlete run 100 meters in 8.0 seconds, such cyborgisation could become an object of criticism. Hence, cyborg ethics research-ers can recognise another need to draw a moral line with respect to cyborg technol-ogy, that of where on the spectrum between the two extremes of an average human and an ultimate cyborg is the line that should never be crossed.

3   Japanese sociocultural environment surrounding body modification

Attitudes to currently available (or in late development) body modification tech-nologies can provide useful guides to the likely reception of further developments: piercing; tattoos; external sensory correctives such as eyeglasses, contact lenses or hearing aids; organ transplants (from recently deceased humans, from genetically modified animals, or vat-grown); implanted sensory correctives such as artificial cochlear; exoskeletons for weakened or paralysed bodies or to enhance strength beyond human norms. In Japan, external sensory correctives as well as implanted ones are widely accepted. It seems to be socially approved that science and tech-nology can correct congenital or acquired disabilities. On the other hand, however, in the Japanese tradition, it has been believed that one's body is not one's own and, therefore, an average Japanese person hesitates to engage in unnecessary body modifications. The attitudes of Japanese people towards body modification seem to reflect the Confucian ethic of filial piety prevalent among the Japanese, which is well described by the quote from the *Classic of Filial Piety*: "Our bodies—to every hair and bit of skin—are received by us from our parents, and we must not presume to injure or wound them. This is the beginning of filial piety" (Legge, 2010). Con-fucianism was introduced to Japan from Korea (Baekje at the time) in the 5th to 6th centuries. Since then, the concepts of Confucianism have been Japanised through syncretisation with Shinto and Buddhist concepts, and remain influential on Japa-nese people's moral standards (Bellah, 1985; Kaji, 2011; Murata et al., 2017a).

In Japan, pierced earrings are widely accepted as an ordinary accessory, whereas nose-, tongue- or body-piercing usually create a negative image of those who have them. Tattoos tend to be seen as symbols of an antisocial person or *yakuza*—a gangster. Tattoos are considered to promote people's sense of fear, and thus those who have a tattoo are often barred from public baths, public swimming pools and bathing beaches. However, Japan has a long tradition of tattooing as a form of body modification. The Report on the People of *Wa* (Japan) of the Book of the Wei, the first Chinese official history book that mentioned Japan (written at the end of the third century), describes that "every Japanese man has tattoos on his face and body, regardless of his status". However, in the mid-7th century, Japan's aesthetics started to change with the development of an aristocratic culture in which physical beauty was considered less important, and the beauty represented by clothes, aromas and so on became emphasised. In addition to this, the social penetration of Confucian-ism depressed the tradition of tattooing in Japan. In 1603, the Edo or Tokugawa Era—a 265-year-long period of peace—started, and in the late Tokugawa period, tattoos became popular among town people as a part of popular culture. In particu-lar, those who engaged in jobs working in their bare skin such as carpenters, fire fighters and postmen liked to have colourful tattoos across their entire bodies. Such colourful full-body tattoos were considered the replacement for their clothes. For them, full-body tattoos were symbols of their manhood, endurance and dapperness. Tattooing technology was significantly developed in this period. However, samurai did not like to have tattoos based on their Confucian morality. In addition, tattoos were used as criminal punishment in the Tokugawa Era from 1720. Those who committed certain crimes had unicolour tattoos placed on their foreheads or arms. In marked contrast to townsfolks' colourful full-body tattoos, these unicolour tattoos were a mark of their shame. In 1872, during the time of the Meiji Restoration, the Japanese government banned tattooing. They considered this ban was necessary for Japan's modernisation or Westernisation. Both applying and having tattoos became illegal, and tattooing became an underground activity, leading to the development of a negative image of tattoos among ordinary Japanese people. The ban was lifted in 1948 during the United States-led Allied occupation of Japan (Yamamoto, 1997, 2017). However, as described above, the average Japanese person's negative attitudes towards tattoos continue to the present day.

Organ transplantation is often criticised in Japan as the act of stealing others' body parts due to a rampant egoism in society. Imamichi (2009: pp. ix, 68–69) described those who wait for the opportunities of receiving organ donations (and their families) as desiring the earliest death of potential donors and that this is morally unacceptable. He strongly recommended the development and use of mechanical organs, supporting human cyborgisation. Actually, although brain death is recognised as legal death to encourage organ transplantation in Japan, based on the Act of Organ Transplantation (Act No. 104 of 16 July 1997), this is not broadly accepted by ordinary Japanese people. At the end of September 2021, 24 years after the act came into force, the number of declarations of brain death under the act reached only 773. Organ donations from people under age 15, which was enabled by the revision of the act in July 2010, had been made only 56 times as of the end of September 2021 (Japan Organ Transplant Network, 2021: pp. 2, 4). Given Japan's population (over 120 million), these low figures seem to demonstrate Japanese people's hesitation in organ donation, borne out by empirical research on attitudes such as that of Okita et al. (2018). Such criticism and hesitation may reflect Japanese religious traditions that prohibit harming dead bodies based on the Confucian view of life and death (Kaji, 2011) and Japanese aesthetics of death that is succinctly described in the phrase "the cherry among flowers, the samurai among men", which praises the graceful death of samurai by association with the graceful fall of cherry blossoms.

On the other hand, regenerative medical techniques and prosthetics are expected to contribute to developing humane treatment for a number of intractable diseases. In particular, the technique of using induced pluripotent stem (iPS) cells is considered as an ethically acceptable medical practice compared with one using embryo stem (ES) cells, because iPS cells are created from a patient's own existing body whereas ES cells, which are created from others' fertilised eggs, involve killing in the process of their creation (Noe, 2013; Takada, 2015). A Japanese research group's attempts to grow human organs such as the pancreas for transplantation within the body of a pig but using human iPS cells (https://muiibr.com) seem to be generally acceptable in Japan. The development, production and rental (not sale, to prevent military use) of Hybrid Assistive Limb® (HAL®), the world's first usable exoskeleton, to improve, support and enhance wearer's body functions, by Cyberdyne Inc. (http://www.cyberdyne.jp/english/) based in City of Tsukuba, Ibaraki Prefecture is considered a successful university venture as well as a significant social action work in Japan.

In their report, the Fifth Science and Technology Basic Plan announced in January 2016, Japan's Cabinet Office proudly declare the development of a "super smart society" or "society 5.0", where robots along with other advanced information technologies such as big data, Internet of Things (IoT) and artificial intelligence (AI) are expected to play a pivotal role and that the government aims for the coexistence of human beings, robots and AI in order to greatly improve the quality of life for people (Cabinet Office, 2015). Japan is already one of the leading countries in robotics. In 2011, Japan's robotics industry held a 57.3% share of the world's industrial robotics market (Ministry of Economy, Trade and Industry, 2013). Their development of humanoid robots has also been among the most advanced. For example, Honda Motors, the world's largest motorcycle manufacturer as well as the seventh largest automobile manufacturer, introduced their bipedal walking humanoid robot ASIMO in 2000, and developed walking assistance devices based on their knowledge of bipedal motion acquired in that robot's development process. These devices have been leased to hospitals and rehabilitation institutions since November 2015 (https://global.honda/innovation/robotics.html).

Japanese robotics researcher Hiroshi Ishiguro has been engaged in the development of various kinds of android robots including the *geminoid*, an android robot of an actual person's likeness, and the *telenoid*, an android satisfying just the minimum requirements for expressing humanlike appearance and motion (http://www.geminoid.jp/en/index.html). Through these developments, Ishiguro says, he investigates the answers to questions "what is a human being?", "what is human existence?", "what is the human mind?" and "what is the self?", and has come to believe that it is possible to let a robot have a function of what looks like the human mind (Ishiguro, 2009, 2014, 2015) because the human mind does indeed not exist (Ishiguro & Washida, 2011).

It is alleged that both individuals' psychological resistance and social pressure against the development of android robots are not strong in Japan, because Japanese society is religiously tolerant and the Christian kind of ideas that God is the creator and that God created man in his own image have not exerted any influence over Japanese social norms (e.g. Takanishi, 2012; Yamanaka, 2019). In fact, android robots have been depicted as good friends and faithful guardians of human beings in many science-fiction manga stories since Osamu Tezuka's *Astro Boy*, which was serialised in manga magazine Monthly Shōnen starting in 1952. Facing the serious concerns of a rapidly ageing society and very low birth rate, policymakers have started to test the practical effectiveness of humanoid and android robot use in the fields of healthcare and nursing care (in addition to powered exoskeletons, walking support robots and so on), for mental healthcare and prevention of dementia of patients and recipients of care (Kondo, 2019; Otake, 2019).

Although it is government policy to promote the development and use of robot technologies including humanoid and android ones, many elements of which can also be used as cyborg technology, the ethical and social issues regarding them have rarely been examined in Japan. The social awareness of the necessity to consider legal regulations on the development and use of robot technology is still lacking in the country (Shimpo, 2016). Nonetheless, all in all, there is a positive attitude to robots among the Japanese. On the other hand, however, when it comes to cyborgisation, particularly with implantables, Japanese people might be hesitant to accept it because of the Japanese culture and tradition related to body modification.

4   Analysis of the results of interview survey

4.1   Overview of the survey

Semi-structured interviews with ten Japanese experts in disciplines relevant to cyborg ethics were conducted from March to May 2018 (eight interviews), and in October 2018 and October 2021 (one interview each). A set of basic questions that were asked at all the interviews was prepared by the authors based on the facts that wearables and implantables are emerging technologies and taking into account that the future usages are uncertain. These questions include asking the interviewee's (a) knowledge about wearables and implantables, (b) attitudes towards the devices, (c) views and opinions on the use of the devices in different contexts, and (d) understanding of the Japanese sociocultural environment surrounding technology, including cutting-edge ones.

The attributes of the interviewees are described in Table 7.1. Every interviewee had Japanese citizenship and was working for a Japanese university, with one interviewee holding a concurrent research position at a German university. Only one interviewee (09IS) was himself using a wearable (a fitness tracker) at the time of the interview, while another interviewee (03GC) had used wearables (a smartwatch and a fitness tracker) previously. No interviewees had used any implantable device. Wearable devices, especially ones for healthcare purposes, were well known to all of the interviewees, whereas implantable devices, even ones for healthcare, were not, for the majority of the interviewees. Three interviewees did have such knowledge: one whose speciality was embodied informatics (05EI), another who is an audiology expert (08AUD) who engaged in interdisciplinary study on wearable and implantable medical devices for the deaf and hard of hearing, and the last a researcher in the defence industry (01DI) who was interested in bio-chips that could be used for checking soldiers' level of stress through monitoring their respiration and blood pressure.

4.2   Survey results and research agendas for the ethics of body modification

4.2.1   Attitudes towards wearables and implantables and related research agendas

Through interviews, various kinds of ethical and social issues related to the development and use of wearables and implantables were suggested. These provide ideas for developing research agendas for the ethics of body modification using these emerging technologies.

Table 7.1 Attributes of interviewees

Interviewee ID	Position[a]	Gender	Age[a]	Expertise
01DI	Assistant professor	Female	30s	Defence industry; applied economics
02LS	Professor	Male	50s	Life science; science education
03GC	Research fellow	Female	40s	Gender and computing; political science
04ME	Professor	Male	50s	Management engineering; production management
05EI	Associate professor	Male	30s	Embodied informatics
06HS	Associate professor	Male	40s	Health and sports science
07SSL	Professor	Female	40s	Social security law
08AUD	Professor	Male	50s	Audiology
09IS	Professor	Male	40s	Information security
10RBT	Associate professor	Male	40s	Robotics

[a] Position and age are as of the time of interview.

Three interviewees expressed positive attitudes towards wearables and implantables.

02LS: I tend to be positive towards these devices.
05EI: Because wearables, which are light enough to be worn without discomfort, and implantables can avoid wasting individuals' physical and cognitive resources, these devices will necessarily be accepted among a wide range of people.
09IS: I basically have a positive attitude towards these devices. They enhance the convenience of our lives.

Simultaneously, however, they also expressed concerns about the possible impacts of the use of the devices. Two of them pointed out the probable confusion about implantable users felt by people around them, which seemed to reflect the Japanese hesitation in using implantables mentioned above.

02LS: I consider that the usage of these technologies should never be allowed to lead to social turmoil. Those who engage in using these devices need to carefully consider social reactions to their use. My concern about implantables is social approval of certain usages. Prosthetic limbs for amputees, for example, are visible at first sight and disabled athletes using such apparatuses have achieved mainstream recognition. However, this is

not the case with implantables. Thus, those who use implantables would possibly get funny looks from others until they are prevalent in society.
05EI: The problem is not only the discomfort of an individual when using wearables or implantables, but also of the people around him/her.
09IS: There is an issue regarding data protection. Our personal data which the devices collect may be sent to databases developed and deployed by US companies which operate and administer relevant online applications. Our data, the personal data of Japanese people, stored in the databases of US companies may not properly protected by US laws, and may be beyond the reach of Japanese laws. [...] The purposes and ways of using personal data collected by wearables and implantables require that individual users should be informed of these things whenever necessary, and such practices should be audited by an independent third party.

A researcher in gender and computing professed a neutral stance to human cyborgisation.

03GC: As long as the devices are used based on a human-centred perspective, my attitudes to wearables and implantables are neutral. Whether a specific device is evaluated positively or negatively should be determined based on who uses it in what way. The evaluation should be made on a case-by-case basis. Sometimes we should view it positively because of its usefulness to humans, and sometimes negatively because of ethical concerns.

An expert in audiology emphasised self-determination of the user regarding cyborg technology usage.

08AUD: Users must not be forced into using these technologies at all or required to make specific use of them. The decision on whether and how to use the devices has to be based on users' own ethical judgement.

The attitudes of the audiology researcher being based on individual self-determination are not surprising given the long debate about cochlear implants and consent in the audiology community and among individuals with hearing limitations (Christiansen & Leigh, 2002).

The following research agenda on the ethics of body modification can be drawn from their statements. The issues related to individual autonomy to be a cyborg (category a) may have to be discussed with Japanese (and other countries' and regions') culture regarding body modification in mind.

a The autonomy of individuals' decision-making on body modification or becoming cyborgs:

a-1 Should individuals be guaranteed freedom to modify their bodies or morphological freedom (Sandberg, 2013) using wearables and/or implantables?

a-2 Should individuals' right to body modification be established?

   a-2-1 Does the freedom or the right allow individuals to arbitrarily set the objectives of their body modification?

a-3 Should cyborg discrimination or discrimination against those who modify their bodies using wearables and/or implantables be prohibited?

b Personal data protection:

b-1 Who can collect, store and/or use personal data available from wearable or implantable devices?

b-2 For what purpose is the processing of data collected from body modification devices socially acceptable?

A robotics expert expected that an implantable would allow human beings to develop perfect recall.

10RBT: I don't have resistance to these devices. These, especially implantables, will and should be used in various areas. If possible, I'd like to use implantables to strengthen my memory, enabling me to recall necessary information just by considering. This function is exciting for me. The enhancement of human memory using implantables would evolve human beings to a new level, and would obviate risks and problems arising from ignorance. Ultimately, implantables would enable human beings to access shared knowledge databases at will without any external device, and to have the total recall.

In association with the usage of a brain chip to enhance brain power for healthy people, another interviewee expressed his opposition to the potential for total recall that could result, echoing the assertion two of the authors made in their previous work (Murata & Orito, 2011).

08AUD: I feel resistant to this kind of use of a brain chip. I consider that total recall and understanding do not necessarily make human beings happy. The forgetting, particularly of hard or traumatic things, enables a person to enjoy peace of mind.

These views lead to the development of a research agenda about how much such technologies could radically change human nature and whether such evolution is desirable.

c Technology-driven human evolution:

c-1 Would total recall technologies be a good thing for human beings?

c-2 To what extent can memory be suitably supported by technologies?

On the other hand, negative attitudes towards implantables were expressed by an interviewee.

> 04ME: I've used neither wearables nor implantables. But, I'm interested in wearables to keep healthy. However, I don't like the idea of using implantables like brain chips for non-medical purposes. I hope to spend my whole life as a flesh-and-blood person: "a real human being". I feel resistant to implantables from an ethical point of view. If I were to feel that an implantable is a unified part of my body, I think that would be worrying.

His opinions evoke further research questions related to technology-driven human evolution which should be considered taking various cultural and religious circumstances into account, given that our physical and mental abilities have already been enhanced by various carryable or wearable technologies on one level or another.

c   Technology-driven human evolution:
   c-3   What is a *real* human being?
   c-4   Is it unethical for a healthy person to become a cyborg using implantables?
   c-5   What kind of technology-based body modification can be socially acceptable?

4.2.2   Regulation on body modification

Several interviewees insisted on the necessity of limits or regulation on the use of wearables and implantables. This may be associated with Japanese people's negative view on unnecessary body modification.

> 06HS: Generally speaking, if healthy people can improve or enhance their abilities using a technology as they want, without any limitation and regulation, our society could turn out to be miserable. People could lose their ambition, and any type of competition could lose its meaning. In such a society, people could lose any enjoyment of their life.
> 02LS: I think not everyone should be eligible to use these devices. It might be permissible that professionals such as medical doctors and lawyers use them for professional reasons. Those who want to use such devices may need to be licensed.
> 09IS: Wearables and implantables must not be used with a criminal intention. Also, the devices should not be used for monitoring people's behaviour. In addition, the use of these devices by those who don't understand the risks associated with it has to be prohibited.
> 07SSL: If these devices can enhance human abilities far beyond innate ones, we need to carefully consider the ways of using the devices. From my viewpoint of a legal scholar, and from my experiences with the discussion

about Japan's Organ Transplant Act in the mid-1990s, this consideration should focus on what legal rules are formulated for the use of such novel devices, taking philosophical and religious values in a jurisdiction into account. In any context, we need a social debate in advance of their broad usage on the acceptable ways of using wearables and implantables so as to protect human rights.

07SSL also expressed her concern over the tie between the devices and commercialism, suggesting the necessity of regulating companies which deal body modification technologies.

07SSL: When the devices are sold by for-profit companies, negative effects on human physical and/or mental health exerted by the use of them would be hard to notice, or would be ignored by rampant commercialism, and the devices would soon spread widely.

On the other hand, with regard to the use of brain chips to enhance healthy people's intellectual capabilities, a robotics researcher pointed out that such brain chip usage would be unstoppable, and an information security expert supported allowing their use with an eye on competition between countries (a Red Queen hypothesis) although he supposed Japanese people's opposition to it.

10RBT: I mostly support the use of brain chips to enhance healthy people's intellectual capability. It is good for human beings to use these devices to prevent brain deterioration. Some people who want to gain an advantage over others will be the first to use these devices. Sooner or later, however, the devices will be used widely just like other technologies.
09IS: I consider this kind of brain chip usage is acceptable, though I think that a large majority of Japanese people would offer opposition. It is no problem that chips are implanted in brains of those who desire to enhance their memory, mental agility and ability. Even though such usage would probably create economic gaps between implanted people and not-implanted people, the gap can be bridged through, for example, building a fair tax system. If brain chip implantation is banned in Japan, the country would be placed at a significant—mainly economic—disadvantage compared to other countries which press forward with allowing implantation.

Their discussions make us recognise the necessity to consider regulations on, or social norms against, body modification, including those applicable to activities of private sector companies, as part of cyborg ethics research. It is desirable to establish the regulations or social norms in advance of the widespread use of wearables and implantables, to proactively address potential ethical issues. On the other hand, such regulations or norms may conflict with the autonomy of individuals wishing to become cyborgs and their morphological freedom as discussed above.

d Regulation on human body modification:
  d-1 What legal regulations on or social norms against body modification should be established?
  d-2 Should restrictions on the use of body modification technologies be avoided so as not to undermine national competitiveness?
    d-2-1 Is it socially acceptable that central/local governments of a country permit or even encourage citizens to become cyborgs to gain and maintain the country's competitive advantage?

4.2.3   Violation of human dignity

Two interviewees expressed concerns about the social risks and problems entailed in using wearables and implantables, in particular potential violations of human dignity and digital eugenics or a new type of eugenics enabled by the use of wearables and/or implantables.

> 02LS: The line between healthy and disabled people, as well as between normal and abnormal ageing, is intrinsically vague. Physical decay and cognitive degeneration with ageing are nothing special, though most people have a negative image of these. If wearables and implantables can be used to repair such physical and cognitive damages, those who can afford to use the devices would willingly use them. This means that those who cannot afford to do so might no longer be seen as ordinary human beings, and thus this relates to human dignity. People whose physical and/or intellectual functions have decreased due to ageing are, of course, human beings. However, I'm afraid, in the near future, aged people would not be regarded as human beings unless their dotage were repaired using implantables such as brain chips.
>
> 07SSL: I'm concerned that the enhancement of human ability using these devices would easily lead to the justification of eugenics. This would violate human dignity, while supporting meritocratic values. In addition, considering there may be an unpredictable detrimental effect on the mental health of those who implant, say, a brain chip, the discussion about the pros and cons of the implantation should carefully be conducted even when a first-person informed consent is given.

Actually, we may need to endeavour to prevent the situation where those who are considered to have superior genes, or are economically well-off, have priority use of brain chips from occurring in the near future.

In terms of the use of a brain chip to enhance healthy people's brain power, another interviewee raised concerns about challenges to human dignity.

> 04ME: I very strongly oppose to the use of brain chips to enhance brain power for healthy people. The reasons are, first, I'm concerned that once a brain chip is used for any specific purpose, it might then become

permitted for any purpose. Legal regulations on the ways of using it have to be established in advance of the usage, and, secondly, I believe that human beings have to maintain their superiority over machines. Excess device usage could result in a kind of degeneration of human beings, and threaten the basic concept of humanity. This undermines human dignity. If brain chips enhance their users' intellectual capacities to a certain level and standardise users' brain power, the world could lose diversity and individuality, and consequently we would face a poverty of culture as well as of science and technology.

A life scientist also expressed his worry about such brain chip usage, emphasising the economic aspects which can lead to technology-based eugenics, and a social security law scholar presented a similar view.

02LS: Brain chips will be used first for healthcare including treatments for dementia and congenital brain disorder. Once these ways of using brain chips are established, some healthy people would want to use them. Because it is impossible to draw a clear line between healthy and unhealthy people, the usage of brain chips would spread little by little. It is scientifically confirmed that the power of human brain weakens with age. Thus, I consider we need to impose a lowest age restriction for the brain chip use for healthy people. On the other hand, I'm concerned about physical and/ or mental problems caused by the use of a brain chip for healthy people. Such problems would occur even if the parts of the chip which attach to physical objects are created with biomaterials. I also worry about the widening gap between haves and have-nots, that is, between those who can afford to implant brain chips and those who can't. This widened gap would result in the revival of eugenics.

07SSL: I don't support the use, because so little discussion about such usage of a brain chip has been made. We need to carefully discuss about this, and also what a human being is, before allowing usage. Otherwise, companies such as Google or Apple would likely promote the use of brain chips, and many of those labourers who desire to become more capable, and employers who feel in need of more capable employees, would accept the use.

These interviewees' responses suggest the necessity to investigate ethical issues related to the violation of human dignity caused by body modification using wearables and implantables including brain chips. Issues e-2 and e-3 may be important in Japan, in particular, where egalitarian values are respected.

e   Human dignity jeopardised by body modification:
    e-1   What is human existence and dignity in the cyborg age?
    e-2   Who is eligible to get the benefits of body modification?

e-2-1 Should equal opportunity to be a cyborg be guaranteed regardless of gender, age, income level, nationality and so on?

e-3   Is digital eugenics socially acceptable?

e-3-1 Are disparities in economic, political or social power between those who are cyborgised and those who are not socially acceptable?

4.2.4   The use of wearables and implantables in different contexts

The views and opinions of the interviewees on the use of wearables and implantables in various contexts including business, education, research, policing and criminal probe, and warfare were explored. In the context of business, a management engineering researcher pointed out that wearables and implantables were useful to enhance the healthiness of working environment in a factory based on labour monitoring.

04ME: I think wearable and implantable devices for health monitoring can be applied to production management. For example, the vital signs of factory labourers monitored by the devices can be used to recognise any unusual body or mental conditions while at work and help to identify improvements to their work environment, which has become complicated due to the introduction of advanced technologies such as IoT. This would contribute to lowering the turnover rate of factory labourers.

In general, however, it is feared that workplace monitoring would lead to the intensification of labour. In this vein, a life scientist considered that the use of the devices would lead to unfair competition.

02LS: These devices would be used as a tool to make sharp decisions in business settings, where people are required to speed up decision making more and more due to computerisation, especially the introduction of AI systems. However, I'm afraid that the use of these devices would lead to unfair competition in business situations, if the devices enable business people wearing or implanting them to have outstanding cognitive abilities.

Given that business people tend to prioritise getting ahead of the competition, their cyborg technology usage may have to be limited legally or on industry-level self-regulatory basis. For those who work for companies operating in an international business environment, cross-border regulations on wearable and implantable usage may need to be set up. The following research agendas are extracted from the interviews.

a The autonomy of individuals' decision-making on body modification or becoming cyborgs:

a-4   Are employers allowed to encourage their employees to use wearable and/or implantable devices?

    a-4-1   Is workplace monitoring through the processing of data collected from employees' wearables and/or implantables socially acceptable?

f   Fairness in the use of body modification technologies:

    f-1   What is fair usage of wearables and implantables in a business setting?

With regard to education, two interviewees expressed their concern about the use of implantables to boost educational effectiveness.

02LS: Considering that wearables can be worn or attached and removed at will, I expect that a new and effective way of educating students using wearables will be developed. On the other hand, if implantables are used for enhancing students' ability to learn, they might be discouraged from learning hard. Young people, I consider, tend to desire to use these devices. I'm concerned that the social environment in which such devices can be used would undermine the motivation for young people to learn. Thus, an age limit for using these may have to be set.

04ME: The learning and proficiency levels of each student would be able to be measured using these devices. These data would be useful for developing tutoring programmes. On the other hand, I can never agree with the usage of a brain chip for educational purposes. The brain and human capacities are developed by learning and study. The brain chip use with the quest for efficiency in education confounds means with the ends. Human imagination and creativity are developed through a trial and error process. Learning by experiencing failure is important for personality building.

We may soon be able to highly improve the effectiveness of education using implantables like brain chips and moreover may be able to create an artificially gifted child through the use of the devices for young children. However, given that creating genetically engineered babies has been heavily criticised, we should be cautious about such technology usage, taking the moral imperatives and social acceptability of the technology use into account.

g   Morally and socially acceptable usage of body modification technologies:

    g-1   To what extent can learning be supported by the technologies?

    g-2   Is the usage of wearables and/or implantables to enhance efficiency in education morally and socially acceptable?

        g-2-1   What is the meaning of learning for human beings, especially for children and youngsters?

        g-2-2   If knowledge can be directly implanted, does the very concept of learning become obsolete?

An information security expert held an optimistic view of the effects of the prevalent usage of wearables and implantables on research activities.

09IS: Using these devices would make it possible for us to acquire new research findings more and more, because the devices provide us with an unprecedented strong ability to track a wide range of people.

He pointed out the benefit of tracking and monitoring individuals for research, even though high-tech monitoring has been considered by many as potentially harmful to human autonomy and dignity. Another interviewee, on the other hand, expressed his concern about the use of the devices by researchers.

02LS: If someone overtly uses wearables and/or implantables, even brain chips, for his/her research, other researchers would soon follow his/her lead and increasingly demand high-spec devices. However, this would result in the uniformity in research activities, because everyone would tend to rely upon big data and AI systems for their research. In that case, computers, not human beings, would make research findings.

Of course, researchers' covert use of the body modification devices for their work may be problematic. Regardless of whether the device usage by researchers is covert or overt, it should be fair and not undermine the nature of research activities.

b   Personal data protection:
  b-3   For what research purpose are tracking and monitoring of individuals using data collected from the cyborg devices they wear and/or implant socially acceptable?
f   Fairness in the use of body modification technologies:
  f-2   What is the fair usage of wearables and implantables in research?
g   Morally and socially acceptable usage of body modification technologies:
  g-3   To what extent can researchers' abilities be enhanced by body modification?
    g-3-1   What is the nature of research activities and how might this be impacted by these technologies?

In the context of policing and criminal investigations, only one interviewee showed a positive attitude towards the use of wearables and implantables, emphasising the possibility of crime prevention.

09IS: The devices are very useful for policing and criminal probe. Law enforcement agencies would be able to get and utilise the data of those who use the devices. That is, the agencies could ask business operators which collect and store personal data transmitted from the devices to pass those data to them. The data can be used for profiling and persona analyses of individuals because the devices collect individuals' state and behavioural information, and thus for sorting out would-be criminals.

Such issues have already come out with home-based devices such as smart speakers (Heater, 2017) and wearable devices such as fitness monitors (Gill, 2019). However, others worried about the issues such device usage would cause, including the advent of surveillance society.

> 10RBT: The devices will be used for personal identification. Implantables will be implanted in bodies of a wide range of people in the near future. This would be very helpful for crime prevention and counter-terrorism. Simultaneously, however, the world would be placed under government supervision enabled by the devices. This is not science fiction but a frightening reality.
> 07SSL: Many people would easily agree with the use of the devices for policing and criminal probe without deep social debate, because such use seems to contribute safety and security in society. Thus, the government may first propose to use the devices in this context, and then attempt to expand use of the devices. This is scary.
> 02LS: It may be acceptable to use these devices for criminal probe, but this would lead to a situation where criminal probe is indistinguishable from crime prevention. In addition, it's awful and scary that police officers might use these devices, considering their physical and intellectual abilities would greatly be enhanced far beyond ordinary people's ones.

In this context, we would face the trade-off between the benefit of social safety and security and the risks of state surveillance and cyborg cops.

a   The autonomy of individuals' decision-making on body modification or becoming cyborgs:
   a-5   Is it socially acceptable for a superior to encourage or force a subordinate police officer to become a cyborg in order to fulfil his/her duties?
b   Personal data protection:
   b-4   Is it justifiable for police agencies to use the personal data including status and behavioural data collected from body modification devices individuals wear and/or implant in a real-time fashion for crime prevention and counter-terrorism?
g   Morally and socially acceptable usage of body modification technologies:
   g-4   Is it socially acceptable for police officers to be cyborgs using wearables and/or implantables to promote effective policing or ensure the safety of their work environment?
   g-5   To what extent can the physical and intellectual abilities of police officers be enhanced using wearables and implantables?
h   Orwellian surveillance in a cyborg society:
   h-1   To what extent can or should we accept state surveillance of those whose bodies are modified with wearable and/or implantable devices in exchange for social safety and security?
      h-1-1   Is indiscriminate mass surveillance of cyborgised people by government agencies socially acceptable?

The usage of body modification technologies in the field of warfare is one of the most worrying issues. Two interviewees explicitly confirmed their opposition to such use.

04ME: Wearables as well as implantables should not be used for political power, especially for any military purpose.
10RBT: Wearables and implantables should not be used for any military purpose.

However, a robotic researcher and another interviewee presented their perception that the usage of the devices to develop cyborg soldiers would be unstoppable and unlimited.

02LS: Although I oppose any usage of these devices for a military purpose, the military use of the devices is unlikely to be limited. The development race for military-purpose wearables and implantables including brain chips is probably unstoppable. This would widen a gap of war capability between rich, industrial nations and poor, developing countries. It's very hard to establish rules for using the devices for military purposes.
10RBT: These devices can be used to strengthen military capabilities through enhancing soldiers' physical abilities and brain power. This enhancement enables soldiers to effectively use state-of-the-art weapons, including cutting-edge combat fighters whose superior performance sometimes brings about paralysis of the hands of pilots or cause them to faint. Wearables like powered exoskeletons will widely be used by soldiers to help them carry heavy loads. In any country in the world, it is important to decrease the number of its soldiers killed in combat. Thus, there would be no choice but to use these devices for soldiers. This can be characterised as sanity within the insanity of war.

Simultaneously, however, they worried about negative effects on the mind of soldiers the development of cyborg soldiers would bring about, echoed by other two interviewees.

02LS: I've heard that there are many people who enter into military service for money in the US. Given that soldiers have to kill the enemies for their country when necessary, if such US soldiers are forced to use these devices to enhance their abilities and to depend on the devices to do their jobs, the use of the devices would have a detrimental effect on the mental health of them. Enhancing innate physical and intellectual abilities of soldiers by the devices might be an invasion of human rights of them.

10RBT: Soldiers may be made artificially fearless using brain chips. This has to be banned.

07SSL: In the field of warfare, the devices would attempt to be used to mitigate soldiers' fear in battle fields. However, this usage has to be prohibited.

01DI: Cyborgisation reduces the psychological burden on soldiers, making it easier for commanders to make decisions to wage war and thus making war more likely to occur.

Another interviewee, however, expressed his relatively positive view on the usage of wearables and implantables for war.

09IS: These devices can be utilised in the most useful ways for military purposes. Through the devices soldiers wear or implant, each soldier could be monitored in a real time fashion. The monitoring is very useful for health and performance control of soldiers in battlefields. Good practices in war time could be found through analyses of the monitored data.

Based on these discussions, the following research agendas are elicited.

a The autonomy of individuals' decision-making on body modification or becoming cyborgs:

    a-6 Is it socially acceptable to force a soldier to use body modification devices to enhance his/her physical and/or cognitive capabilities by military order?

e Human dignity jeopardised by body modification:

    e-4 To what extent can a soldier be a cyborg using wearables and implantables?

    e-5 What kind of cyborg soldiers is socially acceptable?

    e-6 Is it acceptable that a cyborg soldier kills a human soldier?

        e-6-1 Is a battle between flesh-and-blood soldiers and cyborg soldiers acceptable?

        e-6-2 Is a battle between cyborg soldiers and robot soldiers acceptable?

        e-6-3 Should the law of war be reviewed and revised in accordance with the development of cyborg soldiers?

        e-6-4 Are terrorism or urban guerrilla tactics to beat cyborg troops practiced by irregulars in a poor country acceptable?

4.2.5 Views on Japanese sociocultural environment surrounding technology

Interviewees were asked their views on the Japanese sociocultural environment surrounding technology in general and implantables. Two of the interviewees considered that Japanese people generally had conservative attitudes towards the development and usage of technology and didn't understand the nature of technology.

> 02LS: Japanese people are not good at collaboratively developing new technology due to their poor communication skill. There are inventive people in Japan. But, ordinary Japanese people tend to hesitate to positively evaluate inventive ideas put forward by another Japanese person. Instead, they are prone to work against inventive people. Ordinary people are very conservative on technology, because they don't understand the nature of technology and they dislike any change in living arrangement. Even when Japanese people succeed in developing a technology, they usually fail to construct a system which makes successful use of it.
>
> 09IS: On average, Japanese people don't understand the values underpinning science and engineering. Therefore, they tend to seek perfectionism or a hundred percent performance in the development of technology. Actually, I feel that Japanese culture is not suitable for science and technology which are underpinned by rationalism, reductionism and Christian values. In fact, it is not unusual in Japan that the introduction of technology itself is regarded as an end, not a means. On average, Japanese engineers are very good at developing elemental technology. However, they lack an ability to progress to end products.

An expert in robotics mentioned the Japanese animistic culture.

> 10RBT: Japanese people tend to believe that everything has its own spirit. Thus, instead of changing or enhancing themselves, they seem to like to address the development of a technology to embody that somewhat animistic belief. For example, Japanese engineers don't hesitate to develop autonomous humanoid robots, which behave as if they had their own spirit or emotion. I think this is not the case in Western countries.

According to the opinions of the interviewees, Japanese people seem to tend to have a reluctance to use implantables reflecting their conservative tradition and religious values.

> 10RBT: I think the Japanese tend to accept things as they are. They don't like to enhance the physical and intellectual abilities of human beings in unusual ways.
>
> 09IS: In general, Japanese researchers are conservative, thus they would have feelings of resistance towards implantable usage.
>
> 02LS: In Japan, bioethicists tend to play a role to put a brake on the widespread use of these devices. On the other hand, Western bioethicists, I think, tend to encourage the general public to use these devices. This may

be due to the difference in a religious outlook or a view of life and death between Japan and Western countries. For example, in Japan, the idea that brain death is human death is often criticised and the number of organ transplantations has grown at a sluggish pace. These, I think, reflect the Japanese feelings about bodies of dead people. Traditionally, the idea that bodies should not be injured or wounded at all is drilled into Japanese people. Thus, the devices, especially implantables, will start to be used in Western countries first, and after the widespread use in the West, these will start to be used and gradually spread in Japan.

On the other hand, a gender and computing expert emphasised the importance of appropriate legal regulations on technology developments based on Japanese culture.

03GC: I have an impression that the Japanese are more tolerant of new technologies including robots than people in other countries. This relates to, I think, Japanese culture of shame, not bothering others and sensing the atmosphere. […] However, considering that some of these new technologies are not accepted in particular countries due to social backlash against them for religious, moral or ethical reasons, laws should be enforced to regulate technology development in Japan where such tolerance may weaken the functions of social norms.

These interviewees' views and opinions support the arguments in Section 3 and suggest the necessity to investigate ethical and social issues of body modification with wearables and implantables taking Japanese sociocultural characteristics into consideration, as well as from cross-cultural perspectives.

5   Discussions from cross-cultural perspectives

5.1   *From a British perspective*

While tattoos and piercings have not become stigmatised in the UK to the extent they have in Japan, anything beyond piercing of the earlobe for women (and, more recently, men) has been relatively uncommon until recently. Social class also correlates with piercing: those from more affluent groups are less likely to have piercings (Bone et al., 2008). Organ transplants, on the other hand, are regarded as acceptable by the vast majority of the population (Coad et al., 2013). Attitudes to wearable devices in the UK are, as most places, somewhat under-studied. Moran et al. (2013) provided modest evidence that British people may be more wary of the privacy implications of wearable devices than people in Japan. A 2019 survey carried out by YouGov on behalf of UK Children's charity Barnardo's (2019) reported that 22% of adults and 31% of children believed that within 30 years one of our primary modes of communications would be via implanted technology, although the survey simply reported their expectations and not their desire to do so themselves.

The UK has one of the largest defence budgets in the world, and the UK's military technology research is correspondingly significant and well known around the world. Their early 2000s "Future Infantry Soldier Technology System (FIST)" is now in service and integrates wearable technology for communications and targeting.

The idea of inserting an RFID chip into a pet as a less easily removed ID than a collar has been so well accepted in the UK that the government plans to make it a requirement for cat owners (UK Government, 2021) following a public consultation in which 99% of the responses supported the move. In 1998, when the University of Reading (UK)'s Professor Kevin Warwick implanted such a chip in his arm, it was ridiculed by many as a pointless exercise. However, by 2018, a UK firm had begun offering such chip insertion for physical and computer access authorisation. Reaction to the proposal by both business leaders and trade union leaders was quite negative (Kollewe, 2018), however, with the biggest issue being consent on the worker representative side and employee relations on the management side. Such implants are legally allowed in the UK, but clearly even this simple kind of implant raises questions in the minds of people in the UK. However, in keeping with their willingness to impose chipping of pets on their owners, the UK's government has accepted the inalienable nature of the technology that Neil Harbisson has implanted in his brain, initially to enable the colour-blind man to "listen to" colour, but which he claims has given him broader abilities. The device includes a brain implant but also an external antenna which reaches from near the crown of his head to above his eyes. As reported by Zamfir (2018), the UK government permitted a passport ID photo including the antenna (passport photos are rigidly controlled and do not allow the wearing of glasses for example even if the passport holder is effectively blind without them). Harbisson (a UK national raised in Spain) thus claimed to be the first government-recognised cyborg.

One of the best examples we have so far of public reaction to wearables (which provide something of a guide to how future implantables may be received) was the aborted Google Glass product. Just as the Japanese experts had varied reactions to the acceptability of various technologies, so do different groups in the UK. One UK Google Glass tester found most people interested, and a few concerned about privacy (Hague, 2014) but with commercial organisations such as cinema operators and fitness gym operators expressing concerns about copyright and bodily privacy issues.

Concerns about privacy, inequality and other ethical concerns are raised by UK academics even by such simple implants as a payment chip (similar to that used in contactless credit/debit cards and travel payment systems such as Japan's SUICA and London's Oyster) although some UK academics are also involved in their development (Latham, 2022). This echoes the concerns of the interviewees and suggests that people in the UK agree that regulation is needed to limit the deployment of implantable technology.

5.2   *From a Spanish perspective*

The analysis of ethical and social concerns about the emerging technologies is influenced by cultural environment. Comparing the cultural values in Japan with the Spanish ones, we can find different cultures. Spanish culture is based on Christianity (Floristan, 2003) presenting important differences with Confucian ethical principles. Spanish cultural attitudes to body modification differ considerably from those in Japan. Spain is the European leader in human organ donation, with 38 people donating their organs per million of population (La Moncloa, 2021). Spain is the second country of the world in number of transplants with 4,427 transplants done in 2020, while Japan showed 2,237 according to Global Observatory on Donation and Transplantation (2021). If we consider this data in relation with the total population (Japan has approximately three times the population of Spain, making this a 1:6 ratio per capita), we can see an important difference that could be influenced by the different ethical approach to body modification.

Body modification is an important aspect of cyborg technology development. A cross-cultural study comparing Japanese and Spanish acceptance of cyborg technologies in young people showed that ethical awareness is the variable with the most explanatory capacity concerning the intention to become a cyborg in both countries (Murata et al., 2019). In addition, this study extracted unexpected findings about perceived risk. Based on the aforementioned cultural differences, it should be expected that Japanese society showed a lower level of acceptance of cyborg technologies; however, the survey results demonstrated quite similar levels of acceptance in both countries for this technology.

Another relevant aspect that should be considered is the perceptions of cyborgs in films, comics and literature. Media is a cultural weapon that influences the acceptance of technology. Japanese culture has a long tradition with regard to cyborgs and futuristic technologies, but Spain lacks this cultural heritage. Globalisation is changing this situation. Global streaming services such as Netflix and Apple TV+ create content about these futuristic topics and their ethical and social consequences. "Black Mirror" (2011) and "Severance" (2022) are excellent examples that show the dark side of technology, developing a critical view against the dominant technological determinism. In this global context, Japanese culture is also spreading all over the world. In Spain, 20 of the 100 most read books in November 2021 were manga (Nevado, 2022). This globalisation of the media is influencing the ethical perceptions of people regarding technology.

According to Kunst (2022), 40% of respondents to the Statista Global Consumer Survey used wearables (e.g. smartwatch and fitness/health tracker). With such a penetration, we can consider that wearables have left the experimental phase and become an emerging technology in the market in Spain as in many other countries. In the Spanish wearables market, Xiaomi/Mi is in the first position among the top 10 most used wearables brands (23%), and Huawei is in 4th (13%). Apple (19%) and Samsung (14%) are in the 2nd and 3rd places (Statista, 2022). This market situation is important from the economic, social, and ethical impacts of wearables and cyborg technologies. The fact that Chinese companies, rather than US companies about which an interviewee (09IS) expressed his concern, are among the leaders in this market raises many concerns about the use of the information collected by implantables and wearables. If we consider the ethical problems of information collected by digital technologies, such as the Snowden case (Murata et al., 2017b), we can clearly see the huge challenge with wearables and cyborg technologies geo-economically. China is competing in emerging technology markets, such as social media with TikTok, surpassing 3.5 billion downloads in the first quarter of 2021, becoming the first app not owned by Meta (Facebook's new parent company) to cross this threshold (Sensor Tower, 2022). A study conducted by BuzzFeed News proved that Chinese engineers accessed the data of US clients outside the app, such as calendar information and GPS location among others, leading to speculation about the possible existence of backdoors to access users' data (Baker-White, 2022). TikTok data collection can be considered as more aggressive than any of their competitors. This Chinese global technological leadership, considering its political peculiarities, raises important ethical and social considerations with such sensitive technologies as wearables and implantables.

6   Conclusions

This chapter includes theoretical considerations and qualitative empirical investigations of the social acceptance of body modification via wearable and implantable technology in Japan, where people tend to have negative views of unnecessary body modification. Through semi-structured interviews with ten experts in various fields, their attitudes towards the technologies were explored, and their expectations for and concerns about the body modification technologies in various contexts, which directly relate to ethical and social values such as human autonomy, human dignity, and fairness, were investigated. These offered future prospects for cyborg ethics study. Forty four research agenda points were elicited, which fall into eight categories.

The study in this chapter reveals the necessity of establishing a social consensus as to "what is a human being" in parallel with the study on cyborg ethics. The responsibility of individuals and organisations involved in the development and/or use of wearables and implantables should also be examined. In addition, as suggested by the discussions in Section 4, further survey research in countries other than Japan using a similar semi-structured interview sheet would provide us with fruitful findings in terms of the ethics of body modification using wearables and implantables. In particular, interviews in those countries whose cultures and values are definitely different from Japan, as well as from each other, are of high interest. Interviews with implantable users, whom the authors have not found in Japan, would contribute to progress in the study.

Acknowledgement

This study was supported by the JSPS (Japan Society for the Promotion of Science)/ STINT (Swedish Foundation for International Cooperation in Research and Higher Education) Bilateral Joint Research Program "Information and Communication

Technology for Sustainability and Ethics: Cross-national Studies between Japan and Sweden" (JPJSBP120185411 and JA2017-6999), the JSPS Grant-in-Aid for Scientific Research (C) 20K01920 and (C) 22K02063, the Kurata Grants subsidised by the Hitachi Global Foundation, and the Meiji University Grant-in-Aid for the international collaborative research project "Cyborg Ethics".

Note

1 Wurmser (2022) reported that 21.8% of the US population used a smart wearable at the end of the same year.

References

Adam, N., & Wilkes, W. (2016, September 18). When information storage gets under your skin: Tiny implants can replace keys, store business cards and medical data—and eventually a lot more. *The Wall Street Journal*. http://www.wsj.com/articles/when-information-storage-gets-under-your-skin-1474251062

Baker-White, E. (2022, June 17). Leaked audio from 80 internal TikTok meetings shows that US user data has been repeatedly accessed from China. *BuzzFeed News*. https://www.buzzfeednews.com/article/emilybakerwhite/tiktok-tapes-us-user-data-china-bytedance-access

Barnardo's. (2019, September 21). Body implants and holograms are the future of communication, survey says. *Barnardo's*. https://www.barnardos.org.uk/news/body-implants-and-holograms-are-future-communication-survey-says

BBC. (2020, December 9). France to start research into 'enhanced soldiers'. *BBC*. https://www.bbc.com/news/world-europe-55243014

Bellah, R. N. (1985). *Tokugawa religion: The roots of modern Japan*. Free Press.

Benabid, A. L., Costecalde, T., Eliseyev, A., Charvet, G., Verney, A., Karakas, S., Foerster, M., Lambert, A., Morinière, B., Abroug, N., Schaeffer, M.-C., Moly, A., Sauter-Starace, F., Ratel, D., Moro, C., Torres-Martinez, N., Langar, L., Oddoux, M., Polosan, M., Pezzani, S., Auboiroux, V., Aksenova, T., Mestais, C., & Chabardes, S. (2019). An exoskeleton controlled by an epidural wireless brain-machine interface in a tetraplegic patient: a proof-of-concept demonstration. *The Lancet Neurology*, *18*(12), 1112–1122. https://doi.org/10.1016/S1474-4422(19)30321-7

Black Mirror (Seasons 1–5) [TV series episode]. (2011). Netflix. https://www.netflix.com/es-en/title/70264888

Bone, A., Ncube, F., Nichols, T., & Noah, N. D. (2008). Body piercing in England: A survey of piercing at sites other than earlobe. *BMJ*, *336*(7658), 1426–1428. https://doi.org/10.1136%2Fbmj.39580.497176.25

Browne, R. (2016, March 7). U.S. military spending millions to make cyborgs a reality. *CNN*. http://edition.cnn.com/2016/03/07/politics/pentagon-developing-brain-implants-cyborgs/

Cabinet Office. (2015). *The 5th science and technology basic plan*. http://www8.cao.go.jp/cstp/kihonkeikaku/5basicplan_en.pdf

Christiansen, J. B., & Leigh, I. (2002). *Cochlear implants in children: Ethics and choices*. Gallaudet University Press.

Clark, A. (2003). *Natural-born cyborgs: Minds technologies, and the future of human intelligence*. Oxford University Press.

Clark, L. (2014, November 11). Hand-implanted NFC chips open this man's bitcoin wallet. *Wired*. http://www.wired.co.uk/article/mr-bitcoin-nfc-implant

Clynes, M. E., & Kline, N. S. (1960). Cyborgs and space. *Astronautics*, *5*(9), 26–27; 74–76.

Coad, L., Carter, N., & Ling, J. (2013). Attitudes of young adults from the UK towards organ donation and transplantation. *Transplantation Research*, *2*(1), 1–5. https://doi.org/10.1186%2F2047-1440-2-9

Eveleth, R. (2016, May 24). Why did I implant a chip in my hand? My so-called cyborg life. *Popular Science*. http://www.popsci.com/my-boring-cyborg-implant

Firger J. (2015, December 5). 'Tec tats' usher in new generation of wearables. *Newsweek*. http://europe.newsweek.com/tech-tats-usher-new-generation-wearables-401536?rm=eu

Floristan, C. (2003, May 29). Las raíces cistianas de Europa [The Christian roots of Europe]. *El País*. https://elpais.com/diario/2003/05/29/opinion/1054159209_850215.html

Gill, J. (2019, October 2). Fitbit data provides clues in murder case: eDiscovery & criminal investigation. *IPRO*. https://ipro.com/resources/articles/fitbit-data-ediscovery-criminal-investigation/

Global Observatory on Donation and Transplantation. (2021, December). International report on organ donation and transplantation activities: Executive summary 2020 [PowerPoint slides]. *Global Observatory on Donation and Transplantation*. http://www.transplant-observatory.org/2020-international-activities-report/

Greiner, S. (2014). Cyborg bodies—Self-reflections on sensory augmentations. *NanoEthics*, *8*(3), 299–302. https://doi.org/10.1007/s11569-014-0207-9

Guler, S. D., Gannon, M., & Sicchio, K. (2016). *Crafting wearables*. Apress.

Guy, J. (2020, December 9). French army gets ethical go-ahead for bionic soldiers. *CNN*. https://edition.cnn.com/2020/12/09/europe/french-army-soldiers-technology-ethics-scli-intl-scn/

Haddow, G., King, E., Kunkler, I., & McLaren, D. (2015). Cyborgs in the everyday: Masculinity and biosensing prostate cancer. *Science as Culture*, *24*(4), 484–506. https://doi.org/10.1080/09505431.2015.1063597

Hague, S. (2014, July 2). Google glass: From a British perspective—Review. *KnowTechie*. https://knowtechie.com/google-glass-uk-review/

Heater, B. (2017, February 24). Amazon cites First Amendment protection for Alexa in Arkansas murder case. *TechCrunch*. https://techcrunch.com/2017/02/23/alexa-free-speech/

Imamichi, T. (2009). *An introduction to eco-ethica*. University Press of America.

Ishiguro, H. (2009). ロボットとは何か—人の心を映す鏡 [What is a robot? A mirror of the human mind]. Kodansha.

Ishiguro, H. (2014). どうすれば「人」を創れるか—アンドロイドになった私— [How to create a human being? I become an android]. Shinchosha.

Ishiguro, H. (2015). アンドロイドは人間になれるか [Can an android become a human being?]. Bungeishunju.

Ishiguro, H., & Washida, K. (2011). 生きるってなんやろか？ [What is life?]. The Mainichi Newspapers.

Japan Organ Transplant Network. (2021). *News Letter*, 25. https://www.jotnw.or.jp/files/page/datas/newsletter/doc/nl25.pdf

Kaji, N. (2011). 沈黙の宗教—儒教 [Religion of silence—Confucianism]. Chikumashobo.

Kollewe, J. (2018, November 11). Alarm over talks to implant UK employees with microchips. *The Guardian*. https://www.theguardian.com/technology/2018/nov/11/alarm-over-talks-to-implant-uk-employees-with-microchips

Kondo, I. (2019, February 14). 認知症ケアおよび介護におけるAIおよびロボットの活用 [The use of AI and robots in the fields of dementia and nursing care] [PowerPoint slides]. 5th Meeting of the Consortium for Accelerating AI Development in Health Care. https://www.mhlw.go.jp/content/10601000/000478772.pdf

Kunst, A. (2022, May 18). Consumer electronics usage in Spain 2022. *Statista*. https://www.statista.com/forecasts/1001451/consumer-electronics-usage-in-spain

Labios, L. (2022, May 9). Multi-tasking wearable continuously monitors glucose, alcohol, and lactate. *US San Diego News Center.* https://ucsdnews.ucsd.edu/pressrelease/multi-tasking-wearable-continuously-monitors-glucose-alcohol-and-lactate

La Moncloa. (2021, August 16). España mantiene su liderazgo mundial en donación de órganos en 2020, a pesar de la pandemia [Spain maintains its world leadership in organ donation in 2020, despite the pandemic]. *La Moncloa.* https://www.lamoncloa.gob.es/serviciosdeprensa/notasprensa/sanidad14/Paginas/2021/160821_donacion2020.aspx

Latham, K. (2022, April 11). The microchip implants that let you pay with your hand. *BBC News.* https://www.bbc.com/news/business-61008730

Legge, J. (trans.) (2010). *The Hsiao King or Classic of Filial Piety.* Kessinger Publishing.

Ministry of Economy, Trade and Industry. (2013, July). 2012年 ロボット産業の市場動向 [Trends in the market for the robot industry in 2012] [PowerPoint slides]. https://www.jara.jp/various/report/img/20130718002-3.pdf

Ministry of Internal Affairs and Communications. (2016). *White Paper 2016—Information and Communications in Japan.* https://www.soumu.go.jp/johotsusintokei/whitepaper/eng/WP2016/2016-index.html

Ministry of Internal Affairs and Communications. (2020). 令和2年通信利用動向調査報告書（世帯編）[Communications usage trend survey in 2020 (Household)]. https://www.soumu.go.jp/johotsusintokei/statistics/pdf/HR202000_001.pdf

Monks, K. (2014, April 9). Forget wearable tech, embeddable implants are already here. *CNN.* http://edition.cnn.com/2014/04/08/tech/forget-wearable-tech-embeddable-implants/

Moran, S., Nishida, T., & Nakata, K. (2013). Comparing British and Japanese perceptions of a wearable ubiquitous monitoring device. *IEEE Technology and Society Magazine, 32*(4), 45–49. https://doi.org/10.1109/MTS.2013.2286419

Murata, K., Adams, A. A., Fukuta, Y., Orito, Y., Arias-Oliva, M., & Pelegrín-Borondo, J. (2017a). From a science fiction to reality: Cyborg ethics in Japan. *ACM SIGCAS Computers and Society, 47*(3), 72–85. https://doi.org/10.1145/3144592.3144600

Murata, K., Adams, A. A., & Lara Palma, A. M. (2017b). Following Snowden: A cross-cultural study on the social impact of Snowden's revelations. *Journal of Information, Communication and Ethics in Society, 15*(3), 183–196. https://doi.org/10.1108/JICES-12-2016-0047

Murata, K., Arias-Oliva, M., & Pelegrín-Borondo, J. (2019). Cross-cultural study about cyborg market acceptance: Japan versus Spain. *European Research on Management and Business Economics, 25*(3), 129–137. https://doi.org/10.1016/j.iedeen.2019.07.003

Murata, K., & Orito, Y. (2011). The right to forget/be forgotten. *Proceedings of CEPE 2011,* 192–201.

Nevado, R. (2022, July 31). El boom del manga en España: cerca de 30 editoriales publican cómic japonés en nuestro país [The manga boom in Spain: Nearly 30 publishers are bringing Japanese comics to Spain]. *RTVE.* https://www.rtve.es/noticias/20220731/boom-manga-espana/2386130.shtml

Noe, K. (2013). iPS細胞と生命倫理 [Bioethics concerning iPS cells]. 学術の動向 [*Trends in the Sciences*], *18*(2), 26–29. https://doi.org/10.5363/tits.18.2_26

Nora, D. (2015, March 18). La puce dans la peau [The chip in the skin]. *L'Obs.* https://www.nouvelobs.com/l-obs-du-soir/20150318.OBS4923/la-puce-dans-la-peau.html

Okita, T., Hsu, E., Aizawa, K., Nakada, H., Toya, W., & Matsui, K. (2018). Quantitative survey of laypersons' attitudes toward organ transplantation in Japan. *Transplantation Proceedings, 50*(1), 3–9. https://doi.org/10.1016/j.transproceed.2017.11.011

Otake, M. (2019, February 14). 防ぎうる認知症にならない社会に向けた技術開発を起点とする取り組み [Initiatives towards building the society free of preventable dementia starting with the development of technologies] [PowerPoint slides]. 5th Meeting of the

Consortium for Accelerating AI Development in Health Care. https://www.mhlw.go.jp/content/10601000/000478791.pdf

Park, E. (2014). Ethical issues in cyborg technology: Diversity and inclusion. *NanoEthics*, *8*(3), 303–306. https://doi.org/10.1007/s11569-014-0206-x

Rogers, E. M. (1962). *Diffusion of innovations*. Free Press.

Sandberg, A. (2013). Morphological freedom—Why we not just want it, but need it. In M. More, & N. Vita-More (Eds.), *The transhumanist reader* (pp. 56–64). Wiley-Blackwell.

Sensor Tower. (2022, April). Q1 2022: Store intelligence data digest [PowerPoint slides]. *Sensor Tower*. https://go.sensortower.com/rs/351-RWH-315/images/Sensor-Tower-Q1-2022-Data-Digest.pdf

Severance (Season 1) [TV series episode]. (2022). Apple TV+. https://tv.apple.com/us/show/severance/umc.cmc.1srk2goyh2q2zdxcx605w8vtx

Shimpo, F. (2016). ロボット法学の幕開け [The beginning of robot law]. *Nextcom*, *27*, 22–29.

Starr, M. (2016, May 2). Sony patents contact lens that records what you see: Sony has been awarded a patent for a smart contact lens that would be capable of recording video. *CNET*. https://www.cnet.com/news/sony-patents-contact-lens-that-records-what-you-see/

Statista. (2022, August). Wearables: Xiaomi/Mi users in Spain. *Statista*. https://www.statista.com/study/74047/wearables-xiaomi-mi-in-spain-brand-report/

Takada, H. (2015). 幹細胞を利用した再生医療における法規制と生命倫理 : ES細胞とiPS細胞の利用を例に [Legal regulations and bioethics on regenerative medicine using stem cell—Focusing on utilization of ES cell and iPS cell]. *富山大学紀要. 富大経済論集* [*Journal of Economic Studies, University of Toyama*], *61*(1), 1–29. https://doi.org/10.15099/00002028

Takanishi, A. (2012, July 18). Leading the world in humanoid robotics (interviewed by J. Harano). *Nippon.com*. https://www.nippon.com/en/views/b00901/

UK Government. (2021, December 4) Cat microchipping to be made mandatory. https://www.gov.uk/government/news/cat-microchipping-to-be-made-mandatory

Wurmser. (2022, January 13). Wearables: Steady growth in users foreshadows bigger changes to come. *eMarketer*. https://www.insiderintelligence.com/content/spotlight-wearables

Yamamoto, Y. (1997).「文身禁止令」の成立と終焉—イレズミからみた日本近代史—[The enforcement and lifting of the "ban on tattooing": Modern Japanese history from the perspective of tattoos]. *政治学研究論集* [*Journal of Politics (Graduate School of Political Science and Economics, Meiji University)*], *5*, 87–99.

Yamamoto, Y. (2017, January 30). "Irezumi": The Japanese tattoo unveiled. *Nippon.com*. https://www.nippon.com/en/views/b06701/

Yamanaka, T. (2019, August 23). 科学が「神の領域」に近づいた今、改めて「宗教の役割」が見直される時代がくる [As science approaches the "realm of the divine", the time is coming when the "role of religion" will be re-evaluated]. *Diamond Online*. https://diamond.jp/articles/-/211972

Yano Research Institute. (2016, May 16). ウェアラブルデバイス世界市場に関する調査を実施（2016年）[Press release: A survey on the global wearable device market in 2016]. https://www.yano.co.jp/press/press.php/001535

Zamfir, G. (2018, February 14). 'World's first cyborg' with antenna in skull who says 'government recognises him as half-robot'—Claims he's first of many. *The Mirror*. https://www.mirror.co.uk/news/uk-news/worlds-first-cyborg-antenna-skull-12025353

Part II

Creating Ethical and Sustainable Digital Cultures

8 The ascent of memetic movements

Social media, Levinasian ethics and the global spread of Q-anon conspiracy theories

Rickard Grassman, Ryoko Asai and Matthew Davis

1 Introduction

Indeed, the prospects of social media as a source for democracy dimmed already in the aftermath to the Arab spring of 2011 (cf. Kyriakopoulou, 2011). Nevertheless, few could have predicted this seemingly democratized form of information exchange would not just fail to reform authoritarian regimes initially signaled by the Arab spring, but backfire to the extent of the Trump administration taking office in 2017, and culminating in a violent attack on the very house of US democracy itself. It appears social media may be as subversive of democracies as they have been of authoritarian regimes, and it is essentially down to the type of narratives that gain the most traction at certain moments of time (Castells, 2012).

If there are genuine grievances against the government for instance, the democratic argument holds that the various protected rights we enjoy through democratic institutions ensure that whatever story is bound to resonate should be free to do so. Conversely, the more realpolitik logic of authoritarianism relies on the ability to suppress certain stories from ever being voiced in the first place (Kyriakopoulou, 2011). This is especially true for stories that may trigger widespread followings by way of consolidating grievances, which is why social media played such a subversive role in the Middle East of 2011.

Indeed, the Arab spring was essentially the very moment in which the emerging peer-to-peer dynamic of our contemporary social media landscape could no longer be contained through traditional authoritarian tactics of repression (Haas and Lesch, 2017). Under the dense shade of authoritarian government, the increasingly horizontal dynamic of the so-called Web 2.0 was beginning to give voice to one of the perhaps most powerful and long-lasting stories of plight among commoners, consolidating movements and grievances from left to right in the face of endless and concentrated wealth at the top (Castells, 2012).

However, when the common enemies are finally unsettled in and with the dissemination of such stories, and the democratic movements need to work out their own inherent antagonisms along religious and ideological lines (cf. Laclau, 2005), the stories become less resonant. At the end of the day, stories that resonate

DOI: 10.4324/9781003367451-10

in democracies may differ significantly from those under the yoke of authoritarian regimes. It turns out that in the face of latent ethnic and cultural tensions disrupting the smooth running of democratic processes, some may even prefer the authoritarian model to keep things in check. In this view, the memetic logic is harnessed for the very inversion of democratic activism against authoritarian abuses, perhaps never as clearly evidenced as with the storming of the Capitol in 2021.

Even in a mature democracy like that of the US in 2020, there is apparently a very potent story through which people are willing to accept an obvious lie about a free and fair election being stolen, in spite of countless evidence to the contrary. The reason being that this new "reality", widely perceived through a web of narratives on social media of which Q is a case in point, has successfully overshadowed the authoritarian/democracy antagonism with the in-group/out-group divide of the typical racist and populist bent. In other words, this is to say, better to have a leader that is one of us than to keep our democracy going and with it the risk of an out-group candidate prevailing at some point. The Capitol insurrection of January 6, 2021 epitomizes this insidious new dynamic of inherently fear mongering and racist content, structuring an interactive complex of conspiracy-oriented material such as the Q-anon variety transmitted via social media in a form that appears more potent and likely to incite real-world action then hitherto seen.

In this chapter, we explore this new dynamic of an emerging memetic logic in online communications as much as in the real-world action it inspires, which may be as disruptive of democratic institutions of authoritarian regimes. To do so, we will first investigate what exactly it is that online mediation of content offers in this emergent new dynamic. Paralleling McLuhan's prescient dictum that the medium is the message, the aim is here to accentuate the process whereby mediation seems to take precedence over content generation.

In the next section on ethics and conspiracism, we will unpack this process of mediation further, insofar as it may have a significant impact on the phenomenon of conspiracy theories. Furthermore, this in turn may be of critical importance in the appreciation of what the ethical implications of such a phenomenon may be, and how such implications may be fruitfully conceived through the theoretical lens of Levinas. We then turn to a section on memes and movements that explores the memetic structure's impact on social movements in order to make clear, not just the subjective biases of memetic conspiracism, but ultimately to broaden the ethical scope to better account for the intersubjective potency of viral dissemination.

This will concomitantly bring us onto foregrounding our methodology and giving a few illustrative samples of social media posts in Sweden and Japan, which will further testify to the cross-cultural contagion of memetic conspiracism. The ultimate outcome of this journeyed landscape will provide a more nuanced backdrop against which we can embark on a theoretical discussion and hopefully offer some fruitful conclusions as to what democratic societies may face on this front, and how we are to protect ourselves and the freedoms we enjoy, against such an insidious new challenge.

2   The medium is the message: Ethics against the backdrop of conspiracism

While there is nothing new with conspiracy theories or populist messages with classic racist undertones accentuating the particularity of "us" the people standing precariously in the face of an all-conspiring universalized and globally diffused "other". What is new is the intermixing of online and offline interactions in this dynamic, whereby identity impulses are less restrained by factual evidence or dissenting voices, as lived experience can increasingly be backlit by the endless possibilities of the screen. This development, as recent and contemporary events, continuously indicates risks letting confirmation bias impulses go unchecked if this new memetic logic realizes what we take to be inherent tendencies of decoupling itself from more reasoned concerns for the credibility of sources.

In particular, we argue, with the ascent of peer-to-peer communication flows of the social media sphere often referred to as Web 2.0, there is a new interactive element involved in this dynamic whereby users are more radically engulfed in the narrative and the construction thereof by the injunction to help generate spin. It is a bit like a game in which what is verified as legitimate has nothing to do with truth value but with how much traction individual posts seem to stir, and the medium here offers its users the perfect vantage point for precisely appreciating such metrics as opposed to anything having to do with actual quality or real-world accuracy. It is essentially the memetic quality as opposed to veracity that the medium seems to pick up on and often has users confound with the latter, and the unfolding complex of narratives seems to be a direct outcome of such microprocesses.

Someone who believes or refers to traditional editorial news articles in the social media circles we have examined here is often dismissed as having taken the "blue-pill", deriving from the cult-classic 1999 movie *The Matrix* in which the protagonist is faced with the option of "blue-pilling" *qua* continuing life as it "is", or taking the "red-pill" and "waking up to the desert of the real" – as it is articulated in the film. In other words, it's a perfect trope for the conspiracist worldview in which things are not as they seem, and underneath "reality", there are layers and layers to be unearthed. Apart from opening up a vertiginous new dimension in which reality itself is at best mere crumbs of what is taken as truth, it also conveys on the subject a greater sense of agency. This comes through in action and captures the creative dynamic of being an active participant rather than passive recipients in the machinery of content generation. It is an appeal to resist the facticity of historical certainties and established authorities.

This instils a much more potent sense of recognition and ownership of the content they help generate and pass along, for the sense of being quite exclusively part of its "discovery". Furthermore, the consensus appearing experience of broad confirmation is quickly there to reinforce the users conviction that he/she may be on to something. This is achieved through an implied priming of suggestive content coded into certain terms like "deep-state", "Soros"-something or "Illuminati"/"Bilderberg" to name a few, which invites the user to become an active participant in a paranoid type of content generation and validation. This coding essentially amounts to a certain priming of how content is to be read, repro-ducing the spirit of conspiracism when it is used and passed along, be it on screen or in the real world.

However, considering the prevalence of fear mongering and tacit or overt racist tropes setting the tone at the message level of the Q-anon phenomenon, it is dif-ficult not to catch the striking irony in the evermore cross-cultural and even uni-versalized online culture that this fetishization of the particular is being conveyed. Indeed, even this memetic and to some extent gamified format, that is a globalized form of language if ever there was one, is here honed in and perfected as a medium to transcend borders albeit with the somewhat less boundary-crossing message that particularity is under assault from a global cabal of universality and otherness.

The Q-anon complex of conspiracy theories is not just potent enough to incite real-world action as evidenced by more than 15% of the rioters arrested for their participation in the January 6th insurrection, but has declared an explicit affilia-tion therewith. More worryingly is its ability to morph and adapt its assemblage of narratives in ways that overcome falsifying evidence, but also to cross-cultural boundaries by way of users integrating localized and culture-specific grievances, inspiring significant followings in countries as diverse as Sweden and Japan.

While conspiracy theories as previously mentioned are as old as human thought itself, they have historically been culture-specific in the sense that albeit striking affinities and similarities at a meta-narrative analysis, they each have their own myths and stories contingent on historical and cultural particularities in the narra-tives themselves. There are however instances, even before Q-anon when theories and tropes of antagonisms have crossed cultural lines.

Anti-semitism more broadly and in particular the infamous fictional book "The Elders of Zion" is a case in point, originating in Russia and coming to play a sig-nificant role in the anti-semitic lore of Eurasian fascism and communism, as much as in contemporary Islamist extremism in the Middle East and beyond. In other words, the tropes of Q-anon, which by the way have incorporated a great deal from this same anti-semitic source of fiction, are from the good old authoritarian playbook of past and present extremisms alike. So it is certainly not the essential content that makes this movement unique, but it is the universalized form and dynamic through which such content has been repurposed for the Internet age to move across borders that is of interest here, so that hatred's oldest hits can start seeing the light of present-day screens from Stockholm to Tokyo.

Indeed, it may seem a bit unexpected that a US-centered political narrative aimed at white grievances in view of migration would generate a significant fol-lowing in Japan, which is not just a very culturally and ethnically distinct and homogeneous country with little to no immigration, but also an old Second World War adversary to the US, in other words, not exactly the most expected nationality or culture to assimilate a US-centric political discourse on nationalist white griev-ances toward immigration, and make it compelling enough to inspire a significant Japanese following.

Nevertheless, this is precisely what we have seen transpire there, as too in places like Sweden, which albeit certainly a distinct culture from the US, it is perhaps less puzzling considering some historical, contemporary and cultural affinities. Sweden is a Christian majority country of predominantly white ethnicity with euro-centric heritage much like that of the US, and both countries have had a significant influx of other cultures through immigration, which is often portrayed by right-wing extremists in either country as an assault on white European identity.

Therefore, putting these diverse cases together, we can deduce that ethnicity, religion or cultural particularities, however potent they may be in their own right, are not the sole factors at play here. Because had it been so, it would be rather unlikely that this complex of narratives would create such a significant following in Japan. Instead what is at play, as we endeavor to tease out, is that there is something to be found operating at the level of the medium as opposed to the more immediate content level of the message as such (cf. McLuhan and Fiore, 1967).

In our day and age, there is an increasingly universal and global Internet culture that transcends spatial cultures in a great many ways, and one way in particular that becomes strikingly visible with phenomena like Q-anon is when content is epiphenomenal on the mediated process through which it is being generated. This process is one we claim to be structured according to a memetic logic, but one that also takes this gamified logic into habituation, stimulating engagement that becomes self-generative in the affective response it evokes among its users (cf. Kim and Werbach, 2016). No place is this more clear than on social media, with the possible exception of YouTube, where the designers of algorithms have long understood what type of content will keep us clicking and swiping our scrolls.

Furthermore, conspiracy theories are often a logical endpoint of said behavior as fiction ultimately provides more room for calibrating the sought-after memetic affect and engagement, than does fact. The phenomenon instils on the user a sense of being an active agent in the unfolding of the present. There are suddenly layers upon layers of intrigue and conspiracy that have been working in the shadows to keep the user in what he or she perceives as an undeservedly precarious place. Instead of faceless shades-of-gray complexities that are unlikely to arouse much sense of passion or purpose, enters a more crudely black and white world of endless conspiracies that renders evil intentionality upon the many unfavorable circumstances the subject is readily able to discern. Not to mention advancing the prospects of the Other who is always in a relatively better place, whether undeservedly from the periphery *qua* immigrants – or from rigged centers of power *qua* elites.

Against this backdrop, we explore the perils that this new technologically mediated malleability of affect may have on ethics, and in particular with regard to Levinas' constitutive view of ethics as contingent on how we respond to the call of the Other. A call, which for Levinas, precedes even the subject formation itself. In other words, the subject does not proceed from the commonly held process typified in the Cartesian cogito, by which subjectivity comes to know itself through putting all else into question.

In Levinas' view, there is never such abstract foundation of subjective autonomy, before working out its obligation to the world or its inhabitants. Instead, responding to the Other is inherent to this very subject formation in the first place. Moreover, acknowledging this continuous call of the Other, rather than trying to discern more dispassionate and neutral obligations from such autonomous found-ings of subjectivity, is where ethics with Levinas takes a rather different route to that of much of Western Enlightenment thinking.

To further tease out what this means in view of the subject matter at hand, we suggest a tripartite taxonomy of subjectification vis-a-vis this constitutive call of the Other, and our ethical agency founded thereby. Firstly, the *memetic-solipsist subjectification* is one of failure to heed this call by way of radically simplifying the Other through the binary prism of good and evil, us and them, etc. This is to say sameness and difference, as in the ultimate building blocks of identity formation, are here made radically malleable through narration and technological mediation in order to calibrate affect for an agential sense of significance as epitomized in social media-driven conspiracies.

This we counterpose against the *commonsensical rationalist subjectification* that have underpinned much of Enlightenment thinking in the West. According to this view, the face of the Other is largely absent on account of more or less "neu-tral" and "dispassionate" complexities of reciprocity, such as inhabiting the mun-dane shades-of-gray world in which the self and other is virtually the same, or at least on an equal footing morally speaking. Where rationality is the dis-passionate realm in which the sentiments and desires of the ego are supposedly kept at bay and thus enabling us to appreciate our ethical responsibilities toward others, and what we should be able to expect from them. Vastly resonant in a broad range of Western philosophy from Cartesian subjectivity to Spinozism, Hegel and Kant espousing or building on from such foundational conceptions.

Finally, the *subjectification in face of the Other* is what we will foreground as an alternative approach more attuned to the ethical theories of Levinas (2011), and one where our ethical obligations go much further. This identity formation builds on the Other's call having been genuinely appreciated by the subject, and our ethi-cal responsibility transcends even such dis-passions and complexities associated with what we here have termed "*commonsensical rationalist subjectification*". In *subjectification in face of the Other,* we cast off the last vestiges of moral sym-metry and instrumental reason in our attempts to appreciate the obligations we have toward others, and we will come back to this point when some of the ethical problems and limitations of the other two subjectifications have been properly laid bare in view of the phenomenon at hand.

3   From memes to movements

Social media platforms in the sense of user-generated content sites have for better or worse empowered the public to post messages that have the potential to reach and impact large audiences (Curnutt, 2012). Memes are a particularly salient phe-nomenon in this as they provide visual graphics that if successful evoke some kind of emotional response that is often in the form of amusement, that sets a certain tone on textual content that may or may not be overtly political (cf. Shifman, 2014). In other words, memes have the inherent potential of nudging people's mindsets in one political direction or other, without the subjects even realizing the extent of such influence because it may appear in the seemingly unbiased guise of humor (cf. Sunstein, 2021). Moreover, as we will see in action throughout our empirical accounts, this memetic dynamic clearly ensnares subjects further through the sense of ownership and active participation, may also significantly increase the potential for nudging its addressees into real-world action be it individually or collectively. What is more, and what is becoming increasingly visible with phenomena like the Q-anon movement is that domestic and regional controversies can morph and migrate so as to become relevant on a broader global scale.

One of the first instances in which the power of memes was effectively harnessed into real-world political action was the rather ingenious mobilizing of the so-called umbrella movement in Hong Kong of 2014 (Xiao Mina, 2021). The name emerged out of the protests against a new law that gave the Chinese government greater control over who would be eligible to be part of Hong Kong's chief executive council, effectively undermining any real sense of democracy playing its course. This in turn is setting off a tidal wave of pro-democracy demonstrations that would use umbrellas to resist the pepper spray counter-protest tactics undertaken by the Hong Kong police.

What made the umbrella meme particularly powerful is attaching political significance to the essentially defensive symbolic meaning it already carries in shielding the individual against the natural elements, which nearly anyone can access and identify with, and yet when the elements translate into the agency of gas-infused police aggression, it becomes the perfect symbol of non-violent struggle against oppression. The cross-fertilization of online virality and real-world enactments became powerful enough as a symbol of democracy in Hong Kong to make the Chinese government terrified of umbrellas for a considerable time to come, prompting a complete ban on the item in the proximity of the Chinese leader Xi Jinping, and a range of drenched press conferences for an extended period were not so unexpectedly the outcome.

Another powerful meme-to-movement phenomenon that we have seen in the face of police brutality is of course that of Black Lives Matter, with the aim of accentuating normalized racism not just in police conduct but in a great many facets that make up contemporary American society. Unlike the umbrella movement and its simple memetic symbol of nonviolence and democratic struggle against the heavy hand of authoritarianism, the political field surrounding race relations in America as well as the memes emerging in its wake are often more fraught and ambiguous, and consequently contingent on the presuppositions and biases of the recipients of the messages brought forth.

As a little personal anecdote from one of the authors to illustrate this point, the native Swedish author in question once had a very close professional relationship with a black woman from the south of the US, at a time he was living in New York. It became extremely clear from countless conversations, that perhaps not so unexpectedly the working pair had completely different experiences of race relations from their respective upbringings. But also concomitantly that unlike for the author, race played such a critical and omnipresent part in terms of identity for this woman in question. However, the point we aim to make here is not the obvious one on the concept of *intersectionality* and how this concept captures how something can be normalizing and thus barely noticeable on one side of these diverging experiences, nor the point of its counter mirror image being deeply oppressive and thus inescapably salient in signaling deviancy of the norm on the other side (cf. Crenshaw, 2017). Rather what we would like to accentuate with this anecdote is that these intrinsically diverging experiences are in one way even more incommensurable against the backdrop of New York that albeit undoubtedly a part of the US and its history, is a significantly multicultural city and one in which race relations occupy a much more nuanced and ambiguous space than that of the deep south. Consequently, just as the author reflecting back on this is convinced that he may have overlooked instances of real racism taking place in the space of their shared experiences that his old colleague may have been more attuned to pick up on, he is equally convinced that there were less warranted instances when racism did not factor into those shared experiences, yet she still experienced it as such. In other words, the filters we acquire throughout our lives go both ways, so to speak. Experiences may exacerbate as much as downplay political tensions, failing to see antagonisms when they exist or seeing them when they do not.

Moreover, this mutual precarity in the interplay between the spoken significance and its reception is particularly incommensurate in spaces of greater flux in signification where taken for granted identities and boundaries may paradoxically be more fluid and yet over-determining at the same time, and Internet culture exceeds even New York in this sense. This is also why memes may not just contain hidden offensive signals or cues embedded in less provocative surface content, *qua* "dog-whistling". Like the apparent metaphor of dogs picking up certain sounds that appear mute to the rest of us, dog-whistling is a way of targeting some audiences that pick up on controversial content that go undetected by a great many others. In other words, only a select few that are either targeted by or sympathetic to the controversial sentiments conveyed will be able to discern the latent significance thereof, which to external parties is "muted", in the sense of seemingly unremarkable. But also here, the filters go both ways, and genuinely non-provocative content may be ascribed to latent performative significance upon reception of a more controversial variety. In other words, the recipients themselves become co-producers to the content of the message.

This inherent asymmetry of content, significance, intentionality and reception is one part of the complex terrain that makes up the memetic field of contemporary Internet culture. It is also a space in which the memetic new dynamic of conspiratorial narrativizing becomes strikingly fruitful, because if the good old difference of the other is harnessed as a basic premise mirroring the racist impulse more broadly, then what better playing field could there possibly be for *confirmation bias* than the endless memetic landscape wherein we can largely co-create a world of meaning that fits and differs in just the right amount. This caters to our various injunctions of self-affirmation, as Michel Houellebecq (2004) captures with his cynical assertion that the world is medium-sized.

4   Methodology

In light of the phenomenon, we investigate in this chapter it is pretty clear that social media is a strong source of identity formation, especially for the disenfranchized. What is more, so-called "fake news" incubated or evolved in such communities online evidently influence the behaviors of many people, be it online or offline, despite the availability of evidence to contradict the narratives in question (Allcott and Gentzkow, 2017; Jaques et al., 2019). Verifiable facts can be rendered irrelevant in the face of a deeply compelling story (Sam, 2019), and memetic dissemination or gamification may heighten the speed at which these narratives spread (cf. Jaques et al., 2019).

Stories, narratives, or discourses therefore are powerful tools which can be combined into local (and in our case virtual) meta-narratives, wielded by the powerful, to influence public opinion (cf. Foucault, 1972). How we interpret the world is thus governed by competing actors and institutions that privilege and prioritize some discourses over others (Sam, 2019). In the case of social media, algorithms feed an insatiable public their desired narrative in order to monetize and maintain engagement with the platform and its users (Zuboff, 2019). But the dynamic and spread of these narratives are, as we will see here, even more complex by virtue of users themselves starting to emulate the memetic characteristics, in content generation, to help facilitate virality. Yet, these narratives are not fixed; they continue to evolve, injected with new pieces of real-world happenings which members of the community then mutate and fit to their own grand narrative. This digimodernism (Kirby, 2009), where technology amplifies the interactive relationship people have with their perceived reality, is exemplified by the context of our study, the Q-anon phenomena.

In the next chapter, we analyze the various discourses present and available in the public domain on social media. Discourse analysis is very much an interpretive process; hence, the positionality of the researcher is an important consideration for such a method (Tracy and Mirivel, 2009). Consequently, our research team is diverse, and we each bring our own unique perspectives in an attempt to balance our analysis.

To build a corpus of data, we use the social media scraping platform notified, which is a paid service that connects to the APIs of various social media and saves public posts into a cloud database for subsequent analysis. It has various tools that can assist us in our work, such as sentiment analysis, but we mostly use this service to automate the process of data collection, and filter the results based on keywords. We have three streams of data, one specifically primed to scrape Swedish posts, one for Japanese, and one for English. The keywords for each are presented in Table 8.1, and were chosen based on previously observed narratives from Q-anon supporting tweets.

In the following section, we explore the different discourses and how they have changed over time. We have chosen a 12-month period to study our data starting from the time of the January 6th Capitol attack, where we look for the dominant

Table 8.1 Keywords for each search stream on the Notified platform

Post location	Keywords	Rationale
Sweden	Q, qanon, q-anon, QAnon, Anon, wwg1wga, deep state, bilderberg, illuminati, djupa staten, cbts, konspiration, Soros, yoga, satan, djävulsdyrkande pedofiler, Gates, blue, satanister	These search terms are indicative of the type of posts that we have been looking for, and have been used to identify relevant social media activity. Although some of the posts we picked up use these terms critically, the terms have enabled us to scrape Q-sympathetic content that help disseminate Q-narratives on conspiracies and global elites.
Japan	Jアノン, Qアーミージャパンフリン, Qアノン, ディープ・ステート/闇の政府, 日本サンクチュアリ協会 /統一教会, 反中·反中国, 反共·反共産主義, ゴム人間, レプティリアン	These search terms have also been selected so as to pick up on common themes in Q-sympathetic content in Japanese, being shared on social media. The selection contains a similar range to that of its Swedish and English counterparts with terms such as Q-anon, and deep state.
English	Qanon, Q, Qarmy, 5G, deep-state, Trump, illuminati, soros, antivax, conspirituality, great awakening, storm, wwg1wga, pastels evil vaccine	These are some of the more common terms used by followers and spreaders of the Qanon theories, found through other readings and prior research.

discourses on a month-by-month basis based on engagement and reach, as measured by the platform. We then dive into these dominant discourses, wherever they may lead, looking at how they evolve over time in relation to one another and world events.

5   Crossing boundaries: Scraping the social media scape of Sweden and Japan

In view of our scraped material from a wide array of social media activity in Sweden, Japan, including some English language content in the aftermath of the January 6th assault on the US Capitol, it is clear that Q-anon quickly became an internationally known and much debated phenomenon. There are a great many voices espousing all kinds of positions on the subject matter in both Sweden and Japan, including social media posts of ridicule as much as downright support for the Q-anon content, and so too in view of the spectacular events of January 6th it helped orchestrate.

There are posts that appear to have been subscribing to the Q-anon conspiracist agenda even before this particular event and that in its aftermath started explicitly to doubt the veracity of some of Q's predictions on how Trump is going to remain in office, etc. This seems to indicate that there are at least certain limits to how much the *memetic-solipsist subjectification* can domesticate real-world events, and that certain apparent antagonisms can essentially overturn the spell of conspiracism, and pave the way for somewhat more neutral rationalist subjectifications that counter such conspiracist logic. On the other hand, there are posts that seem to double down on the conspiracist wager, only changing the meaning from the literal to the spirit of Q. In other words, even if Q said Trump would remain in office after the 2020 election, for some, the meaning seems strangely able to morph into the notion that he will return in 2024. This is to say there is something quite nebulous about the specific content that is being spread in these posts, and how it can change so as not to let reality interfere with the narratives being spun other than when it can be harnessed as evidence to confirm the Q-narrative at hand.

This is at the heart of what makes Q-anon's conspiracy content memetic in nature and consequently what makes it so easily exportable, originating in the political terrain of the US, with notions that "American politics, business, and media are controlled by the "deep state" (government of darkness)". This apparently fertile dynamic easily translates into conceptions of "the world being ruled by paedophiles who worship the devil" in Japan, Sweden and elsewhere. This is how a Swedish Twitter post puts it, months after the Capitol riot took place:

> Some of us have woken up realizing that something is not right, but most do not realize how brainwashed we have been by elites in the media and government – #illuminati

Illuminati is here one of those indicative terms of a certain meta-narrative coding at play which signals a spirit of conspiracism, like "deep state", "Bilderberg" and the prefix "Soros", that tend to generate a stream of validating and less fact-prone responses. The tweet is followed by a string of encouraging comments, emojis and retweets including Q-anon parlance like wwg1wga (where we go one, we go all), as well as a more substantial responsive tweet that seems to implicitly suggest an affinity between socialism and satanism. This is for instance how one response helps shape the direction of the stream:

> The deep state has taught us to hate Christianity and to see it as evil, but what greater evil could there be, than trying to cast off the teachings of Christ, that is compassion itself, as evil.

This too is cheered on with the odd dissenting voice pointing out that evil is sometimes perpetrated in the name of Christianity, which instead of balancing the discussion seems if anything to be drowned out by sharpening of the tone on the part of supporters raising all kinds of Q-anon resonant conspiratorial claims of the global satanic cabal or some variety of the alternative tropes or allegations of paedophilia. Another interesting phenomenon that we can pick up in the material scraped from Swedish social media use is the intermingling of conspiratorial narratives with narratives on health and spiritual well-being, perhaps best captured in various streams on Facebook and Instagram. This is for instance how an Instagram user expresses herself:

> How much yoga is required to cleanse ourselves of all that we expose our bodies and minds to? If it is not the GMO chemicals in our food, it is Soros-media, CNN and the likes.

This tendency that we may want to refer to in terms of "conspirituality" is one in which the global conspiratorial tone of Q-anon is successfully integrated into an already existing and thriving wellness movement around health and spirituality. It becomes increasingly clear if we follow the discursive progression to this one post that sets off in the direction of global elites poisoning us in one form or other, bodies or minds, and profiting by so doing. The cure, however, in this particular strand of thinking is not so much storming the Capitol, or for that matter the Swedish Parliament (Riksdagen), as it is turning inward – and apparently to Instagram.

However, that does not mean the discursive progression of many of these streams is necessarily inward-looking and as such less likely to act out. While some are clearly satisfied with just voicing grievances on social media from the secluded comfort of one's home, illustrating the phenomenon of *clicktivism* as in relatively docile "on the screen" type activism that lacks real world off the screen activities (White, 2010). Others are clearly emboldened by the recognition such expressions seem to give in taking it out into the street. We could see a great number of such streams announcing a call to protest at specific events, like anti-vax rallies for instance eventually taking place across Stockholm and elsewhere, with photos in the same stream ultimately indicating user-presence on site.

Consequently, illustrating the intermixing of online and offline movements in this dynamic that commentators around "clicktivism" seem to overlook, when portraying it as rather harmless and cathartic ways of acting out online (White, 2010). Just because some users seem satisfied to voice their grievances behind the screen, doesn't mean that others don't get the precise push and confirmation they need therefrom to take their grievances to the streets, or even the Capitol building as in the case of January 6th.

From the social media landscape in Japan, we can discern that the Q-anon movement had reached Japan way before the Capitol riot took place, but this event nevertheless sets off a cascade of Q-anon inspired tweets that seem to have tipped the scale beyond clicktivism, with a significant number of demonstrations to support previous US President Trump taking to the streets across Japan (cf. Wang, 2021: CNN Business reports).

Moreover, social and individual anxieties due to COVID-19, which were spreading since early spring 2020, merged with the Q-anon movement in both countries. The more regional spin on the narrative in Japan is however unlike Sweden, much more geared toward the pre-existing anticommunism/anti-Chinese sentiment.

Q-anon in Japan is called J Anon, and is also sometimes referred to as Q Anon J or QAJF (aka QArmyJapanFlynn). In addition, a number of new groups (factions) have emerged in response to their claims. Among these new groups, the cognitive faction (Ninshiki-ha, 認識派) and the Yamato Q association (Yamato-Q-kai, 神真都Q会) are currently on hiatus due to the recent arrest of core members.

Internet communities in Japan also have a more pronounced cultural affiliation with some of the platforms in which this new memetic dynamic we explore plays out. 4chan, where Q first appeared and started sending messages in 2017, was originally launched on the October 1, 2003 by a Japanese Internet entrepreneur Hiroyuki Nishimura (西村博之). It is an anonymous English-language imageboard website, and mainly used in English-speaking countries. Although the owner and administrator of the site have changed, 4chan still has a proportionally large number of users in Japan.

Moreover, there are a great many links and references back to this platform as well as 2channel (2ちゃんねる), its successor 5channel (5ちゃんねる) 8chan and 8kun, included in the social media posts we scraped in Japan. We have explored these forums but will only use some anecdotal observations therefrom, as it could complicate our comparison with the type of forums and social media scraped in Sweden by way of raising the possibility that there is something in such fora in question that will have an impact on content, and that we would be unable to explore in Sweden in a comparative way. Instead, by scraping the very same social media platforms in both countries, our analysis can say with a greater degree of certainty that the formatting and interface of the content will have a similar impact in both countries.

This is how one Japanese user of Twitter expresses his loyalty to the Q-anon movement, shortly after the riot took place:

Trump fought for our freedoms and against the deep state communist authoritarian ways that are infiltrating our government from within, we are now left to fend for ourselves, unite behind J-anon – #qajf

Another one chimes in about various aggressive patterns in the movements of the People's Liberation Army and Navy of China, threatening Japan, and supposedly "deep-state" complicity among Japanese politicians being implicitly part of such Chinese expansionist agenda. Also here there are a great many examples of posts that call for taking grievances against politicians and elites to the streets, some of which correspond to actual protests taking place across Japan explicitly to support Q-anon or merging with various forms of anti-vaccine campaigns. This is how one Facebook user that announced a call to participate in one of these protests confirms his presence by photos taken after one of those Q-anon events in Japan had taken place, adding the words:

On January 17th (Sunday 2021), I participated in the "March in support of US President Trump" in Fukuoka to check the site of the Fukuoka tournament, which was declared a state of emergency. (…) At the meeting place, Fukuoka City Kego Park, a group of about 200 to 300 people seemed to be mobilized. The pamphlet states, "Each group will form a hierarchy." More people will be announced by the organizers. …Many spiritual people and Q-Anon supporters have high hopes for Q-Anon, and TRUMP will save the world…

The post seems to suggest a tendency of "conspirituality" that we also picked up from the Swedish accounts, although it should be stated that this confluence of narratives between spirituality and conspiracism is much less clear in Japan compared to Sweden. On the other hand, the Japanese posts seem more concerned with organizational issues and dynamics such as having to do with hierarchy and with spreading the word in a more tactical activist sense. The streams we explored are replete with supporting comments to Trumpism and Q-anon, along with the ever-resonant allegations of rings of paedophilia in the upper echelons of society, as they are in Sweden, but often with a clearer practical encouragement on how to take one's grievances into real-life action. "Human trafficking, and more!" are interspersed with encouragements to "Spread the word, to everyone!" and links of personal videologs captured in real life out in the streets at events, and time schedules for broadcasts to occur at given times. In short, the Q-anon content in Japan seems more organized, explicit and with a somewhat more practical agenda, than what seems to be the case in Sweden where it appears more disorganized and spontaneous. Two other obvious dimensions of difference have to do with language and religion in the dynamic through which these conspiracy theories are translated and reproduced in Sweden and Japan, respectively.

Moreover, Q-anon is closely tied to the anti-liberal and conservative movement in the US, which fully embraces Christian values and politically supports former President Trump. In Japan, most followers initially referred to Q-anon in response to former President Trump's anti-liberal and hardline stance on China, which seemed to be a much greater unifying characteristic than that having to do with Christian values. For obvious reasons, having a religious heritage with a belief in 8 million gods (Yaoyorozu-no-kami, 八百万の神), as in the history of Shintoism and Buddhism in Japan, the Christian connection does not have as much appeal in Japan.

As Q-anon propagates the existence of secret societies involving Satanists, this equally does not hold so much significance to the Japanese as it clearly is a narrative that has been spun against the backdrop of implied Christian values. Nevertheless, as the Q-anon phenomenon illustrates more than anything, there is an abundance of semi-synonymous signifiers that comes out interchangeably and that makes sure that if one fails to attract, there are alternatives and often more universal ones like paedophiles and global child-trafficking rings, that may do the trick. The most versatile one in the accounts we have explored here seems to be the "deep state", followed by the assertion that Trump was the vanguard in fighting them. Therefore, in Japan, there is a slightly different type of Q-anon discourse that has attracted more attention, which is primarily concerned with the question of the legitimacy of the election of Trump.

There is also a rich plethora of idiosyncratic narratives blazing their own trail but with an overall tone that largely resonates with the spirit of Q-anon. For instance, one user named "E.T" frequently appears in the Q-anon thread on 5channel, using generally unfamiliar words such as Earth Alliance (Chikyuu-Doumei, 地球同盟), Space Alliance (Uchuu-Rengou, 宇宙連合), Galactic Federation (Ginga-Rego, 銀河連合), White Hat (ホワイトハット), and Ascension (アセンション). E.T lays out, as it were, an epic tale about confrontation with aliens and a world where only the selected by God would survive. Of course, it is interspersed with characteristic terms commonly used by Q-anon followers, such as "deep state", "cabals" and "reptilians". Based on E.T's long post, users gathered in the thread post comments and questions in response to E.T's post. Some of the comments are "we've been waiting for E.T's post!" Each user gives his/her own interpretation to the unfolding Q-anon story and the ongoing very much open-ended deciphering of what the true meaning of Q's true message is.

Another factor that seemed to consolidate the Q-anon adherence in both Sweden and Japan was that of COVID-19, as the anxiety and distrust of corona vaccination seemed to fit hand in glove with the more paranoid world views around deep-state machinations and global agendas. The personal and social experiences of social distance and the physical and mental isolation among people seemed to exacerbate these tendencies more than anything, and Q-anon apparently spoke a language of disaffection that came to resonate far and wide. For example, in the thread on Q-anon on 5channel, there are numerous comments from people opposed to corona vaccination. Many unsubstantiated discourses, such as "the vaccine is microchipped", "the vaccine is made of salt water", "the vaccine is implanted with a microchip and manipulated by 5G communications," and "the vaccine is a biological weapon that will cause death in five years to reduce the population", apparently resonated with the fears that people have about corona vaccination.

6   Beyond solipsist realities: From the empirical to the ethical

As we have seen, there are undoubtedly differences in tone and dynamic in the various social media scapes of Sweden and Japan, and it would be extremely odd if that wasn't so considering the different religious and cultural differences separating the two countries. For instance, we could pick up a bit more on the clicktivist tendency of keeping things more online even if it clearly seeps out in Sweden too, whereas Japanese posts generally seem more actively determined to instigate real-world activism, etc. However, the far more interesting and bigger story of these social media posts is how small and subtle these same differences are, and how strikingly similar the dynamic that play out in either social media scape is including the mentioned codes of conspiracism like the Illuminati and Bilderberg variety, in view of the profound historical and cultural differences between the countries. In other words, the conspiracy theories we have explored here continue to demonstrate a striking ability to migrate and overcome cultural specificities in the way the basic message can reproduce itself. In this chapter, we have referred to this ability as memetic in nature, by virtue of its meme-like form, whereby it reproduces itself and develops over long streams of social media posts in very distinct cultural contexts yet consistent in tone and spirit. The foremost problem about this emergent phenomenon is that the memetic dynamic at play here has largely derailed any rational sense of being able to assess whether something is true or false, which clearly has profound ethical implications.

When people refuse to seek the truth but rather insists on making up their own, we end up fostering what we have here referred to as a *memetic-solipsist* community, which is not unlike the way we may choose to interact in more overtly unreal places such as in computer games. In *memetic-solipsist subjectification*, subjects are set adrift in a deluge of fake news, where the medium will produce its own experienced sense of truth. In one sense, this resonates with Baudrillard's (2010) notion of the "hyperreal" in that the mediated "real" is experienced as more real than anything such simulations may aspire to represent. Although this is not necessarily done by sophisticated visual representations that replace actual experience, but a web of mostly textual narratives that high-jacks our confirmatory cognitive processes, leading us to enact and co-create the misinformation complex that eventually imprisons us.

In other words, the empirical accounts indicate through what we here understand to be a memetic dynamic at the level of the medium, which has largely short-circuited our engagement with the world and the content we consume and help circulate. Our social media posts indicate how the perceived virality and acquiescence among fellow users interpret the truthfulness of content, ultimately appearing to become more potent in determining truth value than anything else.

This mirrors a phenomenon in the field of game design that concerns "alternate reality games" or so-called ARGs, with its apparent risks of certain dynamics from the game-world slipping back into the real world through the backdoor, so to speak. One vocal voice in this controversy is Sean Stacey, the founder of the website Unfiction, who insists the best way to define this genre is *not* to define it at all. Nevertheless, there are those who are not as reticent about ruminating over some of those boundaries, suggesting that "alternate reality games take the substance of everyday life and weave it into narratives that layer additional meaning, depth, and interaction upon the real world" (Martin and Chatfield, 2006: 6). In popular media, a review summed it up as "an obsession-inspiring genre that blends real-life treasure hunting, interactive storytelling, video games and online community" (Borland, 2005: 4). However, game designer Reed Berkowitz (2020) points to a challenge that haunts his profession and in particular opens space and multimedia game designs like ARGs.

The problem is called Apophenia and that is the tendency to see connections between unrelated things. This is to say players may experience clues from random details that are not supposed to drive the narrative, consequently derailing the gameplay from running its course. Berkowitz (2020) explains from personal experience when one of his game designs went awry because random pieces of wood like symbols in the designed lay out of the floor seemed to make up an arrow and clue that there was something in the wall it pointed to, which the players fruitlessly tried to reveal for hours on end. In a nutshell, this captures the precise moment in which the content producer loses control of the significance it generates.

Apart from strikingly fitting the various characteristics, we can derive from some of the attempts to define what an alternate reality game is, and perhaps not least the insistence to leave it open, remember – it's not a game! The Q-anon movement seems to be a case in point in which the receiver takes precedence in the significance of content, and the one subtle threat of Apophenia is neatly solved by complete inversion. "There are no coincidences" is arguably the Q-drop quote that sets off the movement in the first place by affirming the basic gameplay premise, with the implicit injunction for everyone to become active participants in this endless labyrinthine puzzle within puzzles that in this view make up reality.

However, what we call the *memetic-solipsist subjectification* is ultimately an identification fraught with its own inherent antagonisms as it depends on subjugating off-screen facticity to the more malleable narrativity of online conspiracy peddling. This is not to say that such identification is necessarily bound to fail; indeed, Q-anon illustrates how it is able to sustain itself over significant periods of time by integrating on-screen narratives with off-screen experiences. But, the capitol riot and its aftermath is at least a clear example of a real-world event that threatens to derail and undermine Q's predictions.

Nevertheless, while such antagonisms may have proven sufficiently undermining for some, others managed to entrench their beliefs even further by digging in and faithfully holding out for history to get back in line with this overall complex of narratives. Or even more worryingly, buying into this fraught set of narratives in spite of such seemingly insurmountable antagonisms, not to mention doing so across cultural boundaries as in Sweden and Japan.

This continuous re-calibration of experience to fit the sought-after virtual identification and the affect it evokes, which we here term the *memetic-solipsist subjectification*, parallels what Levinas describes in *Totality and Infinity* the subject being "for itself". Being "for itself" according to Levinas is about the subject taking possession of itself through its own representation, and in so doing "it extends its identity to what of itself comes to refute this identity" (Levinas, 2011: 87). "This imperialism of the same is the whole essence of freedom" is how Levinas (2011: ibid) explains it and understands the Other as metaphysical, as the necessary exteriority without which uninhibited freedom amounts to limitless self-delusion and the impossibility of ethics.

The *rationalist commonsensical subjectification* is one form of moderation of such freedom whereby to justify itself, the subject "can endeavour to apprehend itself within a totality" (Levinas, 2011: ibid).

This seems to us to be the justification of freedom aspired after by the philosophy that, from Spinoza to Hegel, identifies will and reason, that, contrary to Descartes, removes from truth its character of being free work so as to situate it where the opposition between the I and the non-I disappears, in an impersonal reason.

(Levinas, 2011: ibid)

Unlike memetic-solipsist subjectification, the other is not subjugated to the imperialism of the self-image, but rather the *commonsensical modality* places the self and other on an equal footing against what is seemingly a neutral background of justice.

While there is certainly room for ethics deriving out of what we here refer to as *commonsensical modality of subjectification.* Indeed, much of the canon of Western intellectual thought from Antiquity to the Enlightenment springboards out of this one basic premise of symmetric equality between subject and other in legal and moral matters. However, ethics according to Levinas necessarily goes further than such neat symmetry of impersonal reason, and with the *subjectification in the face of the Other,* we endeavor to foreground what an ethics founded on having genuinely appreciated the difference of the Other might entail. It is one in which the totality of impersonal reason is transcended by recognizing that the Other can never fully be subsumed into the same as in memetic-solipsism, nor can it be fully accounted for by way of common sense rationalism, with such tropes of neutrality and universalism. Instead, the ethical begins when we start to accept that otherness and difference evade our every attempt at such totalizations and rationalizations, and it is only when we disarm ourselves such modalities of egotism or rational distancing, that we can truly appreciate the particularity of the Other and the ethics engendered thereby.

7   Discussion

In this study, we have seen the striking and cross-cultural power that conspiratorial narratives like that of Q-anon can achieve when disseminated in a de-centralized memetic logic. The Japanese and Swedish posts alike demonstrate a shift in significance attribution at the content level to receiver as opposed to sender in this type of memetic interaction, which means that the spirit of Q-anon is passed forward be it in the form of outlandish or more moderate forms of conspiracist white grievance. Moreover, it adapts and lashes onto idiosyncratic and culturally specific content so as to consolidate forces with other existing causes or movements.

One such example is the phenomenon of "conspirituality" that was indicated by a number of posts in both Sweden and Japan, where the zeal of wellness and spiritual yearnings may be co-opted by a certain Q-anon logic. In Sweden that demonstrated a somewhat more pronounced tendency in this regard, posts draw parallels between the perceived toxins to our bodily and spiritual well-being from the food we eat to the news we see, which in turn seem to be channeled onto the usual old anti-semitic tropes of elitism and its various personifications of evil.

Moreover, balancing freedom of expression against instances of hate speech has been a lingering conundrum since the dawn of modern societies. What we would like to add to that ongoing discussion is to affirm the inherent problems of a virtually "public square" that is owned and controlled by private companies, that becomes all the more problematic when considering how these spaces become subjected to memetic manipulation whereby social movements can be engineered and orchestrated in new and elaborate ways. In this view, mirroring what McIntyre (2018) has described in terms of post-truth societies, the present chapter may have given us some insight as to what societies are bound to face in order to preserve democratic institutions in the midst of discourses that reproduce themselves in processes that in many ways are antithetical to evidential verification?

This dilemma has become all the more timely with Elon Musk's recent acquisition of Twitter. Elon Musk who is himself a social media agitator that publicly ruminates about the power of the meme in various on-line forums, did not just let Donald Trump back on the platform in spite of his Twitter activity being such an obvious catalyst for the events of January 6th, but it seems Twitter is now becoming the fringe-nest go-to forum for voices shunned by legacy media such as Tucker Carlson that even Fox News are unwilling to support. In a 2020 tweet, Mr. Musk says "Who controls the memes, controls the universe", so his reasons for acquiring Twitter are perhaps not so difficult to grasp, especially after he has been able to send stocks and cryptocurrencies in his possession rallying or plunging in real time depending merely on the phrasings of his tweets. The rather more difficult question in this regard is what we can do in democratic societies to safeguard the so-called fourth power, that is the power of a free and independent media to hold powerful individuals to account, when the most powerful are beginning to discover how media can be used to maximize the very same power they possess.

The existence of Q-anon and the information spread by its supporters in this truth-bending form of memetic engagement evoke very broad and classic philosophical questions. However, with the ascent of real-time social media and memetic virality, such classical questions of ethics and philosophy more broadly are perhaps evermore pertinent through the lens of a present that seems to recalibrate the very processes of what is experienced as right or wrong. What is truth, and what is its relationship with power and justice, a question which has haunted philosophy since at least Plato's *the Republic* wherein Thrasymachus famously asserts that the powerful can do what they may and the weak will accept what they must. In other words, power is self-legitimizing in terms of justice, which of course Socrates vividly refutes in this same dialogue. But is power also self-legitimizing in terms of truth, which in the Republic is given voice to by Gorgias and later to reverberate in much of post-structuralist philosophy from that of Jacques Derrida to Michel Foucault. Given the phenomenon at hand, it certainly appears as if the power of the memetic might be significant in what is to be taken as truth. The question however is how well can this power of the memetic be controlled, which hasn't been the specific aim of this chapter, but something we would like to encourage a wider discussion about and hope that future research might explore more directly in coming years.

What we have done in this chapter rather than looking at control directly, is to tease out what on a more subjective level seems to drive users to believe and engage in narrative reproduction, which largely appears to come through in what we have described in terms of a memetic dynamic. In this view, the question is not so much whether truth exists, or who is the ultimate arbiter of what is taken as true, but rather Q-anon points to a more collective malaise whereby existing "truth"-claims are experienced as alienating. What if "truth" is boring, or experienced as demeaning to some of our sensibilities, and what if we can get a deeper sense of purpose or affective reward by scratching the surface until we see a more significant role in there for us to play? In Levinas' view, this amounts to what he describes as an "imperialism of the same", a form of what we term *memetic-solipsist subjectification* whereby the malleability of the meme helps the subject evade some of the confines inherent to rationalist discourse, without giving up the pretence of veracity that is nonetheless confirmed by its inherent virality.

The *memetic-solipsist subjectification* may at some level mirror Robert Nozick's (2003) polemical thought experiment of an experience machine that is experienced as real life but completely compartmentalized from the outside world. In this view we are faced with assessing whether the authenticity of our experiences may have an ethical value in its own right. Indeed, this may capture an aspect of the ethicality spectrum surrounding a phenomena like that of clicktivism, that we mentioned above in regard to one quoted post apparently voicing grievances, which seemingly had little to no discernible real world effects. Clicktivism in this sense is a form of docile activism that begins and ends behind the screen, much like Nozick's (2003) experience machine it is completely compartmentalized from the outside world, merely triggering emotions for the subject in question and in that sense one could argue for it to be rather harmless – however only up to a point.

The existence of Q-anon and the information spread by its supporters in this truth-bending form of memetic engagement evoke very broad and classic philosophical questions. However, with the ascent of real-time social media and memetic virality such classical questions of ethics and philosophy more broadly is perhaps evermore pertinent through the lens of a present that seems to recalibrate the very processes of what is experienced as right or wrong. What is truth, and what is its relationship with power and justice, a question which has haunted philosophy since at least Plato's *the Republic* wherein Thrasymachus famously asserts that the powerful can do what they may and the weak will accept what they must. In other words, power is self-legitimizing in terms of justice, in which of course Socrates vividly refutes in this same dialogue. But is power also self-legitimizing in terms of truth, which in the Republic is given voice to by Gorgias and later to reverberate in much of post-structuralist philosophy from that of Jacques Derrida to Michel Foucault. Given the phenomenon at hand, it certainly appears as if the power of the memetic might be significant in what is to be taken as truth. The question however is how well can this power of the memetic be controlled, which hasn't been the specific aim of this chapter, but something we would like to encourage a wider discussion about and hope that future research might explore more directly in coming years.

It is apparently very comforting to project onto the world a crude good and evil binary logic, on the one hand not having to deal with anxiety arousing nuances in a shades-of-gray world where some things are not so easily cast, and on the other furnishing the subject with a greater sense of significance therein. If what Q-anon followers assume to be an evil government offers them a vaccine, they can either face such apparently puzzling shades-of-gray world in which governments can do good as well as bad things, or they throw out the baby with the dirty water by saying the elites are behind it all and vaccines are just another evil way for them to get us. From the point of view of identification and the constitutive populist tropes of "us" and "them", however gloomy a world it may appear to outsiders, the binary logic is still intact and so are the apparently pleasurable clear cut roles to be played therein.

8   Conclusion

As this web-scraping journey through landscapes unbound by the facticities of real life is coming to an end, we need to remind ourselves of the very real and troubling footprints this ephemerality continues to leave in its wake, and what we may learn from studying its various patterns. In this chapter, we have taken a closer look at one of the most notorious complexes of conspiracy theories as it has moved across the globe and become enmeshed with discourses in Sweden as much as in Japan.

The conspiracy theories developed by Q-anon in the US generally go something like this: The world is ruled by groups of paedophiles who engage in satanic rituals of sexual abuse, cannibalism and/or human trafficking. Q-anon followers also consider such groups to be the "deep state" (shadow government), which they believe includes politicians, wealthy entrepreneurs and individuals, but the term along with other codifications of worldwide elites have proven very resilient and able to migrate across cultural borders.

What is more, it is not so much the real meaning of Q-anon messages that dictates exactly what followers should and should not believe, which testifies to a new emergent dynamic that is significantly enabled through the medium as opposed to the content. McLuhan's famous dictum, the medium is the message referring to the then new medium of television, presupposes a certain passivising one directionality in information consumption, but the moment this soothing personal space of consumption becomes one of co-creations at the content level and further dissemination of a general conspiracist spirit, the dynamic may become both memetic and viral as we have seen with the Q phenomenon.

Q-anon supporters interpret Q's message as they see fit, discover and believe their own "truth", affirming and exacerbating the various leanings and biases of their different identity formations. It begs the question, is such radical world view of good and evil and memetic malleability even reconcilable with institutions based on a certain level of rational consensus as to the process and framework on account of which we can express our individual views and hope to verify what we believe to be true? Furthermore, what can we do to counter this emerging dynamic that in exacerbating the disdain for the other, may well see fit to undermine any rational sense of truth, and possibly democracy itself in doing so?

Indeed, the *memetic-solipsist subjectification* is not only a failure of ethics to Levinas, which is clear, but arguably a failure on a much more fundamental level. After all, even what we have here termed *commonsensical rationalist subjectification* that underpins much of Enlightenment thinking and the very democratic institutions we here reflect on whether we can sustain in face of challenges brought forth by memetic solipsism also fails in view of Levinasian ethics. Although we should here remind ourselves of Samuel Beckett's (1989) lesson that one can fail, then again one can fail better. The commonsensical rationalist subjectification fails insofar as the call of the Other is not fully appreciated as particularities are subsumed under a totalizing framework of equality, and it is perhaps inevitable that ethics of particularity does not withstand systematizing processes, which Levinas reflects on in view of what he terms the problem of the third party. Nevertheless, we shouldn't take this problem of systematicity at the societal level to prevent us from appreciating the particularity of the Other and the call it evokes on us as individuals, which we here describe in terms of *subjectification in face of the Other*, and is how Levinasian ethics could hold up in this regard.

This would amount to not just rationalizing away difference by universal maxims on how one ought to act, but to affirm the insurmountable difference that undermines all such attempts at domestication and totalization. Nevertheless, a far-worse failure of Levinasian ethics is that of the *memetic-solipsist* variety exemplified here by Q-anon, where we not only fail to see insurmountable particularities of others inside or outside our chosen "in-group", but rather project and dramatize such demarcations further to achieve an even greater sense of significance and superiority. In other words, people don't share values and norms with others that are genuinely other in that they differ and/or resemble us in the somewhat unruly way that confronting others has entailed since time immemorial. Instead with the ascent of this, emergent memetic dynamic sameness and otherness is largely produced at the level of the medium as opposed to in the direct realm of content, whereby algorithms are complicit in extensively shielding the subject from this more unruly dimension of the other. The artificial collapse of otherness into a less friction-ridden horizon of the same, that the Q-anon phenomenon epitomizes, is what makes society unstable and fragile.

In our digitalized society, where Q-anon supporters can easily access their favorite news media and social media sites, they look at convenient information, and find and contact the like-minded without meeting each other in person. In a situation where one only accesses the information convenient to him/her/them and meet only the like-minded, the accuracy of information is not ensured. Q-anon followers are often interested in finding the existence of the deep state and the meaning of Q's messages from their own experiences, whether online or offline. Q-anon followers find signs in their own experiences that support the Q-anon's conspiracy theories and use them as the basis for creating their own "truth", and there is obviously a wide palette here to choose from. For instance, one may believe in anything from more subtle deep-state machinations to the more full-out reptilian world view of E.T. mentioned above, and interestingly this inherent width and diversity in narrative tone within Q-anon do not seem to dissuade its believers as much as offering a rather broad onramp from various levels of radicalization.

On the website of QAJF, the following statement appears in the prologue:

> to those of you who feel that there is something strange going on in the world today, it's time to wake up! Anons around the world are united in their hopes to make the world better. There are plenty of clues on social networking sites and the Internet.

The appeal of Q-anon be it in the US, Sweden or Japan seems to be finding answers to our vague fears and frustrations about the state of the world and of our own lives.

To summarize these characteristics, Q-anon's message is about accepting a premise about all the pervasive power of elites, which can be detected and illustrated by a limitless array of memes and postings that seem to confirm this one accepted "truth" over and over again. Q-anon responds to people's vague questions, frustrations and anxieties based on their real lives and experiences, and provides answers that seem plausible and appealing even without evidence. They also camouflage concrete context and propaganda with abstract words and expressions that can be interpreted in any way, and they use a method of information dissemination that makes people feel as if they have found the answers themselves. At times, they use ambiguous language to develop epic stories, and stimulate people's expectations of something unknown, and these expectations even have a calming effect that eases frustrations and anxieties. The memetic dynamic that has made Q-anon a cross-cultural social movement with profound and lasting implications illustrates a new form of power embedded in online meditation, which offers people a sense of purpose and significance.

Acknowledgment

This study was supported by the JSPS/STINT Bilateral Joint Research Project, "Information and Communication Technology for Sustainability and Ethics: Cross-national Studies between Japan and Sweden" (JPJSBP120185411) and Kakenhi (19K12528).

References

Allcott, H. and Gentzkow, M. (2017). Social media and fake news in the 2016 election. *Journal of Economic Perspectives*, 31(2), 211–236.
Baudrillard, J. (2010). *Simulacra and Simulation*, Ann Arbor, MI: University of Michigan Press.
Beckett, S. (1989). *Nohow On*, New York: Grove Press.
Berkowitz, R. (2020). A game designer's analysis of Q-anon: Playing with reality. *Medium*, September 30.
Castells, M. (2012). *Networks of Outrage and Hope: Social Movements in the Internet Age*, Cambridge: Polity.
Crenshaw, K. (2017). *On Intersectionality*, New York: The New Press.
Curnutt, H. (2012). Flashing your phone: Sexting and the remediation of teen sexuality. *Communication Quarterly*, 60(3), 353–369.
Foucault, M. (1972). *The Archaeology of Knowledge* (A. M. Sheridan Smith, Trans.), New York: Pantheon.
Haas, M. L. and Lesch, D. W. (2017). *The Arab Spring: The Hope and Reality of the Uprisings*, New York: Routledge.
Houellebecq, M. (2004). *Lanzarote*, London: Vintage.
Jaques, C., Islar, M. and Lord, G. (2019). Post-truth: Hegemony on social media and implications for sustainability communication. *Sustainability*, 11(7), 2120.

Kim, T. W. and Werbach, K. (2016). More than just a game: Ethical issues in gamification. *Ethics Information Technology*, 18, 157–173.

Kirby, A. (2009). *Digimodernism*, New York: Bloomsbury Academic.

Kyriakopoulou, K. (2011). Authoritarian states and internet social media: Instruments of democratisation or instruments of control. *Human Affairs*, 21, 18–26.

Laclau, E. (2005). *On Populist Reason*, London: Verso.

Levinas, E. (2011). *Totality and Infinity*, Pittsburgh, PA: Duquesne University Press.

Martin, A. and Chatfield, T. (2006). *Alternate reality games white paper—IGDA ARG SIG*. Mt. Royal, NJ: International Game Developers Association.

McIntyre, A. (2018). *A Short History of Ethics*, London: Routledge.

McLuhan, M. and Fiore, Q. (1967). *The Medium Is the Message: An Inventory of Effects*, New York: Random House.

Nozick, R. (2003). *Anarchy, State and Utopia*, Boston, MA: Harvard University Press.

Sam, C. (2019). Shaping discourse through social media: Using foucauldian discourse analysis to explore the narratives that influence educational policy. *American Behavioral Scientist*, 63(3), 333–350.

Shifman, L. (2014). *Memes in Digital Culture*, Cambridge, MA: MIT Press.

Sunstein, C. R. (2021). *Liars: Falsehoods and Free Speech in an Age of Deception*, Oxford: Oxford University Press.

Tracy, K. and Mirivel, J. C. (2009). Discourse analysis: The practice and practical value of taping, transcribing, and analyzing talk. In L. R. Frey & K. N. Cissna (Eds.), *Routledge Handbook of Applied Communication Research* (pp. 153–178), London: Routledge.

White, M. (2010). Clicktivism is ruining left wing activism. *Guardian*, 12 August. Retrieved from http://www.guardian.co.uk/commentisfree/2010/aug/12/clicktivism-ruining-leftist-activis.

Xiao Mina, A. (2021). *Memes to Movements: How the World's most Viral Media Is Changing Social Protest and Power*, Boston, MA: Beacon Press.

Zuboff, S. (2019). *The Age of Surveillance Capitalism: The Fight for a Human Future at the New Frontier of Power*. London: Profile books.

Other sources

Borland, R. (2005). Blurring the line between games and life. *CNET News*. Retrieved from http://news.cnet.com/Blurring-the-line-between-games-and-life/2100-1024_3-5590956.html

Wang, S. (2021). 'Q-anon in Japan': A CNN Business Report by Selina Wang, CNN Tokyo.

Appendix A

Table 8A An example of the month-by-month view of the dominant discourses from the English posts, and their metrics

Month	Language	Dominant discourses	Post stats
April 2021	English	Covid conspiracies – Blood-harvesting and adrenochroming of children.	Brazil, 17k reach, 5.2k comments
May 2021	English	Trump election fraud – Michael Flynn comments on Myanmar coup "There's no reason it shouldn't happen here". Military liberating the nation from the deep state.	Canada, 200k reach, 3k comments
	English	Some kind of storm coming? Sweeping of elites from power. From a suspended Twitter account.	55k Reach, 2.6k comments
	English	Epstein is still alive, and the person in prison who "hung" himself was a body double.	56k Reach, 2k comments
June 2021	English	Trump election fraud. More discussion on Michael Flynn saying a US military coup should happen.	1.3m Reach, 20k comments
July 2021	English	More Michael Flynn discussion, his family recorded taking a Qanon oath. Some discussion on him claiming it's a family motto that just happens to be exactly the same.	US, 300k Reach, 12k comments
August 2021	English	No single obvious dominant discourse, but still a lot of mentions of Michael Flynn, and his influence.	
September 2021	English	Covid conspiracies – Michael Flynn says covid vaccines are being added to salad dressings. Some minor discussions about Trump's election fraud and Biden's impeachment.	US, 43k comments
October 2021	English	Judge dismisses Flynn family lawsuit over Qanon claims. Qanon turns on Flynn.	US, 56k Reach, 9.3k comments US, 200k Reach, 5.8k comments

(Continued)

Table 8A (Continued)

Month	Language	Dominant discourses	Post stats
November 2021	English	JFK Jr. making a big announcement by the grassy knoll. Ends up not happening. Qanon Shaman getting 41months of prison time. Flynn caught saying Qanon is "nonsense", causing Qanon civil war.	
December 2021	English	Nothing dominant. A few discussions claiming that Anthrax pumped through fog machines at Flynn's Qanon conference in a church. Not much other discourses.	
January 2022	English	Mostly random discussions of various protests. However, one interesting Facebook post from Tommy Truthful TV. "88 is a white supremacist symbol HH which stands for hail Hitler, Trump Real name Drumph he's hide your manic crypto Jewish bloodline, direct descendant to Adolf Hitler. Qanon=666, in the Fibonacci cipher, Maga the highest level in the Church of Satan! So keep worshiping him and putting your energy into the beast, Trump, Biden, Obama they all are 33rd degree Freemason Jesuits born into witchcraft bloodlines! #thebeast #antichrist #trump #88 #hh #gematria #numbergod #maga"	10k Reach, 90 comments
February 2022	English	No real clear dominant narrative. But a few of counter narratives suggesting that Qanon is a Russian psyop, and the Flynn is a Russian operative.	
March 2022	English	Some reference to a "US biolabs in Ukraine" conspiracy, that appears to originate as Russian propaganda, with one suggesting it will be used as a false flag to hide chemical weapons attacks. Not much other identifiable discourses.	300k Reach, 4.9k comments
April 2022	English	"Watch the water" discourse, a new right wing "documentary" that suggests snake blood was put in the vaccines to inject us with Satan's DNA. A previous month talked about "watch the water" as Trump watermarking ballots to try and identify voter fraud, so perhaps that is a discourse that has evolved?	

9 Cultural frictions in the ethics of smartphone games

The example of Pokémon GO in Japan and Poland

Akira Ide and Paweł Pachciarek

1   Introduction

In recent years, the gaming market has undergone major changes. Up until 2010, the market was dominated by home consoles such as Nintendo and PlayStation, while PC games had an unignorable market share. According to Newzoo (2022), however, mobile games, including smartphone applications, have accounted for almost half of the $200-billion market and have completely been changing the ethical situation. Before the rise of mobile gaming, home consoles, along with PCs, were the mainstay of the gaming industry, and software was supplied on physical media. For example, in the early 21st century, Nintendo consoles used cartridges, and Sony PlayStation software was sold on CD-ROMs. The regulation of software distribution was, in a sense, simple; if the distribution of the software of a game was physically stopped by, say, a regulating authority, it was impossible to play. It was easy for a regulator, as well as for companies in the video game industry, to suspend the release of games that were deemed to be harmful to players' (especially children's and youths') mental growth or society and to control the distribution of such games.

However, because games can since the late 1990s be downloaded from or played via the Internet, it has become ever more difficult to control their distribution. This is especially true in liberal countries, where constitutional commitments to freedom of expression make it difficult for regulating authorities to restrict people's use of such games. Therefore, the video game industry is often encouraged to establish and implement self-regulations by pressure from governments and society. Although there are plenty of games targeted to local and regional markets, many games are developed to be sold in several countries, or ideally – for the game developers – in all countries. Given that there are cultural differences in the reception of games as one form of digital technology, particularly related to ethical issues, gaming companies carefully attempt to ensure that issues like racism, gender discrimination and violence, which are ethically unacceptable issues across borders, are not involved in a game they produce, given that it may breach local taboos and cultural norms. To explore these issues, we take up Pokémon GO, a location-based mobile game, as a subject of investigation, and compare how this game is viewed in Japan and Poland. We will explain the extensive literature on

DOI: 10.4324/9781003367451-11

ethical aspects of Pokémon Go, but in this chapter, we will zoom in on one particular ethical issue – playing at sites of genocide.

Pokémon GO was first released in Japan in July 2016 and a major social debate arose when it was played in a park near the Atomic Bomb Dome in City of Hiroshima (hereafter simply referred to as Hiroshima). Every year on August 6 (Atomic Bomb Memorial Day in Hiroshima), a large event is held in the park to commemorate the victims. In 2016, the solemnity of the ceremony was thought to be disturbed by an influx of Pokémon GO players, who were widely viewed as 'impolite', because the Peace Memorial Park is seen as a place of solemn prayers (D'Anastasio, 2016). In Poland, playing Pokémon GO in the Auschwitz concentration camp (hereafter simply referred to as Auschwitz) has been considered inappropriate; 'fun' is not to be enjoyed in the place where millions died (Ntelia, 2020). Both the Atomic Bomb Dome and Auschwitz are well-known World Heritage Sites. As we will explain in more detail, the main reason for exploring these countries is that they both hold sites of genocide, where there have been strong reactions to playing Pokémon Go, but we also identify cultural differences which can help us understand the responses in the two countries.

In the following section, we explain what Pokémon Go is and why a cross-national study between Japan and Poland is suitable for our research. In the third section, we review relevant previous studies and confirm the novelty of the present work. In Section 4, we describe the cultural friction caused by Pokémon GO in various places in the world and situations in which ethical issues arise from the game being played. The first half of Section 5 discusses cultural issues in Japan and the second half examines those issues in Poland. In Section 6, we summarise the work from a comparative cultural perspective and draw conclusions.

2   The impact of Pokémon GO

2.1   What is Pokémon GO?

Pokémon GO is a location-based game from Niantic for smartphones. The characters and motifs in the game are based on those in a series of popular pocket monster games, role-playing video games for handheld game consoles like the Game Boy, released since the mid-1990s. However, the design of the game is completely new; the screen background and features are unique to the age of smartphones (Craddock, 2017). The use of smartphones as communication devices made it possible for the game to handle large amounts of data, in turn allowing the operations and management of games incorporating location information in real time. Originally, Niantic was an internal Google venture; the founders were deeply involved in improving Google Maps according to the Niantic Story (n.d.). Before the advent of Pokémon GO, many games used location information. 'Ingress' developed by Niantic and released in December 2013 was an extremely popular 'camping game' that was downloaded more than 20 million times worldwide as of November 2018 (The Niantic Team, 2018). However, the social impact of Ingress was limited compared to that of Pokémon GO, which has been downloaded more than 500 million times by the end of 2021 (Iqbal, 2023). The launch of Pokémon GO marked a new epoch. Whereas smartphone games prior to it were regarded as child-oriented ones, many adult users have enjoyed playing Pokémon GO (Rasche et al., 2017). It was even reported that a reporter played the game during a briefing at the US Department of Defense (Matyszczyk, 2016). However, as this game becomes popular, social problems appeared.

Pokémon GO is an epochal video game in terms of its implementing augmented reality (AR) technology so that fictional characters are projected onto the map on a smartphone screen. Nevertheless, up to the present date, AR has rarely come up in conversation among gamers including Pokémon GO players, presumably because they assume that AR technology is non-problematic for them and society. However, from an academic viewpoint, in particular, investigating the ethical and social aspects of AR is important, given that this technology has potentially been incorporated into every video game.

Usually, Pokémon GO is described as using AR technology. However, what problem would be caused by the use of the technology? On the initial screen of the video game, PokéStops, where players can collect various items such as Pokémon Eggs and Poké Balls, characters, monsters, etc., appear. At this stage, however, the real world is not depicted; the game is played only in a virtual world. When the 'AR mode' is activated, the game combines the outside world (as seen by the smartphone camera) with on-screen stuff such as monsters and items. The point is that this mode is not necessarily inactivated in areas supposedly unsuitable for playing this kind of game. The combination of the meanings of such areas and the presence of a monster could bring about a sense of aversion to the video game. We focus primarily on the 'place' rather than the game per se to address the ethics of smartphone games.

2.2   Why compare Japan and Poland?

We view Japan and Poland from a comparative perspective or 'gaze', a technical term in sociology, for three reasons. The main reason for comparing these two countries is that both have sites of genocide. As pointed out by Margalit (2014), in Auschwitz, extremely ill-advised photographers took pictures that lacked respect for the victims to get 'likes' on Instagram. Similarly, in Japan, a video of a dance performance against a backdrop of the Atomic Bomb Dome caused a firestorm of criticism when it was uploaded. The question of how to behave in such sites of tragedy in the era of advanced information-sharing has become a major social issue. As mentioned, we focus on the ethics of playing Pokémon GO in the vicinity of the Atomic Bomb Dome and Auschwitz. Through investigating the ways of local residents processing the memory of genocide and the nature of the game, we will reveal how citizens in each country view life and death. Although death is common when playing, say, a shooting game, it is not unusual for thousands of people to die during a single game session. However, the deaths in Hiroshima and Auschwitz were historically true, being irrevocably tied to the memory of the places. The ethical implications of playing games in such places of mass death have not been considered. Because both Japan and Poland experienced modern forms of mass death, the cross-cultural study between the two countries is meaningful and unique results would be produced.

Second, video games are positioned differently in the two countries. In Japan, although the video game industry is one of the major industrial sectors and many adults have played video games for decades, they have tended to be viewed as toys for children until recently because a big toy maker, Nintendo, first supplied home video game consoles (Wolf, 2015). On the other hand, in Europe, games are increasingly being used for educational purposes and are regarded as products for adults. Poland (and Eastern Europe in general) is an exporter of games; the video game industry may become a major revenue source in the (near) future (Kamosiński, 2019). As the number of children in the developed world declines, the business model of the Japanese video game industry which has targeted at children will soon lose its effectiveness. It is important to consider these differences when we look toward the future development of the game industry. In Poland and the rest of the former Communist bloc, economic constraints rendered it difficult to develop advanced industrial technologies. Instead, mathematics and computer science, which do not require high research expenses, progressed (*"Elektronika, komputery w PRL"*, 2015). The Polish video game industry has lessons for Japan, showing how national strength is maintained even without large investments.

The third reason is that the situation regarding religion is contrasting between the two countries. Poland is a devoutly Christian country, whereas Japanese are largely indifferent to religion. In fact, Główny Urząd Statystyczny (2022) reported that over 90% of Polish people believed in Catholicism. On the other hand, more than 70% of the Japanese had no religious belief according to the survey of ISSP Research Group (2022).

Given the similarities regarding having sites of genocide in addition to the differences regarding attitudes toward games and religion, a cross-national study on ethics of the smartphone game at sites of genocide is relevant.

3   Overview of previous studies

In July 2022, 6 years had already passed since Pokémon GO was released, and the game was played in over 150 countries and had been downloaded over 1 billion times as already shown on Iqbal (2023). The literature dealing with the video game is thus extensive. Tang (2017) sought to understand why the game was so attractive. Player demographics have been studied; Rasche et al. (2017) attempted to find what kind of people would be addicted to Pokémon GO. The technical aspects have also received attention; this was the first game to fully embrace artificial intelligence (AI) (Rauschnabel et al., 2017). When playing Pokémon GO, one walks in real space to capture monsters that appear in AR; long distances must be walked to obtain high scores. Pokémon GO has thus been used to enhance health (Althoff et al., 2016). As movement is essential, movement per se has been studied. Colley et al. (2017) and Evans and Saker (2019) analysed spatial behaviour; the game changed human trajectories. From the social science perspective, the economics have been studied. Although Pokémon GO is free, many players are willing to pay for 'special' items. The economic impacts of the game on local economies are significant. A study showed that an event of the game held in Dortmund had a direct economic impact of €19 million, including money spent on food, drink and accommodation (Kautz, 2019). Aluri (2017) explored the potential of Pokémon GO as a travel guide. Carbonell Carrera et al. (2018) described how Pokémon GO encouraged a new form of place perception using 'spatial orientation'. Pokémon has always been based on a journey into unknown situations. Because various kinds of monsters appear as the game progresses, a 'travel atmosphere' is an integral part of the game, combined with AR. The Harry Potter Wizarding World (which ended in February 2022), which was developed based on the Pokémon GO technology, placed even more emphasis on visiting strange places. In a similar vein, Pokémon GO has been used to aid disaster recovery in Japan: visitors are attracted to disaster areas where economic recovery is lagging (including those affected by the Great East Japan Earthquake), and spend money at these sites (Newman, 2016). Japan suffers from many disasters, and this way of using the game is unique to Japan. Psychological phenomena associated with such behaviour have been intensely studied. Orosz et al. (2018) and others asked: 'Why do we play this game?' and, 'What is the motivation?' Zach and Tussyadiah (2017) found that playing this game increased happiness, which was not a specific intention of the creators.

Several studies have reported that Pokémon GO triggers behavioural changes. Kari (2017) found that the game enhanced sociability. Tateno et al. (2016) reported that Pokémon GO brought stay-at-home people outside. However, every phenomenon has both bright sides and downsides. Some have reported on the 'negative effects' of Pokémon GO. According to Dworak et al. (2007), children who played Pokémon video games (pre-Pokémon GO) lost sleep and had trouble learning.

The danger of looking at mobile phones during driving has been pointed out by many (e.g. Madden & Rainie, 2010). Smartphone games require players to take action, and they are required to make complex decisions and manipulations while playing Pokémon GO. Thus, the game would enhance the risk of traffic accidents further (Faccio & McConnell, 2019).

Many analyses of cultural friction (the topic of this paper) have appeared; here, we focus on Hiroshima (e.g. Judge and Brown, 2017) and Auschwitz (e.g. Ntelia, 2020). Although there have been many previous studies in this field, the aim here is not simply to add another stone to the pile; this work has special significance. Hiroshima and Auschwitz are similar in that the massacres occurred at about the same time. The Auschwitz deaths were based on racism but those of Hiroshima on a military offensive; thus, the massacres differ in nature. The atomic bombings were carried out as part of the Manhattan Project, which was originally intended to target Nazi Germany. However, Germany surrendered in May 1945, and as a result, the atomic bombs were instead dropped on Japan; there was no racial element, as Reed (2015) has shown.

As strong cultural taboos severely challenged playing Pokémon GO in these places, we believe that a comparison of the two may reveal cultural differences between Japan and Poland that have not yet been explicitly identified. Judge and

Brown (2017) and Ntelia (2020) focussed primarily on the ethics of gaming in general, while we focus on Pokémon GO in order to show the cultural differences between the two countries.

4   The social problems caused by Pokémon GO

Globally, friction caused by Pokémon GO is common. Thailand (a Buddhist country) has imposed restrictions on Pokémon GO use in royal palaces and temples (Chuwiruch, 2016). Neighbouring Vietnam, also a Buddhist country, has requested removal of PokéStops from religious and historical sites. At the PokéStops, players can get items. It is natural that users would come together around PokéStops. However, as Vietnam is a socialist country, it may be that the government is additionally interested in avoiding traffic confusion caused by crowds, along with ensuring an air of religious tranquillity at sacred sites ("Vietnam wants pokemon", 2016). Pokémon GO regulation is an important issue in the Islamic world, where idolatry is banned. Brunei (a strict Muslim country) has removed PokéStops and Pokémon gyms from religious facilities (Hab, 2016). At the gym, a player can have a battle between his/her own monster and one that another player has. Pokémon GO has caused cultural friction in relatively secular Muslim countries such as Turkey and Indonesia. In Turkey, imams have advocated for a Pokémon GO ban (Akkoc, 2016); in Indonesia, religious leaders have declared a Haram Fatwa (religious ban) ("Ulema council", 2016). The game is considered incompatible with Islamic values. In Indonesia, the principal concern is that the game will distract from daily prayers. This problem is not unique to Pokémon GO; any addictive pastime will yield comparable results.

There are many examples of friction in the Christian world where religious condemnation of Pokémon GO is evident. An Italian Catholic bishop calls for a ban; he argues that the sight of people walking around with smartphones is seen as 'diabolical' (Skinner, 2016). One Pokémon GO player was arrested for playing in a church in Russia (Yin-Poole, 2017). The US has raised a number of issues pertaining to spiritual matters (such as the inappropriateness of playing in a Holocaust memorial or in churches) and management issues (the potential for accidents) if play is allowed in churches (White, 2016) ("US Holocaust Museum", 2016). PokéStops has been removed from indigenous cemeteries in Canada (Trumpener, 2016).

The high court in Gujarat state was asked by Hindus and Jains to ban Pokémon GO since it showed eggs in places of worship, thus hurting religious feelings ("Pokemon go in", 2016). Pokémon GO has also been of concern to indigenous cultures such as the Maori of Oceania, who thought religious observances might be affected were PokéStops to be placed in places of religious significance or if monsters appeared (Taiuru, 2019). Notably, Iran has banned the game for security reasons ("Iran bans Pokémon GO", 2016). Similarly, as the game uses geographical positional information, it was not available in South Korea for some time after release (Shead, 2017). In China, the world's most populous country, Pokémon GO cannot be played since Google Maps does not work in China (Hill, 2022).

Thus, Pokémon GO restrictions are aimed at:

1 preventing traffic and other accidents;
2 protecting religious conventions and sites;
3 protecting tranquillity of mind, which has little to do with any doctrine;
4 protecting national security.

Sometimes, several of the above are combined.

The next section focuses on the areas around the Atomic Bomb Dome in Hiroshima and Auschwitz as concrete examples.

5 Hiroshima and Auschwitz

Here, we introduce some of the perceived problems associated with Pokémon GO use in Hiroshima and Auschwitz. A comparative analysis follows.

5.1 *Hiroshima*

The game was released in Japan on 22 July 2016, thus about two weeks later than in the US (6 July). Gameplay exploded; the 'download' servers were often down (Third, 2016). After its release in Japan, the dangers quickly became public knowledge: injuries when walking and accidents when driving ("'Pokemon go' behind", 2016). In Hiroshima, large numbers of players walked in the Peace Memorial Park which hosts the Atomic Bomb Dome, as Atomic Bomb Memorial Day (August 6) which commemorates the victims was approaching. The more serious social impact of Pokémon GO was close to occurring. Hiroshima citizens repeatedly claimed that the park was 'a place of prayer' and the city asked Niantic to remove the gameplay locations, the PokéStops; Niantic rapidly did so. Officials recognized that the city had requested removal; players confirmed that the PokéStops had disappeared ("Hiroshima", 2016). If it was not considered that the area around the Atomic Bomb Dome was a 'place of prayer', this could be challenged. Before the war, Hiroshima was a large military city and the area around the Atomic Bomb Dome a bustling commercial district. The Dome was then a product exhibition hall, and thus a place of business. Therefore, if the Dome was to be remembered as a 'bustling place where many people came and went', there might be little need to ban games.

Although many players did not use the AR function as Stalberg (2022) pointed out, the game initially attracted a lot of attention because of the AR capability. For location-based games lacking AR, the main ethical debate is on whether it is acceptable to play them in certain locations. In the case of games with AR functionality, however, a completely new question emerges: namely in what way it is acceptable to project a game's worldview onto the real world. In fact, in Milwaukee County (WI, US), the development of AR games used to require permission; however, this was eventually declared illegal on the grounds that it excessively restricted freedom of expression in the case of *Candy Lab Inc. v. Milwaukee Cnty.* (2017). As Purzcyki (2019) noted, it is difficult to determine what should actually be subject to regulation. Pokémon GO features many ghostly monsters such as Haunter and Gastly that can be projected onto the real world (as revealed by the smartphone camera). Many would consider it inappropriate to superimpose ghosts on the Atomic Bomb Dome. Thus, the ethical issues expand to include the intentions of gamers. The request to ban Pokémon GO in the vicinity of the Atomic Bomb Dome in July 2016 was based on the real difficulty of holding a memorial service if people played there; however, the problem of AR is different fundamentally. If an AR game played in, say, a deserted park without causing any actual harm is regulated, such regulation is no longer content-neutral; as such, this amounts to censorship and is akin to a ban on reading books with certain ideological content. Of course, such a ban should be avoided.

5.2   *Auschwitz: Founding trauma and moral panic*

Here, we recall and juxtapose two places that serve as mythologised monuments of post-war, collective historical trauma, and that bear traces of founding traumas. As noted by LaCapra (1999), trauma typically plays a tendentious ideological role, for example, in terms of the concept of a chosen people or a belief in one's privileged status as a victim. Thus, when the identity and memory of a nation or a group are based on locations such as Auschwitz or Hiroshima, such places will be protected, even when an alleged action is accidental, or not potentially dangerous, or (as in the case of Pokémon GO) is dependent solely on the algorithm used (even if the camera is on, this may be unintentional). In other words, although people do not necessarily intend to do harm by playing in memorial sites, reactions to such behaviour can be strong, which is evident in the Auschwitz example. We argue that media's as well as general public's responses to the perceived problem of playing Pokémon GO within the former Nazi concentration camp were mostly exaggerated. Those who condemned such video game play might assume that historical memory was being defiled. In reality, a few individuals erred and the game developer blocked gaming as soon as they were told of the problem (Lucas, 2016). However, in an effort to understand the overreactions to a few cases that people's playing the game led to the national and international momentum to protect the sanctuaries, we need to trace the perception of Pokémon over the last two decades and explore the alleged 'anti-biblical' and 'mystical' dangers posed to the social development of children. As Łuczaj and Hoły-Łuczaj (2017) showed, Pokémon GO poses an apparently irrational threat to many people.

The early popularity of Pokémon in Poland was indicated by the observation that 'Pokémon' entered Polish slang as a person with an exceptionally low IQ and an exceptionally high level of pink in (and on) their entire bodies (Miejski słownik slangu i mowy potocznej, 2008). An Internet Pokémon language developed among the young; this featured the visualisation of speech to underline one's own creativity as Wikeczek (2011) stated; this was perceived as negative. The language is characterised by the use of 'alternating caps' (alternate upper- and lower-case letters), abbreviations of words and 'loan' words from other languages, for example pokEMON w aushCHWiTZ nIe JEEst CoOl! (Pokémon in Auschwitz is not cool!).

In Poland, the first Pokémon boom coincided with the Pokémon craze in the West. Manga and anime became extremely popular among young people, being the first generation raised after 1989 and the fall of communism in Poland and Europe. This can be partly explained by the popularity of other manga animations, *Sailor Moon* and *Dragon Ball*, in the 1990s, whose huge popularity paved the way for subsequent animations including Pokémon. This generation constitutes what may be called the caesura (the poetic pause). For some of the older generation who were never exposed to Japanese anime (or, because of dubbing, were not aware that it was anime), the phenomenon was often misunderstood and provoked much anxiety. Among the Catholic Church hierarchy, Pokémons and the entire anime phenomenon became synonymous with pagan practices that posed risks to Christian values (e.g. Skawińska, 2003). In the (Catholic) nationalist press of the time, Pokémon was linked to brutality, eastern mysticism and Shintoism, especially the worship of natural elements, such as water, fire or earth (Bielecki, 2001), and was thought to threaten childhood development. Pokémon evolution and transformation into later forms were taken to be evidence of the dangerous influence of Eastern philosophy, in which re-incarnation was a 'suspicious' idea. This issue was thoroughly described in Skawińska (2003). The alleged occult content (in contrast to the supposedly appropriate content of the Decalogue) of anime was (reportedly) clear; the creators 'do not cover up their intention to take over the souls of young people' as Bielecki (2001) stated.

As reported by Łuczaj and Hoły-Łuczaj (2017), more than 15 years later, the critique of Pokémon in Poland persists. An analysis of the bodies of 63 texts revealed that articles on Pokémon GO commonly contain the words/phrase 'danger', 'religion' and 'adverse consequences'. Unforgettable newspaper headlines stated that 'Pokémon is a game full of dangers' and 'A law blocking Pokémon? This game poses risk to social stability'.

The disproportionate and exaggerated reactions to the small number of events involving personal injuries, the supposed abuse of historical memory and the so-called abuse of sacred symbols caused by Pokémon GO demonstrate the range of irrational phobias raging in society. One of the most vivid examples could be the case of a famous YouTuber who faced five years sentence in prison for playing Pokémon Go on his phone in a church in Yekaterinburg (Amnesty International UK, 2020). A key term used to understand such phobias is 'moral panic' as Cohen (2011) mentioned. Extreme social reactions and moral outrage are directed to an object perceived as a potential risk to societal well-being and existing moral values. According to Cohen, this object can be new, or an old object returned to the limelight. The media play key roles in targeting alleged 'social pathologies', fuelling the sense of threat and reinforcing fear when the situation is in fact rather harmless.

AR, which blurs the border between the real and the virtual, initially distinguished Pokémon GO from other 'games', capturing the attention of smartphone users. In other words, the fear of new entertainment based on AR triggered a social need to seek a rational hazard. At the same time, the appearance of Pokémon GO exposed a fear of 'diabolic action'. Reflecting events at the turn of the century, the need to insert the Pokémon discourse into the realm of black magic was resurrected principally by the far-right, Catholic and nationalist media, and priests according to Salwawski (2016). Suchecka and Szostak (2016) argued that 'plenty of priests claim that the game features occult and demonic elements'. 'Pokémon prepares them [children] to accept humanist "wisdom" and "occult spirituality"' according to Brown's report (Brown, n.d.). The pseudo-scientific and intellectually derelict ramblings on the relationship between Pokémon and dark forces and comparisons of the laws of Pokémons to the structure of Hell (based on the hierarchy of weaker and more powerful demons) rendered social outrage incandescent when it was found possible to play Pokémon GO within the Auschwitz-Birkenau Museum (which of course features geographical co-ordinates and was thus not automatically excluded from the realm of the game). A few players deliberately violated the site of Auschwitz-Birkenau taken by many to be sacred. Article 196 of the Polish Penal Code regulates questions of insulting religious feelings, prohibiting insults to places or objects of worship in any religion. The penalty for such a crime is a fine or up to two years in prison. Actually, LGBTQ+ cause supporters who painted the images of the Virgin Mary and baby Jesus with rainbow halos as a reaction to hateful statements of the catholic church hierarchs in Poland were accused (but, were finally acquitted) (Deutsche Welle, 2021). Therefore, in a country where the criminal code features offenses against religious feelings, the return of moral panic became very possible. Again, the Museum administration asked the game creator to exclude 'the Site of Memory of Auschwitz and any similar places. Now and in the future.' as a Dziennik Gazeta Prawna (newspaper) article reported (Dziennik Gazeta Prawna, 2016); this request was immediately granted. An analogous situation arose in the Holocaust Memorial Museum in Washington DC (US); a player uploaded a picture with Koffing (a Pokémon emitting toxic fumes) next to a signpost for the Helena Rubinstein Auditorium, which hosts the testimonies of gas chamber survivors.

However, connecting Pokémon to the above-described 'dangers', encouragement of moral panic, and expressions of neophobia, technophobia, or fear of the mystic-occult (without any discussion of validity) in the context of the Auschwitz-Birkenau Museum is inappropriate because the violation of the sacred space is indirect. The catching of Pokémons on premises of former death camps, which was considered morally wrong, was indeed possible but was not the focus of the game (the game co-ordinates spanned the globe). In that sense, site violation by a lack of sensitivity was not forced by the game; it was an individual decision. Theoretically, the user could commit a similar act by playing any other game in the Auschwitz-Birkenau Museum, or in the Holocaust Memorial Museum. It is difficult to see malevolence in the developers; they immediately blocked the game responding to the requests. The moral panic that exploded in Poland after the Pokémon games, anime, and, later, Pokémon GO appeared, and after the violation of Auschwitz, was initially very turbulent and then quickly disappeared; again, the developers promptly blocked the game.

6   Conclusion and future work

What differences are apparent when comparing Hiroshima and Auschwitz, two sites that experienced mass deaths? These sites initially appear to be similar, but in fact, differ in terms of 'religious conviction'. For Hiroshima, the fact that 'a lot of people died there' is important; the content of Pokémon GO per se is not. In other words, the fact that the game was played at the memorial site and interfered with memorialisation matters. In the Polish case, on the other hand, the act of playing the game at Auschwitz violated religious convictions. The difference is attributable to a difference in acceptance of the game per se in the two countries. The manga animation Pokémon (on which Pokémon GO is based) is Japanese content produced by a Japanese TV station. Therefore, even though the game is played worldwide, it does not create cultural friction in Japan; it is simply perceived as annoying. Today, Japanese people tend not to be religious, although Japan has a long religious tradition of Shinto and Buddhism. As Kawano (2005) showed, it is difficult to state that religious convictions inform everyday Japanese behaviour.

On the other hand, in the case of Auschwitz, Pokémon GO is not respected by the Polish Catholic culture, and gaming at the site of the massacre offends national sentiment. In this sense, the emotions raised by Pokémon GO play in Auschwitz differ qualitatively between Poles and foreigners.

In studies on cultural friction caused by Pokémon GO, 'religion' is often a key concept. We thus explored real-world views on Pokémon GO between its country of origin and those to which it was subsequently exported.

There are two possible future challenges. As noted in Section 3, cultural friction has been observed in the Middle East and Asia; a comparison of these cases with that of Poland (a Catholic country) may afford new insights. Additionally, it may also be possible to understand cultural friction more generally by comparing the country of production with countries of export. For example, the American film 'Joker' of 2019 deals with murder in a manner that seeks to be funny. In fact, murders in Japan have mimicked those of the film and have had a significant social impact (Wingfield-Hayes, 2021). Similar sequelae occurred in the US (Gross, 2019). Thus, a comparison of the impact of US content on the rest of the world with the impact within the US may reveal cultural differences. Extension of the method could afford new and useful insights into cultural anthropology.

Acknowledgement

Special Thanks to Masatoshi Tokuoka, Dai Tanaka & Sho Sato (LUDiMUS Inc.)

This study was supported by the JSPS/STINT Bilateral Joint Research Project "Information and Communication Technology for Sustainability and Ethics: Cross-national Studies between Japan and Sweden" (JPJSBP120185411) and by JSPS KAKENHI Grant Numbers JP20H01220 and JP 21K18396.

References

Akkoc, R. (2016, July 14). Pokemon Go Mania Sweeps Turkey, but some want it banned. *Phys.org*. https://phys.org/news/2016-07-pokemon-mania-turkey.html

Althoff, T., White, R. W., & Horvitz, E. (2016). Influence of Pokémon GO on physical activity: Study and implications. *Journal of Medical Internet Research*, *18*(12), e315. https://doi.org/10.2196/jmir.6759

Aluri, A. (2017). Mobile augmented reality (MAR) game as a travel guide: Insights from Pokémon GO. *Journal of Hospitality and Tourism Technology*, *8*(1), 55–72. https://doi.org/10.1108/JHTT-12-2016-0087

Amnesty International UK. (2020, May 18). Russian youtuber facing prison for playing Pokémon go. *Amnesty International UK*. https://www.amnesty.org.uk/russia-youtube-blogger-prison-playing-pokemon-go-church-ruslan-sokolovsky

Bielecki, A. (2001). Harry Potter i Kieszonkowy Potwór [Harry Potter and Pocket Monster]. *Miłujcie Się!*, *41*, 11–12.

Brown, D. L. (n.d.). Problem Z pokemonem. *Czytelnia Chrześcijanin*. http://czytelnia.chrzescijanin.pl/problem-z-pokemonem/

Candy Lab Inc. v. Milwaukee Cnty. (2017). 266 F. Supp. 3d 1139.

Carbonell Carrera, C., Saorín, J. L., & Hess Medler, S. (2018). Pokémon GO and improvement in spatial orientation skills. *Journal of Geography*, *117*(6), 245–253. https://doi.org/10.1080/00221341.2018.1470663

Chuwiruch, N. (2016, August 8). Royal Palace among places Pokemon shouldn't go in Thailand. *New Jersey Herald*. https://www.njherald.com/story/news/nation-world/2016/08/08/royal-palace-among-places-pokemon/2989814007/

Cohen, S. (2011). *Folk devils and moral panics*. Routledge.

Colley, A., Thebault-Spieker, J., Lin, A. Y., Degraen, D., Fischman, B., Häkkilä, J., … Schöning, J. (2017, May). The geography of Pokémon GO: beneficial and problematic effects on places and movement. In *Proceedings of the 2017 CHI conference on human factors in computing systems* (pp. 1179–1192). https://doi.org/10.1145/3025453.3025495

Craddock, D. (2017, July 6). Happy accident: The making of pokemon go. *Shacknews*. https://www.shacknews.com/article/100513/happy-accident-the-making-of-pokemon-go

D'Anastasio, C. (2016, August 8). You can no longer catch Pokémon at Hiroshima's Memorial or the Holocaust Museum. *KOTAKU*. https://kotaku.com/you-can-no-longer-catch-pokemon-at-hiroshima-1784985508.

Deutsche Welle. (2021, February 3). Polish activists acquitted in Rainbow Virgin Mary case. *dw.com*. https://www.dw.com/en/polish-court-acquits-lgbt-activists-in-rainbow-virgin-mary-case/a-56749372

Dworak, M., Schierl, T., Bruns, T., & Strüder, H. K. (2007). Impact of singular excessive computer game and television exposure on sleep patterns and memory performance of school-aged children. *Pediatrics*, *120*(5), 978–985. https://doi.org/10.1542/peds.2007-0476

Dziennik Gazeta Prawna. (2016, July 14). Rzecznik Muzeum: Działanie Gry Pokemon go W miejscu pamięci auschwitz narusza pamięć ofiar. *GazetaPrawna.pl*. https://www.gazetaprawna.pl/wiadomosci/artykuly/960087,rzecznik-muzeum-dzialanie-gry-pokemon-go-w-miejscu-pamieci-auschwitz-narusza-pamiec-ofiar.html.amp

Elektronika, komputery w PRL – konferencja [Electronics, computers in the People's Republic of Poland – conference] (2015, June 14). *AKTUALIZACJA*. https://dzieje.pl/dzieje-sie/elektronika-komputery-w-prl-konferencja

Evans, L., & Saker, M. (2019). The playeur and Pokémon GO: Examining the effects of locative play on spatiality and sociability. *Mobile Media & Communication*, *7*(2), 232–247. https://doi.org/10.1177/20501579187988

Faccio, M., & McConnell, J. J. (2019). Death by pokémon go: The economic and human cost of using apps while driving. *Journal of Risk and Insurance*, *87*(3), 815–849. https://doi.org/10.1111/jori.12301

Główny Urząd Statystyczny. (2022). *Religious denominations in Poland 2019–2021*. https://stat.gov.pl/en/topics/other-studies/religious-denominations/religious-denominations-in-poland-2019-2021,1,3.html

Gross, S. (2019, November 14). Man accused of fatal shooting during zombie crawl claimed to be 'Arthur Fleck' from 'joker'. *Reno Gazette Journal*. https://www.rgj.com/story/news/crime/2019/11/14/reno-man-accused-murder-dressed-like-joker-arthur-fleck/4182735002/

Hab, R. (2016, August 16). Mosques versus Pokestop/Gym. *Justina Baharun*. https://justinabaharunblog.wordpress.com/2016/08/12/mosques-versus-pokestopgym/

Hill, S. (2022, March 10). Banned in China, pokemon go still takes the internet by storm. *iMyFone*. https://www.imyfone.com/change-location/pokemon-go-china/

Hiroshima: Keep Pokemon away from Atom Bomb Memorial. (2016, July 28). *BBC News*. https://www.bbc.com/news/world-asia-36891787

Iqbal, M. (2023, January 9). Pokémon go revenue and usage statistics (2023). *Business of Apps*. https://www.businessofapps.com/data/pokemon-go-statistics/

Iran bans Pokémon GO. (2016, August 8). *Guardian News and Media*. https://www.theguardian.com/world/2016/aug/08/iran-bans-pokemon-go

ISSP Research Group. (2022). *International Social Survey Programme: Environment IV – ISSP 2020*. https://doi.org/10.4232/1.13921

Judge, E. F., & Brown, T. E. (2017). Pokémorials: placing norms in augmented reality. *UBCL Rev., 50*, 971.

Kamosiński, S. (2019). Birth and development of the computer game industry in Poland. Analysis of selected cases. *UR Journal of Humanities and Social Sciences, 3*(12), 101–116. https://doi.org/10.15584/johass.2019.3.6

Kari, T., Arjoranta, J., & Salo, M. (2017, August). Behavior change types with Pokémon GO. In *Proceedings of the 12th international conference on the foundations of digital games* (pp. 1–10). https://doi.org/10.1145/3102071.3102074

Kautz, P. (2019, September 9). *Pokémon GO Fest: Economic impact study* [Power Point Slides]. VP Research & Analysis, Statista. https://www.statista.com/study/70046/studie-pokemon-go-fest/

Kawano, S. (2005). *Ritual practice in modern Japan: Ordering place, people, and action.* University of Hawaii Press.

LaCapra, D. (1999). Trauma, absence, loss. *Critical Inquiry, 25*(4), 696–727. https://www.businessinsider.in/a-lot-of-people-had-police-run-ins-when-they-played-the-game-pokemon-go-is-based-on/articleshow/52440848.cms

Lucas, E. (2016, August 16). Pokemon No: The battle to declare Auschwitz off-limits. *The Times of Israel*. https://www.timesofisrael.com/pokemon-no-the-battle-to-declare-auschwitz-off-limits/

Łuczaj, K., & Hoły-Łuczaj, M. (2017). Pokemon to demon?: analiza dyskursu strachu przed grą "Pokemon go" w polskim internecie [Pokemon is a demon?: an analysis of the discourse of fear of the "Pokemon go" game on the Polish Internet]. *Zeszyty Prasoznawcze Kraków: Wydawnictwo RSW Prasa, 4*(232), 964–984.

Madden, M., & Rainie, L. (2010, June 18). Adults and cell phone distractions. *Pew Research Center*. https://www.pewresearch.org/internet/2010/06/18/adults-and-cell-phone-distractions/

Margalit, R. (2014, June 26). Should Auschwitz be a site for selfies. *The New Yorker, 26*, 2014. https://www.newyorker.com/culture/culture-desk/should-auschwitz-be-a-site-for-selfies

Matyszczyk, C. (2016, July 24). Reporter plays Pokemon go during state department isis briefing. *CNET*. https://www.cnet.com/culture/reporter-plays-pokemon-go-during-state-department-isis-briefing/

Pokemon. (2008, April 17). *In Miejski słownik slangu i mowy potocznej.*□https://www.miejski.pl/slowo-Pokemon

Newman, N. (2016, September 21). *Even better than the real thing.* Gavekal Research. http://newsfeed.sovereignwp.com/wp-content/uploads/2016/10/Even-Better-Than-The-Real-Thing.pdf

Newzoo. (2022). *Global games market report.* https://newzoo.com/insights/trend-reports/newzoo-global-games-market-report-2022-free-version

Ntelia, R. E. (2020). Play everywhere: Can we play (in) Auschwitz? *DiGRA '20 – Proceedings of the 2020 DiGRA international conference* (pp. 1–4).

Orosz, G., Zsila, Á., Vallerand, R. J., & Böthe, B. (2018). On the determinants and outcomes of passion for playing Pokémon GO. *Frontiers in Psychology, 9*, 316. https://doi.org/10.3389/fpsyg.2018.00316

'Pokemon go' behind second fatal traffic accident in Japan, Kyodo reports. (2016, August 28). *CNBC.* https://www.cnbc.com/2016/08/28/pokemon-go-behind-second-fatal-traffic-accident-in-japan-kyodo-reports.html

Pokemon go in Indian Court for 'hurting religious sentiments'. (2016, September 7). *BBC News.* https://www.bbc.com/news/world-asia-india-37294286

Purzycki, K. (2019). For atatopistic places. In H. Jamie, A. Kulak, & K. Purzycki (Eds.), *The Pokémon GO phenomenon: Essays on public play in contested spaces.* McFarland.

Rasche, P., Schlomann, A., & Mertens, A. (2017). Who is still playing Pokémon GO? A web-based survey. *JMIR Serious Games, 5*(2), e7197. https://doi.org/10.2196/games.7197

Rauschnabel, P. A., Rossmann, A., & tom Dieck, M. C. (2017). An adoption framework for mobile augmented reality games: The case of Pokémon GO. *Computers in Human Behavior, 76*, 276–286. https://doi.org/10.1016/j.chb.2017.07.030

Reed, B. C. (2015). The Manhattan Project. In *Atomic Bomb: The Story of the Manhattan Project: How nuclear physics became a global geopolitical game-changer.* Morgan & Claypool Publishers.

Salwowski, M. (2001). Pokemony wciąż atakują. *Niedziela.pl.* https://www.niedziela.pl/wydruk/67775/nd

Shead, S. (2017, January 24). Pokemon go has finally landed in Asia's third biggest gaming market. *Business Insider.* https://www.businessinsider.com/r-pokemon-go-unleashed-on-game-mad-south-korea-six-months-late-2017-1

Skawińska, M. (2003). Zniekształcony obraz rzeczywistości lansowany przez katolickie pisma młodzieżowe: analiza treści periodyków "Miłujcie się!", "Droga" i "Źródło" [Malformed image of reality reflected in the letters of catholic young people.] *Occasional paper (Monash University. Slavic Section. Polish Studies), 18*, 101–103.

Skinner, D. (2016, August 19). An Italian bishop ties pokemon go to Nazism, wants to go to court to ban it. *Twinfinite.* https://twinfinite.net/2016/08/italian-bishop-pokemon-go-nazism-court/

Stalberg, A. (2022, July 26). Pokemon go: How to turn off ar mode. *Game Rant.* https://gamerant.com/pokemon-go-turn-off-ar-mode/

Suchecka, J., & Szostak, N. (2016, July 19). "Pokemon GO". Inwazja potworów na Polskę. Kościoły, lotniska, wieże ciśnień... *wyborcza.pl.* https://www.wysokieobcasy.pl/wysokie-obcasy/7,115167,20419725,inwazja-potworow-zalewa-kraj.html

Taiuru, K. (2019, January 21). Pokemon go on the Urupa, marae – tapu? *Dr Karaitiana Taiuru PhD, JP, ACG, MInstD, RSNZ.* https://www.taiuru.maori.nz/pokemon-go-on-urupa-marae/

Tang, A. K. (2017). Key factors in the triumph of Pokémon GO. *Business Horizons, 60*(5), 725–728. https://doi.org/10.1016/j.bushor.2017.05.016

Tateno, M., Skokauskas, N., Kato, T. A., Teo, A. R., & Guerrero, A. P. (2016). New game software (Pokémon GO) may help youth with severe social withdrawal, hikikomori. *Psychiatry Research, 246*, 848–849. https://doi.org/10.1016/j.psychres.2016.10.038

The Niantic Story. (n.d.). *Niantic.* https://nianticlabs.com/about?hl=en

The Niantic Team. (2018, November 5). Welcome to ingress prime. *News.* https://nianticlabs.com/news/ingress-prime?hl=en

Trumpener, B. (2016, September 22). PokeStopped! Indigenous Cemetery's PokeStop removed|. *CBCnews.* https://www.cbc.ca/news/canada/british-columbia/pokemon-go-pokestop-removed-from-indigenous-cemetary-1.3774237

Ulema council in lebak issues haram fatwa against pokemon go as players could neglect prayer for game. (2016, July 25). *Coconuts.* https://coconuts.co/jakarta/news/ulema-council-lebak-issues-haram-fatwa-against-pokemon-go-players-could-neglect-prayer/

US Holocaust Museum asks pokemon go players to stop. (2016, July 13). *BBC News.* https://www.bbc.com/news/world-us-canada-36780610

Vietnam wants pokemon out of religious, historic sites. (2016, August 17). *VnExpress International.* https://e.vnexpress.net/news/news/vietnam-wants-pokemon-out-of-religious-historic-sites-3454191.html

White, K. (2016, August 5). Some Arizona pokestops vanish from pokemon go. *The Arizona Republic.* https://www.azcentral.com/story/news/local/arizona/2016/08/05/pokemon-go-remove-pokestops-arizona/88076174/

Wileczek, A. (2011). O języku młodego Internetu [About the language of the young Internet]. *Zeszyty Prasoznawcze, 210*(3+4), 82–99.

Wingfield-Hayes, R. (2021, November 14). What the 'Joker attack' revealed about Japanese society. *BBC News.* https://www.bbc.com/news/world-asia-59257736

Wolf, M. J. P. (Ed.). (2015). *Video games around the world.* MIT Press. https://doi.org/https://doi.org/10.7551/mitpress/9780262527163.001.0001

Yin-Poole, W. (2017, May 15). Russian youtuber slapped with three-and-a-half year suspended sentence for playing Pokémon go in a Church. *Eurogamer.net.* https://www.eurogamer.net/russian-youtuber-given-three-and-a-half-year-suspended-sentence-for-playing-pokemon-go-in-a-church

Zach, F. J., & Tussyadiah, I. P. (2017). To catch them all—the (un) intended consequences of Pokémon GO on mobility, consumption, and wellbeing. In R. Schegg & B. Stangl (Eds.), *Information and communication technologies in tourism 2017* (pp. 217–227). Springer. https://doi.org/10.1007/978-3-319-51168-9_16

10 From strangers to neighbours

How the sharing economy can help building and maintaining local communities

Per Fors and Tina Ringenson

1 Introduction

In the late 2000s, the emergence of the sharing economy coincided with the widespread adoption of information and communication technologies (ICTs) and digitalisation, according to Schor (2020). This period was also characterised by the financial crisis, which prompted discussions on how society and the economy could be reorganised in a more sustainable and equitable way. An emergent discourse questioned conventional forms of production, consumption and work under global capitalism. In early utopian visions, digital sharing and collaboration were seen as critical, laying the groundwork for what would later be known as the sharing economy.

In previous studies, we have concluded that in Sweden, the sharing economy today is mainly understood and promoted as a potential contributor to environmental sustainability through increased resource efficiency (Naturskyddsföreningen, 2015; Svenfelt et al., 2019). While early proponents of the concept often emphasised the potential for a more fair and socially sustainable society, possible benefits of conventional sharing platforms for social sustainability are today overlooked in research and policy in Sweden (Fors et al., 2021). Such advantages are typically associated with collectively owned, not-for-profit (NFP) platforms, as stated in studies conducted by Schor and Thompson (2014), Light and Seravalli (2019), Schor and Vallas (2021), Fedosov et al. (2021) and Hagbert et al. (2018). However, while such platforms do indeed exist, they constitute only a small fraction of the overall number of sharing platforms. Research on the social issues of conventional platforms tends to instead focus on how dominant actors promote social *unsustainability* through tax evasion, gentrification, social isolation, exploitation and more (Martin, 2016; Wachsmuth and Weisler, 2018; Balding et al., 2019; Schaller, 2021). While we agree that there are many social sustainability-related problems with how the sharing economy is currently constituted, these problems are pointed out and sufficiently understood by the research community. We are also sympathetic towards utopian visions of a more collaborative sharing economy as it was first envisioned by Schor and Thompson (2014), Belk (2007) and others. Despite this, we argue that there are potential social benefits of digitally enabled sharing through conventional platforms that have not been properly highlighted in research, at least not in a Swedish context (Fors et al., 2021).

DOI: 10.4324/9781003367451-12

The objective of this chapter is therefore to emphasise the existing sharing economy's potential to contribute to various facets of social sustainability, such as fostering social cohesion, overcoming segregation between people of different background and culture, and re-establishing individuals' ties to their communities. We are inspired by how the sharing economy is promoted in Japan, where it is seen as a driving force for cohesion and social sustainability (Majima et al., 2021). Our main concern is urban social sustainable development, which is defined by Polèse and Stren (2000) as the development that promotes a harmonious evolution of civil society that supports the coexistence of culturally and socially diverse groups and encourages social integration while enhancing the quality of life for all segments of the population. Our argumentation is mainly theoretical but illustrated with an empirical case of an existing sharing economy platform operated in Sweden called Hygglo, which was established in 2016. In this case, we highlight how many sharing platforms are highly dependent on physical proximity between users and embedded in local communities, despite being digital and thus essentially accessible anywhere in the world.

2   Visions of a sharing society

As previously mentioned, the platform-based, digital economies started to emerge in the late 2000s. The American researcher Belk (2007) asked in a research article *Why not share, rather than own?* In this article, he stated that sharing was a much more resource-efficient mode of consumption than permanent ownership and that sharing can foster community. Botsman and Rogers (2010), in their influential book *What's Mine Is Yours: The Rise of Collaborative Consumption*, present sharing as a new cultural and economic force that would transform businesses and how we consume goods and services. Schor and White were also working on a book at the time, called *Plenitude: The New Economics of True Wealth,* published in 2010. In their book, they envisioned a future in which people would do more for themselves and their peers with the help of digital platforms, and that these platforms would transform the way people related to work and free time. People would work less, but the feeling would be more rewarding, satisfying and productive. According to Schor (2020), young adults who were reaching economic independence during the financial crisis turned to sharing as an alternative to global capitalism dominated by multinational corporations (MNCs). They believed that sharing would not only address the issue of work but also alleviate social disconnection, inequality and environmental degradation.

While it did not have a name at the time, what these authors were envisioning is now often referred to as the sharing economy. The sharing economy, as briefly mentioned in the introduction, is often understood as an economic model based around digital platforms where individuals interact with the intention to share products and services, often for a fee. This type of transaction is what we will refer to as "sharing" henceforth. The sharing economy is one of many emerging economies of the late 2000s. Related economic concepts include the platform economy, the P2P economy, the gig economy and more. As we have argued elsewhere, there are no clear distinctions between many of these concepts, and despite there being collectively agreed-on definitions, these definitions are sufficiently flexible for different actors to claim that they are indeed part of these economies (Majima et al., 2021). As these ideas began to develop, Airbnb and Uber also emerged. While often claimed not to be part of the sharing economy, these companies would heavily impact and shape the sharing economy and the discourse surrounding the concept (Laurell and Sandström, 2017).

3   Sustainability within the sharing economy

Sustainability is a key concept in the sharing economy. Among the various sustainability discourses that exist today, sustainable development is the most well-known. Sustainable development is described by the World Commission on Environment and Development (WCED) (1987) as development that meets the needs of the present without compromising the ability of future generations to meet their own needs. The contemporary understanding of sustainable development is based on the idea that there are economic, environmental and social challenges that should all be considered simultaneously.

A recent literature review by Boar et al. (2020) concluded that the majority of research papers on sustainability of the sharing economy focused on its potential environmental benefits, mainly reduction of resource use and greenhouse gas emissions. It is not unexpected, as Belk emphasised the circular potential of the digitally enabled sharing economy already in 2007: individuals who possess infrequently used items, such as clothing, tools and hiking gear, can share them with strangers online. In theory, sharing has the potential to increase the efficiency of product usage, resulting in reduced production needs and consequently lower waste and emissions (Leismann et al., 2013; Hamari et al., 2016; Seegebarth et al., 2016; Plewnia and Guenther, 2018; Laukkanen and Tura, 2020). Thus, sharing can contribute significantly to Sustainable Development Goal (SDG) 12, which concerns responsible production and consumption. While the sharing economy is promoted for many different reasons by various actors, this is by far the most commonly used argument for expanding the sharing economy model (Heinrichs, 2013; Boar et al., 2020).

Although most researchers agree that there is much potential for environmental sustainability in the sharing economy [Boar et al. (2020), for example, argue that it helps to achieve *all* SDGs], it is unclear whether this potential has been at all realised. Instead, some researchers argue that sharing can instead lead to increased environmental unsustainability (Verboven and Vanhereck, 2016; Clewlow and Mishra, 2017; Balding et al., 2019; IBISWorld, 2021; Schaller, 2021). While early proponents of the sharing economy often used social arguments to promote the emerging digital economies, these aspects are generally less emphasised today (Boar et al., 2020; Fors et al., 2021). Instead, many researchers tend to have a less optimistic outlook when considering the social implications of such platforms, not least due to the growth of the related (and to some extent overlapping) gig economy (Martin, 2016). They claim that conventional platforms can contribute to gentrification and lack of long-term rental contracts (Thoem, 2015; Mermet, 2017; Wachsmuth and Weisler, 2018), racism (Curran, 2023), inequalities (Schor, 2017; Frenken and Schor, 2019), problems related to tax evasion, precarious work and limited possibility to unionise (Plenter, 2017). Light and Miskelly (2015) claim that there is a difference between the sharing economy and a sharing culture of generosity and collaboration. They understand the sharing culture as one where neighbours and friends share between themselves for altruistic and social reasons, without economic transactions, and suggest that the sharing economy undermines such practices. To summarise, we can conclude that while the sharing economy emerged with hopes of a better and more sustainable, fair and altruistic society and culture, nowadays the emphasis in a Western context is often put on its (hitherto largely unrealised) potential to contribute to environmental sustainability while accusing conventional digital platforms of contributing to various negative social impacts.

While there are certainly many negative social impacts related to these new digital economies, we have chosen in this chapter to explore a different vista, inspired by our previous research on how researchers and policymakers in Japan are more inclined to promote conventional sharing as also beneficial for social sustainability (Majima et al., 2021). For instance, it is emphasised by the Japanese government that the sharing economy has the potential to increase social sustainability in rural areas that find it difficult to sustain their local communities and that the sharing economy is expected to facilitate new collaborative communities (see Noda et al., 2018). In the Japanese policy document *Dynamic Engagement of All Citizens* (一億総活躍社会), it is stated that the sharing economy should contribute to an inclusive society where everyone – youngsters and older adults, women and men, people with deficiencies or intractable diseases, and those who have experienced failure – can play an active part and where everyone feels worthy of living at home, in local communities (Tors et al., 2021).

4   The decline of social cohesion in the wake of crises

In the very beginning of this chapter, we described how the sharing economy emerged as a response to the financial crisis, and at the time of writing, we are experiencing the effects of another global crisis (apart from the ongoing climate crisis), namely the COVID-19 pandemic. This dramatic event has had major impacts on the digital economies, as Hossain (2021) shows: As most people refrained from travelling during the COVID-19 pandemic, digital platforms mainly used by tourists experienced much less activity. At the same time, due to lockdowns and restrictions, the market for digital food delivery platforms such as Ubereats and Foodora has doubled in value (Ahuja et al., 2021). More importantly, Müller et al. (2021) and Banerjee and Rai (2020) among others have shown how the lockdown in many countries, deployed by governments in order to stop the spread of the COVID-19 virus, have had serious implications on societal development and individual well-being. In an essay in the New Yorker, Masha Gessen (2020) focuses on the political consequences of loneliness and isolation during the pandemic, drawing on Hannah Arendt's (1973) claim that that isolation – understood as the inability to act together with others – renders people impotent. Gessen's main argument is that democracy cannot thrive in a society consisting of isolated individuals, and indeed, according to IPU (2021), 75 percent of the world's population live in countries that face democratic backsliding.

Christopher Lasch's book *The Revolt of the Elites* (1996) discusses the decline of democracy in mid-1990s America and highlights the importance of self-governing local communities for the well-being of American society. He believes that local communities thrive when people take responsibility for their immediate surroundings and work together with friends and neighbours. This relies on the collective efforts of individuals within these communities. Lasch (1996) noticed that this mentality, which had previously been the norm, was rapidly eroding, not least due to the increased mobility of capital and the emergence of global markets giving rise to a new managerial elite. The livelihoods of the new managerial elites differ from those of their predecessors, as they rely less on owning property and more on their ability to manipulate information and their professional expertise. Their high degree of cosmopolitanism contributes significantly to the reduction of local and regional loyalties. Lasch argues that success in modern business and professions requires individuals to be willing to pursue opportunities wherever they may arise. According to the author, those who remain rooted in one specific place forfeit their chances for upward mobility. In fact, success has never been as closely linked to mobility as it is today. Those who covet membership in this new aristocracy turn their back on their local communities and instead cultivate ties with other actors on international markets. They are home only in transit and are unlikely to contribute to and participate in the development of a specific, local community. They rather end up in specialised geographical pockets, zones or networks (Reich, 1992) such as Silicon Valley, Cambridge or Hollywood. Such places, although physical in nature, differ greatly from communities in a more traditional sense: They are inhabited by transients and lack the sense of continuity that comes from a shared sense of place and self-consciously cultivated standards of conduct passed down from generation to generation (Lasch, 1996).

While the managerial elites broke loose from local communities in favour of globalisation and cosmopolitanism, other demographics have stopped being active participants in local communities for completely different reasons. In a report from the ONS (2019), it is established that young people in the UK between the age 18 and 24 often feel like they do not belong in their local community and report lower degrees of trust towards and willingness to interact with their neighbours. There are many explanations to this. According to the EC (2015), a majority of people living in European cities consider it difficult to find a good place to live at a reasonable price. All over Europe, the number of people aged between 25 and 29 still staying home with their parents has increased since 2009 (Eurostat, 2021), with severe effects on their transition to independence (FEANTSA, 2021; Worth, 2021). Many young people who want to live in or near cities, for work or other reasons, never get the chance to settle down for an extended period of time, which hinders them from forming deep and long-lasting relations to people in their immediate surroundings.

Furthermore, according to Sweden's rent tribunals, the number of tenants with precarious contracts has doubled in the last ten years (SR, 2019).

Oldenburg (1989) argues that community building often takes place in informal hangouts that exist between the home and the workplace, such as bars, cafés and restaurants. He calls these hangouts "third places". He sees third places as physical places designed first and foremost around casual meetings and conversations unrelated to work and family life. For the individual, these places can bring well-being and belongingness, and they help strengthen local communities. The third place, much like the larger neighbourhood it serves, brings together individuals who are united by physical proximity rather than choice (Lasch, 1996). While many of us might prefer to hang out with our closest friends or co-workers – people who generally share our worldviews, perspectives and values – a third place hosts people who share nothing more than the local community in which they live. Rather than mutual admiration and unchallenged agreements, these places can host lively conversations. It is not surprising, Lasch (1996) argues, that third places have traditionally been the resort of pamphleteers, politicos and revolutionaries. Without such places, according to Oldenburg (1989), local communities become populated by people who are concerned merely with getting through the day. Today, many of these physical third places have either disappeared or transformed into places where the social interaction is held at a minimum, such as malls or fast-food restaurants. In fact, acquiring the things we need to get through our lives now only requires a bare minimum of social interaction by design.

Much has happened since Oldenburg (1989) introduced the concept of the third place, especially when it comes to how we connect with strangers. Nowadays, interactions between strangers mainly take place online, and researchers have argued that online platforms have replaced traditional third places (e.g. Rheingold, 1993; Soukup, 2006; Parkinson et al., 2021). However, such digital third places rarely emphasise the role of any particular local community (although there are exceptions, such as local Facebook groups). There are also examples of hybrid third places. Battista (2018) looks into the mobile game Pokémon Go and argues that its digital infrastructure fulfils the basic criteria for a third place according to Oldenburg's original formulation of the concept and that its function of bringing people together both in a digital and a physical environment makes it a third place.

5   Trust among community members

One major concern for Lasch (1996) is the lack of trust among individuals in local communities. He cites Jacobs (2016, originally published in 1961), who contends that communities are the means by which individuals and families are connected to the wider world and that it is the informal trust among members of the public that sustains these local communities. High levels of trust lead to an inclination to care about how the community develops. Yamagishi and Yamagishi (1994) define trust as an expectation that someone has goodwill and benign intent in their dealing with us, even when it is difficult to detect another's true intentions. This differs from the expectancy of benign actions from someone that one has extensive information about, which would rather be an assurance. Trust can generally be divided into two types of trust: particular trust, which is trust in people that you know and interact with often, such as family and friends; and general trust, which is trust in unfamiliar or "most people" (Delhey et al., 2011). However, who are included in the idea of the unfamiliar "most people" can still be blurry, as this is understood differently in different contexts, especially how different these unfamiliar people can be from the respondent (ibid.). There are indications that trust tends to be higher in more homogeneous environments, although not solidly proved that this is always the case (Dinesen and Sønderskov, 2017; Newton et al., 2017) and that this may have more to do with economic status than with homogeneity (Tolsma et al., 2009). For example, Wang (2016) found that intergroup trust was higher among Shanghai residents in more ethnically diverse areas, surmising that this may be because increased contact with out-group neighbours can foster tolerance. A study by Finseraas et al. (2019) on Norwegian soldiers showed that ethnic majority members from diverse neighbourhoods had lower trust towards members of ethnic minorities but that close personal contact between members of majority and minority ethnicities removed this negative correlation. In Sweden, socioeconomic segregation has increased significantly since the 1990s (Delmos, 2021). Income, education and employment – factors which to a large extent define social class – are often closely related to where people live (Hedström, 2015). Furthermore, Sweden has a high rate of ethnic residential and social segregation (Rokem and Vaughan, 2019). This means that it is perhaps not always possible to meet people outside your own social class in your local community.

We see the connection between individuals in physical, local communities as both an important aspect of, and a prerequisite for, social sustainability (Polèse and Stren, 2000). According to Lasch (1996) and Jacobs (2016), the connection between individuals and their local communities depends on personal interactions and discussions, which both require and reinforce trust. Many argue that the success of the sharing economy is dependent on the level of trust among users on the platforms, as well as towards the sharing economy companies (ter Huurne et al., 2017; Möhlmann and Geissinger, 2018; Yang et al., 2019; Akhmedova et al., 2021; Räisänen et al., 2021). This chapter includes examinations of the opposite relationship: If the sharing economy can increase contact between society members, it may influence levels of trust in society, which could improve local social sustainability.

6　　The case of Hygglo

This study, which was conducted between 2020 and 2021, examines how four cases of for-profit companies in the P2P economy, including both entrepreneurial ventures and incumbent companies, work for sustainable development. Qualitative data was collected through semi-structured interviews, observations and analysis of social media posts and news articles. Due to the COVID-19 pandemic, most interviews were conducted via Zoom, but participant observations were also utilised whenever possible. A total of 21 interviews were conducted, including three with Ola, one of the founders of Hygglo and its current CEO. Additionally, the authors of the chapter have personal experience using the service.

Hygglo ("Decent" in Swedish) is a Swedish sharing economy platform that was launched in 2016 by Swedish entrepreneurs from Stockholm. On the platform, people can share tools, hiking equipment and more for a fee. The business idea is not by any means unique; rather, it epitomizes the very essence of the sharing economy. However, previous attempts to launch similar sharing platforms in Sweden had failed, mainly since the sharing economy was still very niche before 2016. The Hygglo founders realised, however, that the market was growing rapidly and were eager to capitalise on it. While they did indeed launch their platform almost right away, they soon realised that they were not in a hurry, since they had a relatively experienced team and could easily attract funding. This meant that they could focus on developing their team and product further. Although they could theoretically start growing right away, they did not want to take the risks associated with growth without knowing that the time was right and without knowing that their team was ideal.

A first challenge for Hygglo was to make sure that users could find what they were looking for in their immediate surroundings. They assumed that few would travel for several hours just to borrow a tool or a pair of ice skates. This meant that they had to reach a critical volume of available products on their platforms in a certain area, and to do so, they had two main options: either to focus on a subset of products – such as camera equipment – or to open up for many different product categories but limit themselves to a specific geographical area. The assumption that physical proximity would be a central aspect for most users turned out to be correct, and therefore they decided to go with the second option. Thus, they began marketing the platform in Hammarby Sjöstad in southern Stockholm.

Focusing on a single geographical area turned out to have both short- and long-term consequences for Hygglo. The short-term effects were that Hygglo could promote the platform among people who lived nearby. Word of mouth spread quickly in the neighbourhood, and the platform grew organically within this area without any marketing efforts. However, the decision to let Hygglo grow from Hammarby Sjöstad also had long-term consequences on the user demographic. Today, the typical user on Hygglo is a white male and between 26 and 51 years old, and most of them have a higher education and a relatively high income (Hygglo, 2020). Ola argues that this might be connected to the fact that they decided to let the platform grow from a rather gentrified part of Stockholm where this is a dominant demographic, in combination with the founders also being part of this demographic:

> One challenge is that we are few and that we are three men who started the company. ... We have an overrepresentation of highly educated and highly paid people among our users. It hasn't really taken off among other demographics.
>
> Our long-term ambition is to grow, and we have focused on growing where we've met the least resistance, without thinking that 'hey, this might

be a problem for us in the future'. ... That we started in Hammarby Sjöstad is definitely one cause for this. We have not been able to grow as quickly in other areas.

While their platform could theoretically facilitate sharing among complete strangers all over Sweden, Ola argues that proximity is the single most important factor deciding whether a product will be shared or not. This means that most of the time the people who interact on the platforms are in fact neighbours, although they might not know each other. In order to use the platform in the intended way you need to share your exact location, and search results on the platform are ranked only based solely on proximity rather than price or rating. Today, if you search for a product on their platform, half of the time you will be able to find the product you are looking for within two kilometres of your current position but, according to Ola, this figure should be closer to 75% or 80% in order for people to find the service convenient enough to be a relevant alternative to conventional shopping or rental services (Hygglo, 2022). How far people are willing to travel varies a bit depending on the product. Roof boxes, for example, can typically be picked up further away than for example drilling machines or ice skates. Typically, however, the majority of Hygglo's rentals take place within one kilometre.

7   Sustainability at Hygglo

Sharing is currently generally more environmentally friendly than shopping, and this is to some extent reflected in the way Hygglo is marketed. Ola speaks of sustainability as one of their core values as a company. When they are invited to speak at conferences, they naturally focus on the environmental sustainability of their service, and when they are hiring, they look for people who express sustainability-related values. In addition to promoting environmental sustainability, Hygglo has implemented measures to prevent tax evasion, racism, sexism, fraud, harassment, and to keep the platform inclusive in a more general sense. Despite their emphasis on sustainability, Ola argues that there are other values that are more important for most users, not least accessibility. Sustainability is often viewed more as a positive side effect.

Furthermore, he argues, the focus on sustainability in public discourse and research can potentially create problems for the sharing economy in general. While it is expected from conventional companies to mainly strive for economic profit and growth, these issues are controversial for sharing economy companies, especially those with an outspoken focus on sustainability. Ola:

Some users may find it problematic that we strive towards making money through our platform. P2P economies first emerged in an era where profits were not enough of a motivation – instead there had to be a collective that ran the platform and profited from it. Those who accept that you can make a profit within the sharing economy often instead talk about the 'platform economy', which I feel is a concept with mainly negative connotations.

Another issue that Ola raises is the diversity of the user base, which we have previously described. At the moment, the company is not actively working towards increasing the diversity, but rather to scale up on a more general level. However, they have identified the current limitations in what user groups they appealed to as both a problem from a sustainability standpoint and from a business standpoint. Despite offering a relatively cheap service that would theoretically appeal to the low-income demographics, Ola speculates that it could have to do with the fact that they launched the platform in Hammarby Sjöstad and the fact that they created a platform that they themselves would use, which in turn could have excluded other kinds of users.

8   Social interactions among Hygglo users

Ola anticipated that the challenge would be to attract people who were willing to share things they own, but it turned out that the challenge was to find people who were willing to rent instead of buying new. According to him, Hygglo is often cheaper and more convenient than other rental services both in urban and rural areas. Additionally, individuals can access goods during evenings and weekends. Still, people think of sharing as more complicated and troublesome, according to Ola, not least since they feel uncomfortable with the social role that sharing afforded, compared to traditional business-to-consumer (B2C) services where people know what is expected from them:

> When you go to the grocery store, you can be a friendly customer and greet the cashier, or you can [quietly] take your groceries and leave. It is clear that you are the customer and the other person is the cashier. [In P2P] you find yourself in a situation where your role is unclear. This makes you confused.

It seems some people who use the argument that buying new is more convenient than sharing are rationalising in order to avoid uncomfortable, unwanted or awkward social situations. On many P2P platforms, there are technical solutions that aim to simplify the social interactions that using the service entails, not only for those who feel uncomfortable with these kinds of situations but also to increase trust. For example, Hygglo is using the electronic identification service BankID[1] so no one can be truly anonymous on their platform. While this system excludes people that do not have a Swedish bank account, such as some exchange students or other people who otherwise could benefit from using their service, it makes it more difficult to commit fraud. They are also using their own payment system to make sure that no money is paid upfront to further prevent fraud and to make sure that the transaction is insured. This also means no bartering, which is something that some users find awkward. In comparison with platforms that do not have these systems in place, in the UK for example, problems related to fraud and harassment are very rare according to Ola. To further promote trust, Hygglo is using review systems for users and products, and they carry out credit checks on all users who want to rent out items on their platform.

9   Sharing to promote neighbourhood belonging

As previously discussed, the sharing economy is commonly referred to as a digital and worldwide phenomenon (Revenko and Revenko, 2019), which is largely dominated by MNCs (Felländer et al., 2017). Sharing platforms are typically perceived as being digital in nature, implying that their offerings are accessible from any location at any time (Parguel et al., 2017). However, as demonstrated in the preceding section of this chapter, global and digital platforms still encourage traditional sharing practices of physical objects in physical spaces among individuals. The sharing economy is furthermore frequently described as enabling sharing between strangers (Thoem, 2015; Slee, 2017). While the sharing economy technically encourages sharing among individuals across the globe, geographical proximity is viewed as a crucial factor for those who participate in the sharing economy. While individuals who interact on sharing economy platforms such as Hygglo may be strangers, they frequently happen to be neighbours or "local strangers". In the following pages, we will show how sharing among local strangers can improve the sense of belonging to a certain community, and the willingness to contribute to the development of that community, drawing on the previously presented literature together with the Hygglo case.

10   Sharing among local strangers

Our initial inquiry is centred around the following question: If physical third places have indeed experienced a decline, as Oldenburg (1989) started discussing some time ago, can digital platforms serve as a substitute for physical third places? In the case of Hygglo, using the platform is free of charge and it is accessible for everyone with a BankID, and it is welcoming people irrespective of social class, age and background. A prime example of this is that although young people may be more likely to damage or mistreat shared products, leading to an increase in insurance costs, the insurance fee is the same for all users regardless of age, thereby levelling out some inequalities. In addition, Hygglo can be accessed from any location within Sweden and Norway, but the platform lists search results based on proximity. The platform enables browsing of the most popular categories in one's specific area, as well as viewing the most recent search results, reviews and comments in that area.

In the digital environment, the users decide where to meet up physically and when, but the sharing itself takes place in a physical environment. An overwhelming majority of all meetings happen within a couple of kilometres from where the users live. Meetups can take place in homes or stairwells, in suburban centres or in parking lots. Therefore, the digital infrastructure that facilitates interaction, payment and reviews is tightly connected to a local, physical place. Sharing platforms do not explicitly facilitate or promote spaces for more general discussions and conversations between people. They clearly facilitate meetings between local strangers both in a digital and a physical environment; however, this is more seen as a necessity rather than a purpose of the service. Local strangers might not be meeting to discuss politics, but they are led to some kind of interaction and are invited to recognise each other's existence with or without the pretext of sharing. Without participating in a sharing practice, these local strangers might not have met.

A third place should ideally gather people from different socioeconomic backgrounds in a single, physical place. While we have already argued that sharing practices should theoretically appeal to people from many different demographics, we have shown that the demographic of Hygglo users is quite homogeneous. This may be a challenge for sharing platforms to act as social levellers and, calling back to Polèse and Stren's (2000) definition of social urban sustainability, encourage the compatible cohabitation of culturally and socially diverse groups while promoting social integration. Lasch (1996) and Oldenburg (1989) question whether a third place that only attracts people from a certain social class can bring understanding and cohesion to a community or if such a place further solidifies existing clusters. Considering how important physical proximity is when sharing, the diversity of the platform (who you interact with) is intrinsically tied to the demographics of the area where the users reside. In other words, Hygglo and similar platforms cannot easily overcome the challenges of segregation, which limits their potential as a social leveller. On the other hand, although many European cities are segregated with different people living in different areas, at least in smaller cities, this segregation is often not so spatially extreme that people of different groups would not show up within, for example, a two kilometres radius. Hygglo admits that they have had trouble recruiting users within certain demographics, and this is presented as a sustainability problem. However, as Hygglo is still focused on growth, they are actively working to expand in areas where the service is currently underused, and since there are so many different motivations to participate in sharing (e.g. environmental, economic and social), they may be able to reach a broader audience in the future. In that case, they may connect strangers of different groups to a larger extent, and thus increase their third-place-like functions.

11   Trust in the sharing economy

Much research has focused on the role of trust in the sharing economy. Such research often argues that lack of trust between individuals is affecting the growth of the sharing economy negatively (ter Huurne et al., 2017; Möhlmann et al., 2018; Yang et al., 2019; Akhmedova et al., 2021). Therefore, sharing platforms have implemented features that aim to promote trust, such as the previously mentioned BankID. While sometimes seen as infringing privacy and integrity, the general willingness in Sweden to be affected by such systems seems to be high (Husz, 2018).

Less emphasised in research is the claim that the sharing economy may produce increased trust towards other people in society. Drawing on Yamagishi and Yamagishi (1994) – who assume that trust is characterised by the expectation of benevolence without information or guarantees of such benevolence – one can argue that electronic identification is a way of *bypassing* the need for trust rather than *building* it. By thinking that we have outsourced trust to a digital platform, we might ignore building trust between people without technological mediation.

Several of the studies reviewed focusing on diversity and trust suggest that people tend to be less trusting towards out-groups (although it remains unclear whether that is magnified or mitigated by proximity to them in one's neighbourhood), but close actual contact with out-group individuals seems to strengthen trust (Tolsma et al., 2009; Dinesen and Sønderskov, 2017; Newton et al., 2017; Finseraas et al, 2019). Hygglo's electronic identification system may reduce the need for trust between individuals from different groups participating in exchanges, thereby creating a point of contact from which a more general relationship of trust may develop. A notable difference between this platform and any gig economy platform is that most users meet as equals – regardless of whether they are looking to rent or renting out. Gig work can theoretically bring together people of different classes or ethnicities, but tends to result in very unequal relationships (Schor, 2017). As Ola describes it, the more equal relationships developed in P2P sharing may bring with it a certain amount of social awkwardness. Nonetheless, we argue that it is still beneficial and possibly imperative to challenge conventional norms associated with these kinds of transactions to establish more sustainable consumption practices. More generally, Chen et al. (2009) show how social interactions between members of a community tend to affect trust in the community as a whole positively when studying a Chinese P2P community. As we have argued that there is a tight connection between the sharing platform (Hygglo) and the local communities in which the platform is used, it is reasonable to assume that social interaction among members of the digital platform can spill over to the local, physical community.

It is possible to bridge the trust gap between local inhabitants through sharing that facilitates physical meetings between users and provides assurance of trust. As a best-case scenario, it can strengthen a sense of belonging in the local community, as well as support interactions between its members and a neighbourly culture of sharing. However, its levelling effects and third-place-like qualities would be stronger if the meetings were between people of culturally and socially diverse groups (Polèse and Stren, 2000). In its current form, these meetings seem to be mainly between people that are similar to each other, both because of a homogeneous user base of the service and because of segregated housing patterns.

12   Conclusions and future research

Throughout this chapter, we have discussed how ordinary people are becoming more separated from the neighbourhoods in which they live. We traced this discussion back first to Lasch (1996) who showed how a managerial elite turned its back against local communities and formed exclusive, globalised networks with no connection to specific physical places. Then, we presented other societal developments that have resulted in voluntary and involuntary withdrawal by people from their local communities. We argued that an unwillingness or inability to invest oneself in one's local neighbourhood can be a severe problem for social sustainability, not least since self-governing communities are the basic units of democratic society (Lasch, 1996).

Digital economies, including the sharing economy, have been praised for their potential to promote environmental sustainability since their emergence. However, a general consensus seems to be that conventional sharing economy platforms risk contributing to social *unsustainability* (Martin, 2016; Schor, 2020). Home sharing platforms such as Airbnb are sometimes criticised for contributing to gentrification and the lack of affordable long-term rental contracts, and gig economy platforms have been criticised for producing a new kind of lower class or precariat consisting mainly of immigrants that are in many ways alienated from society (Peticca-Harris et al., 2020). While acknowledging that the sharing economy has not developed as early proponents of the concept hoped, we here explore possible benefits of sharing economy platforms for urban sustainable development. Our main findings are presented in this section followed by a brief section of potentially fruitful vistas for future research on the sharing economy and its effect on social sustainability.

One of the main selling points of the sharing economy is that it allows sharing among complete strangers, in contrast to traditional sharing, which often occurs among friends and neighbours. Light and Miskelly (2015) argue that the kind of sharing that occurs on most digital platforms is very different to how they understand a sharing culture. They claim this is because the sharing economy does not foster egalitarian values and relationships in a local setting, since it focuses on increasing resource efficiency rather than caring for the local community. We argue that there is some merit to those claims, and indeed the concept of sharing has been hijacked by profit-maximising MNCs with little to no interest in promoting increased social cohesion but rather the opposite (Martin, 2016). Our study has shown that Hygglo, a conventional for-profit sharing economy company, appears to have different priorities that could be beneficial to well-functioning and cohesive local communities and may be part of fostering a sharing culture. Sharing economy platforms provide access to local products through local strangers, thus complementing the return of local shoppers to neighbourhood stores following the COVID-19 pandemic (Accenture, 2020). While perhaps not functioning as a meeting spot where conversing is the main activity (Oldenburg, 1989; Battista, 2018), such as a (hybrid) third space, it inarguably facilitates meetings and interaction between local strangers. Therefore, platforms such as Hygglo hold the potential to strengthen connections between local strangers, and thereby support the recreation and upkeep of their community. In conclusion, Hygglo and similar platforms fulfil some of the original characteristics of a third place as it is accessible, inclusive and encourages local participation (Soukup, 2006).

According to research on trust, close contact with out-group people can lead to a generally higher trust towards others that are different from oneself under the right conditions. A lack of civic trust casts a heavy pall on society's ability to enable people to live with their differences (Lasch, 1996). Research on trust in the sharing economy is well established. The sharing economy places people and human interactions at its core, enabling strangers to build relationships through digital platforms, thus mitigating our inherent stranger danger bias (Möhlmann et al., 2018).

As we have seen in our case, Hygglo "outsources" traditional person-to-person trust to technological artefacts such as identification processes, secure payment systems and other anti-fraud measures. Such technology-mediated trust does not replace or erode person-to-person trust. Instead, research has shown that trust-enhancing digital cues can contribute to increased trust between strangers (ibid.). Therefore, we conclude that technology-mediated trust, as practised on online platforms like Hygglo, may lead to higher civic trust (Lasch, 1996), primarily because it enables users to interact with people they would normally not interact with, allowing them to develop trust with local strangers who may be different from themselves.

However, as previously mentioned, sharing services of this type face several challenges in realising their full potential to promote social sustainability. Despite this potential to bridge the gap between people of different genders, ethnicities, and class, it is unclear whether this potential is actually realised on the investigated platform. This can be attributed to two main factors, both of which reinforce each other: the homogeneity of users on the platform under study and widespread segregation in its primary usage areas. The reason for this can be attributed to the focus on physical proximity, as demonstrated in this chapter, when engaging in sharing practices. This same principle is what is projected to generate the social benefits of the sharing economy. In addition, the interactions facilitated by Hygglo are not necessarily explicitly social in their nature: As previously mentioned, they create an unclear relationship that is neither clearly customer and provider, nor fully casual. These platforms do not automatically encourage people to engage in conversation outside of what is necessary for participating in the sharing practice.

In future research, it would be interesting to see how sharing economy companies, but also policymakers, could work towards promoting higher levels of out-group trust among community members through sharing. Furthermore, the resituating of the daily life into the local neighbourhood during and after the COVID-19 pandemic is of interest for all aspects of sustainability, and it would be useful to learn more about what role sharing economy platforms may serve in that development – especially if it has a role to play in its continued trend. Lastly, the unclear social roles in P2P in general urge for a deeper understanding through empirical studies of users in order to better understand the reported reluctance to participate in such – essentially more sustainable – practices.

Acknowledgement

This study was supported by the JSPS/STINT Bilateral Joint Research Project, "Information and Communication Technology for Sustainability and Ethics: Cross-national Studies between Japan and Sweden" (JPJSBP120185411) and Kakenhi (19K12528).

Note

1 According to Statistics Sweden (2020), 85 percent of the Swedish population uses electronic identification services such as BankID.

References

Accenture. (2020). COVID-19: New habits are here to stay for retail consumers. Available at: https://www.accenture.com/us-en/insights/retail/coronavirus-consumer-habits. Accessed: 2022-11-29.

Ahuja, K., Chandra, V., Lord, V., & Peens, C. (2021). Ordering in: The rapid evolution of food delivery. McKinsey. Available at: https://www.mckinsey.com/industries/technology-media-and-telecommunications/our-insights/ordering-in-the-rapid-evolution-of-food-delivery. Accessed: 2022-03-22.

Akhmedova, A., Vila-Brunet, N., & Mas-Machuca, M. (2021). Building trust in sharing economy platforms: Trust antecedents and their configurations. *Internet Research*.

Arendt, H. (1973). *The origins of totalitarianism*. New York.

Balding, M., Whinery, T., Leshner, E., & Womeldorff, E. (2019). *Estimated percent of total driving by Lyft and Uber*. Fehr & Peers.

Banerjee, D., & Rai, M. (2020). Social isolation in Covid-19: The impact of loneliness. *International Journal of Social Psychiatry*, 66(6), 525–527.

Battista, C. (2018). We're going on a Pokéhunt: The community behind Pokémon Go. *Debating Communities and Networks IX Conference 2018*.

Belk, R. (2007). Why not share rather than own?. *The Annals of the American Academy of Political and Social Science*, 611(1), 126–140.

Boar, A., Bastida, R., & Marimon, F. (2020). A systematic literature review. Relationships between the sharing economy, sustainability and sustainable development goals. *Sustainability*, 12(17), 6744.

Botsman, R., & Rogers, R. (2010). What's mine is yours. *The Rise of Collaborative Consumption*, 1.

Chen, J., Zhang, C., & Xu, Y. (2009). The role of mutual trust in building members' loyalty to a C2C platform provider. *International Journal of Electronic Commerce*, 14(1), 147–171.

Clewlow, R. R., & Mishra, G. S. (2017). Disruptive transportation: The adoption, utilization, and impacts of ride-hailing in the United States. Institute of Transportation Studies, University of California, Davis, Research Report UCD-ITS-RR-17-07

Curran, N. M. (2023). Discrimination in the gig economy: The experiences of black online English teachers. *Language and Education*, 37(2), 171–185.

Delhey, J., Newton, K., & Welzel, C. (2011). How general is trust in "most people"? Solving the radius of trust problem. *American Sociological Review*, 76(5), 786–807.

Delmos. (2021). Segregation i Sverige – Årsrapport 2021 om den socioekonomiska boendesegregationens utveckling. Available at: https://www.delmos.se/wp-content/uploads/2021/07/Segregation-i-Sverige.pdf. Accessed: 2022-03-31.

Dinesen, P. T., & Sønderskov, K. M. (2017). Ethnic diversity and social trust: A critical review of the literature and suggestions for a research agenda. In Uslaner, E. M. (Ed.), *The Oxford handbook of social and political trust*. Oxford University Press.

European Commission (EC). (2015). Quality of life in European cities. Available at: https://ec.europa.eu/regional_policy/sources/docgener/studies/pdf/urban/survey2015_en.pdf. Accessed: 2022-03-24.

Eurostat. (2021). Living conditions in Europe. Available at: https://ec.europa.eu/eurostat/statistics-explained/index.php?title=Living_conditions_in_Europe. Accessed: 2022-03-24.

FEANTSA. (2021). The 6th overview of housing exclusion in Europe 2021. Available at: https://www.feantsa.org/en/report/2021/05/12/the-6th-overview-of-housing-exclusion-in-europe-2021.

Fedosov, A., Lampinen, A., Odom, W., & Huang, E. M. (2021). A dozen stickers on a mailbox: Physical encounters and digital interactions in a local sharing community. *Proceedings of the ACM on Human-Computer Interaction*, 4(CSCW3), 1–23.

Felländer, A., Ingram, C., & Teigland, R. (2017). The sharing economy – Embracing change with caution. Näringspolitiskt forum. Available at: https://entreprenorskapsforum.se/wp-content/uploads/2015/06/Sharing-Economy_webb.pdf. Accessed: 2022-03-31.

Finseraas, H., Hanson, T., Johnsen, Å. A., Kotsadam, A., & Torsvik, G. (2019). Trust, ethnic diversity, and personal contact: A field experiment. *Journal of Public Economics*, 173, 72–84.

Fors, P., Inutsuka, Y., Majima, T., & Orito, Y. (2021). Is the meaning of the "sharing economy" shared among us? Comparing the perspectives of Japanese and Swedish policymakers and politicians. *The Review of Socionetwork Strategies*, 15(1), 107–121.

Frenken, K., & Schor, J. (2019). Putting the sharing economy into perspective. In *A research agenda for sustainable consumption governance*. Edward Elgar Publishing.

Gessen, M. (2020). The political consequences of loneliness and isolation during the pandemic. The New Yorker. Available at: https://www.newyorker.com/news/our-columnists/the-political-consequences-of-loneliness-and-isolation-during-the-pandemic. Accessed: 2022-03-30.

Hagbert, P., Finnveden, G., Fuehrer, P., Svenfelt, Å., Alfredsson, E., Aretun, Å., … Öhlund, E. (2018). *Framtider bortom BNP-tillväxt: slutrapport från forskningsprogrammet "Bortom BNP-tillväxt: scenarier för hållbart samhällsbyggande"*. KTH Royal Institute of Technology.

Hamari, J., Sjöklint, M., & Ukkonen, A. (2016). The sharing economy: Why people participate in collaborative consumption. *Journal of the Association for Information Science and Technology*, 67(9), 2047–2059.

Hedström, J. (2015). *Residential mobility and ethnic segregation in Stockholm*. Stockholm: Stockholm University.

Heinrichs, H. (2013). Sharing economy: A potential new pathway to sustainability. *Gaia*, 22(4), 228.

Hossain, M. (2021). The effect of the Covid-19 on sharing economy activities. *Journal of Cleaner Production*, 280, 124782.

Husz, O. (2018). Bank identity: Banks, ID cards, and the emergence of a financial identification society in Sweden. *Enterprise & Society*, 19(2), 391–429.

Hygglo. (2020). Statistik över 2020. Available at: https://www.hygglo.se/blog/statistik-oever-2020. Accessed: 2022-03-24.

Hygglo. (2022). Prylar nära dig. Available at: https://help.hygglo.info/sv/articles/5210537-prylar-nara-dig. Accessed: 2022-03-31.

IBISWorld. (2021). Car sharing providers in the United States. Available at: https://www.ibisworld.com/industry-statistics/market-size/car-sharing-providers-united-states/ https://www.valuepenguin.com/auto-insurance/car-ownership-statistics. Accessed: 2022-03-24.

Inter-Parliamentary Union (IPU). (2021). Is democracy really in crisis? Available at: https://www.ipu.org/event/democracy-really-in-crisis. Accessed: 2022-03-24.

Jacobs, J. (2016). *The death and life of great American cities*. Vintage.

Lasch, C. (1996). *The revolt of the elites and the betrayal of democracy*. WW Norton & Company.

Laukkanen, M., & Tura, N. (2020). The potential of sharing economy business models for sustainable value creation. *Journal of Cleaner Production*, 253, 120004.

Laurell, C., & Sandström, C. (2017). The sharing economy in social media: Analyzing tensions between market and non-market logics. *Technological Forecasting and Social Change*, 125, 58–65.

Leismann, K., Schmitt, M., Rohn, H., & Baedeker, C. (2013). Collaborative consumption: Towards a resource-saving consumption culture. *Resources*, 2(3), 184–203.

Light, A., & Miskelly, C. (2015). Sharing economy vs sharing cultures? Designing for social, economic and environmental good. *IxD&A*, 24, 49–62.

Light, A., & Seravalli, A. (2019). The breakdown of the municipality as caring platform: lessons for co-design and co-learning in the age of platform capitalism. *CoDesign*, 15(3), 192–211.

Majima, T., Fors, P., Inutsuka, Y., & Orito, Y. (2021). Is the meaning of the "sharing economy" shared among us? Comparing the perspectives of Japanese and Swedish researchers. *The Review of Socionetwork Strategies*, 15(1), 87–106.

Martin, C. J. (2016). The sharing economy: A pathway to sustainability or a nightmarish form of neoliberal capitalism? *Ecological Economics*, 121, 149–159.

Mermet, A. C. (2017). Airbnb and tourism gentrification: critical insights from the exploratory analysis of the 'Airbnb syndrome' in Reykjavik. In *Tourism and gentrification in contemporary metropolises* (pp. 52–74). Routledge.

Möhlmann, M., & Geissinger, A. (2018). Trust in the sharing economy: Platform-mediated peer trust. *Cambridge Handbook of the Law of the Sharing Economy*, 27–37.

Müller, F., Röhr, S., Reininghaus, U., & Riedel-Heller, S. G. (2021). Social isolation and loneliness during COVID-19 lockdown: Associations with depressive symptoms in the German old-age population. *International Journal of Environmental Research and Public Health*, 18(7), 3615.

Naturskyddsföreningen. (2015). *Ägodela: Köp mindre – få tillgång till mer*. Bonnier fakta.

Newton, K., Stolle, D., & Zmerli, S. (2017). Social and political trust. In Uslaner, E. M. (Ed.), *The Oxford handbook of social and political trust*. Oxford University Press.

Noda, T., Tanaka, H., Wang, Y., Izumi, H., Sunami, H., & Nozawa, K. (2018). Progresses and issues of sharing-economy policy in local areas. *Journal of Economics*, 45, 1–29.

Office for National Statistics (ONS). (2019). Are young people detached from their neighbourhoods? Available at: https://www.ons.gov.uk/peoplepopulationandcommunity/wellbeing/articles/areyoungpeopledetachedfromtheirneighbourhoods/2019-07-24. Accessed: 2022-03-24.

Oldenburg, R. (1989). *The great good place: Cafés, coffee shops, community centers, beauty parlors, general stores, bars, hangouts, and how they get you through the day*. Paragon House Publishers.

Parguel, B., Lunardo, R., & Benoit-Moreau, F. (2017). Sustainability of the sharing economy in question: When second-hand peer-to-peer platforms stimulate indulgent consumption. *Technological Forecasting and Social Change*, 125, 48–57.

Parkinson, J., Schuster, L., & Mulcahy, R. (2021). Online third places: Supporting well-being through identifying and managing unintended consequences. *Journal of Service Research*, 25(1), 108–125.

Peticca-Harris, A., DeGAMA, N., & Ravishankar, M. N. (2020). Postcapitalist precarious work and those in the 'drivers' seat: Exploring the motivations and lived experiences of Uber drivers in Canada. *Organization*, 27(1), 36–59.

Plenter, F. (2017). Sharing economy or skimming economy? A review on the sharing economy's impact, *Twenty-third Americas Conference on Information Systems*, Boston. https://core.ac.uk/reader/301372060

Plewnia, F., & Guenther, E. (2018). Mapping the sharing economy for sustainability research. *Management Decision*, 56(3), 570–583.

Polèse, M., & Stren, R. (Eds.). (2000). *The social sustainability of cities: Diversity and the management of change.* University of Toronto Press.

Räisänen, J., Ojala, A., & Tuovinen, T. (2021). Building trust in the sharing economy: Current approaches and future considerations. *Journal of Cleaner Production*, 279, 123724.

Reich, R. B. (1992). *The work of nations.* New York.

Revenko, L. S., & Revenko, N. S. (2019). Sharing economy phenomenon in the digitization era. Журнал Сибирского федерального университета. Гуманитарные науки, 12(4), 678–700.

Rheingold, H. (1993). *The virtual community: Homesteading on the electronic frontier.* New York: HarperCollins.

Rokem, J., & Vaughan, L. (2019). Geographies of ethnic segregation in Stockholm: The role of mobility and co-presence in shaping the 'diverse' city. *Urban Studies*, 56(12), 2426–2446.

Schaller, B. (2021). Can sharing a ride make for less traffic? Evidence from Uber and Lyft and implications for cities. *Transport Policy*, 102, 1–10.

Schor, J. B. (2017). Does the sharing economy increase inequality within the eighty percent?: Findings from a qualitative study of platform providers. *Cambridge Journal of Regions, Economy and Society*, 10(2), 263–279.

Schor, J. (2020). *After the gig.* University of California Press.

Schor, J. B., & Thompson, C. J. (Eds.). (2014). *Sustainable lifestyles and the quest for plenitude: Case studies of the new economy.* Yale University Press.

Schor, J. B., & Vallas, S. P. (2021). The sharing economy: Rhetoric and reality. *Annual Review of Sociology*, 47, 369–389.

Seegebarth, B., Peyer, M., Balderjahn, I., & Wiedmann, K. P. (2016). The sustainability roots of anticonsumption lifestyles and initial insights regarding their effects on consumers' well-being. *Journal of Consumer Affairs*, 50(1), 68–99.

Slee, T. (2017). *What's yours is mine: Against the sharing economy* (Vol. 10). New York: Or Books.

Soukup, C. (2006). Computer-mediated communication as a virtual third place: Building Oldenburg's great good places on the world wide web. *New Media & Society*, 8(3), 421–440.

Statistics Sweden. (2020). Befolkningens it-användning 2020 [ICT usage in households and by individuals 2020], Solna: SCB.

Svenfelt, Å., Alfredsson, E. C., Bradley, K., Fauré, E., Finnveden, G., Fuehrer, P., … Öhlund, E. (2019). Scenarios for sustainable futures beyond GDP growth 2050. *Futures*, 111, 1–14.

Sveriges Radio (SR). (2019). Fler bor med osäkra hyreskontrakt. Available at: https://sverigesradio.se/artikel/7166666. Accessed: 2022-03-31.

ter Huurne, M., Ronteltap, A., Corten, R., & Buskens, V. (2017). Antecedents of trust in the sharing economy: A systematic review. *Journal of Consumer Behaviour*, 16(6), 485–498.

Thoem, J. (2015). Belong anywhere, commodify everywhere: A critical look into the state of private short-term rentals in Stockholm, Sweden. *Master's dissertation, Department of Urban Planning and Environment*, Royal Institute of Technology.

Tolsma, J., van der Meer, T., & Gesthuizen, M. (2009). The impact of neighbourhood and municipality characteristics on social cohesion in the Netherlands. *Acta Politica*, 44, 286–313.

Verboven, H., & Vanherck, L. (2016). The sustainability paradox of the sharing economy. *uwf UmweltWirtschaftsForum*, 24(4), 303–314.

Wachsmuth, D., & Weisler, A. (2018). Airbnb and the rent gap: Gentrification through the sharing economy. *Environment and Planning A: Economy and Space*, 50(6), 1147–1170.

Wang, Z. (2016). *Neighbourhood social interaction-Implications for the social integration of rural migrants* (Doctoral dissertation, UCL (University College London)).

World Commission on Environment and Development (WCED). (1987). *Our common future*. Oxford University Press.

Worth, N. (2021). Going back to get ahead? Privilege and generational housing wealth. *Geoforum*, 120, 30–37.

Yamagishi, T., & Yamagishi, M. (1994). Trust and commitment in the United States and Japan. *Motivation and Emotion*, 18, 129–166.

Yang, S. B., Lee, K., Lee, H., & Koo, C. (2019). In Airbnb we trust: Understanding consumers' trust-attachment building mechanisms in the sharing economy. *International Journal of Hospitality Management*, 83, 198–209.

11 How does the digitally driven sharing economy promote cultural sustainability? The case of a musical instrument-sharing business in Japan

Yohko Orito and Takashi Majima

1 Introduction

The term 'sharing economy' appeared in the 2000s and gradually gained prominence with the rise of sharing services on online platforms and the spread of mobile networking devices, such as smartphones and tablets, that enable access to such services (Negoro, 2017). Consequently, a wide variety of initiatives in the sharing economy were launched in various industries. Popular examples of sharing businesses worldwide include the home-sharing service Airbnb and the ride-sharing service Uber. Previous research has argued that the sharing economy can contribute not only to economic sustainability but also to environmental and social sustainability (Henry et al., 2021; Ciulli & Kolk, 2019; Mi & Coffman, 2019; Böcker & Meelen, 2017; Schor, 2016; Sundararajan, 2016; Daunorienė et al., 2015; Belk, 2014; Rifkin, 2014). These effects can go hand-in-hand but also be in conflict with each other.

Since the early 2010s, sharing economy practices have also emerged in Japan with different types of sharing businesses being launched in the country. The characteristics and preferences regarding the usage of sharing services have been observed during this period in the Japanese context, even if individual users do not clearly recognise the term or understand whether certain initiatives and services relate to a sharing economy. Based on the findings from our previous studies (Fors et al., 2021; Majima et al., 2021; Orito et al., 2021), which provided a comparative literature review of the academic and governmental viewpoints of Japan and Sweden on the sharing economy, it seems that in Japan a sharing economy can contribute to economic revitalisation and address social issues in the Japanese society, such as excessive vacant housing in local areas, ageing population, depopulation, and labour shortage. This Japanese perception that mainly focuses on business outcomes or economic impact along with the social issues in Japan is referred to as the "Japanised" sharing economy (Orito et al., 2021). It is considered that if a sharing business addresses social issues, it can contribute to the growth of a sustainable society.

On the other hand, several new types of sharing businesses have emerged in Japan. These are somewhat different from the aforementioned traditional Japanese style, and they aim to address social or cultural sustainability. Siraj-Blatchford

DOI: 10.4324/9781003367451-13

(2008, p. 68) explains that social sustainability is a concept that concerns social issues in a broad sense, including cultural and political issues, that affect the quality and continuity of people's lives within and between nations. Thus, cultural sustainability is an aspect of social sustainability that refers to the maintenance and development of traditional values, customary practices, and modality that are rooted or respected in the society or community. Today, there are some cases in which a certain cultural value and its related customs or modalities have not been widely prevalent or have been gradually declining in society for a long time because the enjoyments of the cultural values tended to be limited to certain people or areas only until now, for example, the wealthy people or urbanised area. Then, in cases where the necessity of preservation and development of such culture and its values in more general ways are recognised for those purposes, sharing economy-based businesses are seemingly emerging with aims to share and maintain such cultural values in Japan.

Among them, one musical instrument-sharing company seeks to preserve the work of artisans and professionals in mainly classical/Western music and promote the sustainability of the culture of musical instrument artisans. In Japan, the ownership of classical music instruments, such as violins and flutes, can be expensive or unusual for the ordinary people. Furthermore, they may not be played frequently, if people do not have some interest in classical/Western music and music instruments, except for the people belonging to the music community or related organisations. However, the sharing business in focus in this study has been attempting to provide an opportunity to share or transact music instruments, mainly music instruments for classical/Western music, among ordinary people in Japan,[1] with the aim of sustainability of music artisan culture. Thus, as part of our research on the sharing economy, we conducted a semi-structured interview with the founder and chief operating officer (COO) of this recently launched company that provides musical instrument-sharing services. With this interview, we intend to understand his thoughts on how the music culture and artisans of music instruments can be sustained, the prospects and challenges of running a sharing business, and how the business' aim of boosting cultural sustainability is perceived. In other words, we aim to assess his thoughts on music instrument artisans and the sustenance of the music instrument industry through the operation of his sharing business utilising digital technologies.

As Benjamin (1968) and du Gay et al. (1997) pointed out, the evolution of technology often changes the culture and meaning of art, as in the case of music being programmed using machines. The advancements in music production technology and the resulting machine-made musical creations have deteriorated the aura of musical works and values in some cases. For example, the emergence of the Walkman, which enabled us to listen to music anywhere, has transformed not only the way people enjoy music but also the nature of social space which is described as 'mobile privatisation' (du Gay et al., 1997). However, the sharing business discussed here provides relevant insights on the use of digital technologies to maintain the existing values of the music culture, its meaning, and practices. The sharing business in this case also demonstrates interesting features in these aspects. Then, based on a qualitative analysis, we investigate the founder's perspective on the sustainability of the music and music instrument industry and its culture in a contemporary society where the quality and enjoyment of music itself are changing due to the spread of digital music technology and its culture, and in some cases the value of live musical performances tends to be disregarded.

Some studies have analysed the overall relationship between the sustainability of local culture, community, and sharing businesses (Botsman & Rogers, 2010) as well as in the context of travel or tourism (Paulauskaite et al., 2017; Tussyadiah, 2015). However, to the best of our knowledge, no case studies have explored sharing businesses that aim to promote sharing practices for increasing cultural sustainability related to music artisans. Therefore, the case study discussed in this chapter is novel and represents the new dimension of sharing-economy businesses in Japan, and it can contribute to highlighting new kinds of sharing businesses that aim to promote cultural sustainability. Moreover, this study discusses sharing business in Japanese context and also intends to contribute to the state-of-the-art international research because there are no other studies that deal with the promotion of sharing practices for increasing cultural sustainability on music and music instruments, in Japan and in other countries.

The remainder of this chapter is structured as follows. Section 2 explains the diffusions of sharing economy and its services, and traditional culture of classical/Western music and musical instrument industry/artisans in Japan. Section 3 provides an overview of the semi-structured interview of the COO of the musical instrument-sharing company as well as its findings. Section 4 discusses the specific characteristics and findings of this case. Finally, the conclusions and future research directions are presented in Section 5.

2 The background of the case: Sharing economy and classical/ Western music culture in Japan

2.1 *The diffusion of sharing economy in Japan*

There have been a wide range of services based on the sharing economy concept, also referred to as sharing businesses in Japan. The sharing business model enables individuals to share their tangible and intangible resources with others. While no specific definition of the sharing economy is socially accepted or understood in Japanese society, sharing services are classified as space (i.e., house, room), transportation (i.e., car/bicycle ride-sharing), goods (i.e., books, records), skills (i.e., housework, teaching), and money (i.e., crowdfunding) (Sharing Economy Association Japan, n.d.). Such services are constantly expanding into areas such as clothing, accessories, and agricultural land. According to a report published in Japan, various types of sharing services and their markets are expected to grow in the coming years (Yano Research Institute Ltd, 2018).

However, while developments surrounding sharing businesses and services in Japan have been observed, there is no consensus on a widely accepted definition or classification of sharing services. In fact, the context of the sharing economy varies by country and region. Based on the findings of previous studies (Fors et al., 2021; Majima et al., 2021; Orito et al., 2021), it seems that the nature of the sharing economy has not been vigorously debated in Japan, nor have the interpretations of the economy been closely examined. Even if Japanese researchers' understanding of a sharing economy is vague, broad, or all-encompassing, in many cases, they have introduced the concept and accepted it to a certain extent. In other words, Japanese researchers and policymakers tend to prioritise discussions on how to put the sharing economy to practical use and implement it in society.

These tendencies can be observed in reality. For example, before the COVID-19 pandemic, the use of home-sharing and guesthouse services had been flourishing, with guesthouse intermediary services, such as Airbnb, becoming widely popular in Japan. Although initially promoted to address a possible shortage of lodging facilities due to the Olympic Games in Tokyo, home-sharing services began to be promoted and used during the mid-2010s, and consequently, peer-to-peer (P2P) vacation rental services became more popular. Simultaneously, regulation of lodging services and the facilitation of more appropriate lodging became social issues, and the Private Lodging Business Act (Act No. 65 of 2017), also known as the New *Minpaku* Act (民泊新法), was approved and enacted in June of 2017 and 2018, respectively.

Since early 2020, Japanese people have been asked to refrain from going out due to the COVID-19 pandemic, and restaurants have had to either suspend operations or reduce business hours (Kyodo, 2020). Consequently, the use of home delivery services, such as Uber Eats, has increased in Japan. This implies that skill sharing (delivering food and drinks) became prevalent in Japanese society, in the sense that individuals provide skills (labour) in the form of home delivery. In addition, the promotion of online activities at home and teleworking may have led to an increase in the number of users of online skill-sharing and crowdfunding services and the sharing of goods mediated by online platforms.

As mentioned above, these phenomena could be interpreted as constituting the formation of a practical and flexible sharing economy in the Japanese context that matches the realities of the Japanese society and the social issues that must be addressed at that time. Of course, there are no major changes in the industries where existing interests are protected by strong regulations or self-regulation, as in the case of ride-sharing services. Thus, the level of progress varies considerably in line with the current situation in the Japanese society. Furthermore, while the sharing business case discussed in this chapter is also intended to address the situation in Japanese society, it is also a new case or model that seeks to respect the cultural values of music, cultural sustainability. The next section provides some background to this case.

2.2 Historical background of classical music culture and its sustainability in Japan

How were the classical Western music (henceforth called classical music) and instruments that are discussed in this study, introduced into Japanese society? The classical music was imported to Japan during the *Meiji* period (1868–1912), at first, when classical music performances were played by military bands and the performances were adopted as part of military training. Thus, it was only enjoyed by a small portion of the upper class people in Japan (Homma, 2013). Subsequently, with the spread of phonograph and recording technology in the *Taisho* (1912–1926) and *Showa* (1926–1990) eras and the adoption of music appreciation as a part of school education, the classical music culture has been further popularised within society (Homma, 2013; Nishijima, 2010; Watanabe et al., 2005). In the pre-war period, classical music was spread mainly among people in higher education institutions. After the Second World War, especially in the 1980s, with the opening of Suntory Hall, which is concert hall for classical music funded by Suntory Ltd in 1986, classical music became a fashionable symbol and a part of Japanese music culture (Homma, 2013; Watanabe et al., 2005).

Later, while classical music tended to be regarded as a legitimate culture in Japan and a tasteful hobby, especially among the well-educated (university graduates) and those with high income,[2] the increasing education of society as a whole has made it a hobby of a much larger number of people, though not as large a number as popular music. Today, it is explained that classical music is even more widely enjoyed through eclecticism and collaboration with popular music (Homma, 2013; Kataoka, 2000). According to the Basic Survey of Social Life (Ministry of Internal Affairs and Communications, 2022), which is a survey of people aged ten years and older conducted from 1996 to 2021 and examined the number of people who went to concert halls or live events, an average of 8% of people went to such facilities for listening to classical music and 11% for popular music. The difference is thus not big. Furthermore, the number of professional orchestral groups has also almost not changed between 2004 and 2020. Therefore, it can be said that classical music has become a part of Japanese musical culture and popular music, although it has undergone some changes (becoming a fashionable icon, joining with popular music, and taking a more commercial form).

However, the market for musical instruments in Japan is shrinking rapidly (Akamatsu, 2018; Inoue, 2018). For example, according to Akamatsu (2018, p. 167), the value of shipments by Japanese musical instrument manufacturers declined 33.4% from 1990 to 2014; domestic sales has decreased about 34% in the same time period. Also, household expenditures on musical instruments decreased by about 56%. Additionally, according to the aforementioned Basic Survey of Social Life (Ministry of Internal Affairs and Communications, 2022), the number of people who played musical instruments as a hobby had been constant or slightly declining from 1996 to 2021.

On the other hand, recently, desktop music, or digital music creation, performance and sharing on personal computers, has become increasingly popular as personal computers have become more widespread and their capabilities have improved since the 1990s (Kato, 2012; Kakihara, 2003). This allows us to play (or produce the sound of) any instrument inexpensively and easily in comparison to ever before (e.g., Masuda, 2008). Another fundamental problem is that many classical music performers have to buy or rent instruments, which are expensive, and this prevents many people from enjoying classical music. Consequently, although the culture of listening to classical music has taken root and there is a certain need for it, the number of people who enjoy playing musical instruments seems to slightly decline, and instead, the mechanisation/digitalisation of music, such as playing on a computer, has increased. Due to these factors, at present, the demand for musical instruments may have reached its ceiling. Under these circumstances in Japan, the livelihood of instrument producers, especially craftsmen, is in jeopardy, as is the preservation of the culture of playing musical instruments and enjoying orchestral performances. To put it in a slightly exaggerated way, the sustainability of traditional classical music and music craftsman culture is being threatened by digitalisation and the economically rational behaviour of consumers.

Nevertheless, a new sharing economy-based business has appeared that attempts to address this situation: a musical instrument-sharing business company that this study deals with. It is a start-up company that is attempting to counter or overcome the crisis of classical music in Japan, or to maintain the culture of enjoying classical music in the country, by using a sharing business scheme and by utilising the economically rational behaviour of consumers in reverse, which has a potential to deteriorate real music playing culture in some cases.

3   Case study of a musical instrument-sharing company

In this section, we describe the results of our interview with the founder of a Japanese sharing economy company. The interview was conducted to explore their motivations for the sharing business, their business model, and their awareness of cultural sustainability.

3.1   Methodology

To elucidate the new type of sharing economy businesses, we conducted a semi-structured interview with a person on the management level of a musical instrument-sharing company that emphasises the sustainability of the artisans of music instruments and the industry. The interviewee, Mr A (Male, 20s), was the founder and COO of the company. The online interview was conducted using Zoom in October 2021 and lasted around 90 minutes.[3] The major interview questions are listed in Appendix 1.

Using a thematic analysis approach, a narrative analysis method described by Riessman (2008), this study clarifies the interviewee's understanding of their own business, the sharing economy, and the relationship to sustainability. We attempt to analyse what was considered to be an important part of the company's understanding of the sharing economy. To avoid bias in the authors' understanding of the responses and comments, several discussions were conducted among the authors, and our interpretation was also confirmed by the interviewee. The interview's original transcript in Japanese was translated into English by the authors for the purpose of this study.

3.2   *Findings from the interview survey*

3.2.1   *Background and motivation for starting a sharing business*

This musical instrument-sharing business was launched in 2018 by two young businessmen. One of them, Mr A originally studied architecture at a university in Japan. His father was a violin craftsman, and thus, Mr A decided to follow in his father's footsteps. After graduating from the university, he enrolled in a professional school abroad to learn the art of violin making.

However, during his studies at the violin-making professional school, he learnt more about the current situation and working conditions of musical instrument craftspeople, realising that many of them find it difficult to make a living and are forced to go into musical instrument repairs. This harsh reality has been unchanged for a long time. As each musical instrument is handcrafted by the artisan, they tend to be expensive, getting out of reach for many users who love music and musical instruments. Mr A said that because it takes at least 200 hours to make a violin, the instruments become very expensive. Consequently, only a handful of people could make a living from it, although they would really like to continue making instruments. This is common in Japan and not a favourable situation for craftspeople of musical instruments, and it is not easy for them to make a living, even if they have developed a high level of skill and expertise in creating musical instruments.

Mr A confided in his old friend Mr B, the other founder and the chief executive officer (CEO) of the company, about the severe conditions faced by craftspeople in the musical instrument industry. At that time, Mr B was recognising the revitalising capability of China's sharing economy. Consequently, they came up with the idea of a sharing service for musical instruments, which they later turned into a business.

However, the founders faced several challenges to realise a sharing business. According to the explanation of Mr A, traditionally, musical instruments rental was not structured in a way that it could be widely distributed to people who had little or no personal relationship with it. It was customary for exchanges to take place only within a very narrow circle of trust without money. Therefore, in order to operate the business model of musical instruments-sharing, the problems of the existing musical instrument industry, as well as the concerns and risks regarding lending instruments, were considered by the founders. Mr A explains these points as follows.

A: *'One of the things that scared us was "why was nobody else doing it [musical instruments-sharing service]?" That was the scariest issue; there was no competition.*

There are some instrument manufacturers and dealers who do not want to think about renting out their musical instruments because there is a risk of them breaking. Occasionally, music stores allowed renting out instruments that could be repaired and the value of the instrument was not significantly reduced, but they do not want to rent out instruments worth nearly hundreds of thousands or millions of yen because of the risk of breakage.

However, since I used to work for a music store, I know that sometimes music stores lend instruments to trustworthy customers who are willing to buy them for a week or two as a trial. This is a bit confusing to me. I could understand if the customers actually paid for borrowing the music instruments, but, in many cases, it was a world without such money, where they exchanged only based on mutual trust'.

He also made an analogy of the exchange of instruments and information about music instruments in such a closed and narrow circle of trust to the hair salon system, and points out that this would lead to a narrow range of choices for the consumers, as follows.

A: *'You cannot just go to a musical instrument store for the first time and sud-denly borrow instruments there. We had to visit and buy the instrument several times, and then we had to connect with them as customers. It is a little like a hair salon system, right? Once you go to that store, you will keep coming back to it, and it will be difficult to change to another store. In almost all cases, if you go to that store, you will go there all the time, and you may not change to another store. But, if this is the case, you may still only be able to choose musical instruments from within that store'.*

Moreover, Mr A also pointed out that Japanese school education only taught how to play the vertical flute and pianica, and only a few people joined brass band clubs to learn how to play wind instruments. Consequently, Japanese education on playing classical musical instruments is not very popular for ordinary people relatively, even though listening to classical music is more popular in the Japanese society than ever before. He explained this environment and his desire to change it.

A: *'Especially in the United States, the violin, viola, and cello are instruments that are played by ordinary people, and students in elementary, middle, and high school can all play one of these instruments. In Japan, there is not enough space to play, and, often, violins are not taught in schools because they say violins are too difficult to play. This environment is also not favourable. While this story is mainly about string instruments, I think the environment for playing other musical instruments in Japan is also not so friendly, and so I thought it would be great if we could improve it. In particular, the violin has an image of being very expensive for customers. It would be wonderful if we could change the image that only celebrities can use musical instruments and make the sys-tem more accessible to everyone'.*

Under such influences regarding musical instruments, Mr A's desire to establish a new system to improve these artisans' conditions by utilising the Internet or online networking to his advantage, was the main driving force behind the launch of this musical instrument-sharing service. Then, Mr A addressed the current situation of the industry, and their attempt to turn the situation around.

A: *'What can we do to solve this problem? Artisanal musical instruments, even those that are not well-known, cost at least hundreds of thousands of yen. It is quite expensive to buy one for ordinary people, and so I thought of making it more affordable using the Internet in this age. It was not considered in the past, but now it is normal for people to meet each other on the Internet.*

   At first, almost all people said that it was unlikely that they would buy a musical instrument online, and that they would never buy a musical instrument online without trying it out. However, our goal is to move away from deciding what to buy or not to buy and create a new standard where customers can borrow and then buy using new systems'.

He also considered that an Internet-based musical instrument-sharing service would offer many options for consumers as follows.

A: *'This [above situation such that consumers can borrow music instruments in narrow trust-based circle] led to the idea that if a customer had sufficient trust, they could make use of the Internet and be able to rent out some of the more expensive instruments. I thought that if it was on the Internet, we could create a situation where people could look at a lot of different musical instruments, and borrow different instruments'.*

Mr A wishes to work as a craftsperson in the future, considering that the number of musical instrument craftspeople is decreasing. However, currently, he intends to support them by developing a new business model for the sharing of musical instruments.

A: *'At the moment, we would like to develop a good environment for musical instrument craftspeople. So, although we are aiming to make our instrument sharing service available to individuals so that anyone can use the instrument easily and casually, our starting point is to help the craftspeople or artisans of musical instruments. We focus on artisans by increasing their reach through special features, such as artisan category, and finally users can try out their original musical instruments at a reasonable price.*

   If such a system could be established on the Internet, artisans' music instruments would no longer be kept on the shelf. We hope that this will increase the value of their instruments, and their names will become known to a wider public. Right now, I am trying to create a new environment for such a system, and when that system becomes established as an operation, I think I would like to be able to make musical instruments myself'.

3.2.2   The business model and operation

The operations of the company are basically led by two people, Mr A (COO) and Mr B (CEO). Mr A is also involved in the wholesale of musical instruments, in parallel with his sharing business. The management of the company's website, apps, and systems is supported by external directors of the company. Mr A conveyed their intentions to keep the organisation relatively small as follows:

A: *'The fact that we are not motivated by profit is one of our advantages. We are just two people, and so it does not cost that much. We are working with the aim of setting a new standard for social change'.*

In this musical instrument-sharing service, users can borrow instruments and equipment, such as violins, guitars, cellos, drums, accordions, and wind instruments, from individuals or music instrument retailers for a minimum of one week using a software application. The users also can rent their own music instruments and equipment, while the number of user-to-user (consumer to consumer: C to C) transactions is still lower than that of business to consumer (B to C) transactions until now. The service offers the possibility of buying them, if a customer likes the instrument and wishes to buy it. A 25% commission on each transaction is set to be the company's revenue stream. If the customer who lent or borrowed the instruments does not prefer to hand them over to the person themselves in the transaction process, the company provides an optional service whereby the company is responsible for the entire transaction. The company has also introduced ID registration and a peer-to-peer review system, as well as a compensation system that covers repair costs and insurance. Mr A explains the process of repairing and insuring the instrument as described below.

A: *'I think the most important thing about the sharing economy is the mutual review process. The first transaction alone will inevitably be zero, but when the customers decide to continue, if we do not incorporate mutual evaluation into every transaction, it will never lead to the next opportunity. So, of course, we secure personal licenses, bank accounts, etc. and check all of them. If something goes wrong, we can track it from there.*

Further, since we provide musical instrument-sharing services, it is also important to think about repair compensation. This is one of the reasons why music stores, as well as music shops, have never offered such a service. The insurance companies also did not have such a system. It would be fine if we were insuring our own inventory, but our services are not our own, we intermediate the exchange. The first thing that happened was that all insurance companies refused to cover the items that people had lent to others. Only one insurance company was willing to do it. So, it took a year for the insurance company to build the system, thanks to the person in charge of the company who was trying to find the most suitable insurance policy for us. For the time being, the maximum coverage is 50,000 yen and repairs are limited to 60%, but we hope to be able to cover 80% of the cost. We would like to try it out and change it after seeing how many customers use it.

However, we have decided that we will not pay for all of the work because if we do, there will definitely be people who intentionally break the instruments and try to get just the repair fee, so we have to ask them to pay even if it is a little amount. In order to gain the trust of the customers, we place a lot of importance on such compensation and proof of ID'.

Initially, the target customer group of the business was mainly university students who could not afford to buy expensive instruments and may have become new customers as a result of spending more time at home due to the spread of COVID-19. The following are some of the current perceptions of instrument lenders and borrowers, as told by Mr A.

A: *'When we were making the business plan before we started this service, we thought that university students would be the target. The students have circles or clubs, and if they want to start playing a musical instrument in a music circle or club, they need musical instruments. However, I have also experienced that it is very difficult for university students to purchase instruments that cost tens or hundreds of thousands of yen. For example, if they use our services, they can borrow a guitar for a week for 3,000 yen or 5,000 yen as a trial at a reasonable price and try playing it to see if it suits them.*

Meanwhile, as the lenders of musical instruments, we targeted people in their early and late twenties who had bought guitars or other musical instruments when they were in college and did not have time to play the instrument since they started working. In fact, when we actually launched the business, we found that there were a surprisingly large number of people who retired from a band or from playing music in their 40s and 50s who still had their own collections.

Of course, there were people in their late twenties, but there were also many women and housewives who had more time at home and wanted to find a new hobby. There were also mothers who wanted their children to learn a musical instruments. However, they were not sure if their children would continue to learn; so, they rented it once for about six months'.

The company's promotional campaign has been implemented with major railway companies as partners and broadcasted via mass media and online advertising. However, Mr A says that the most powerful promotional tool is 'word-of-mouth'.

A: *'In fact, we are currently working in partnership with a famous railway company, which operates railways between Tokyo and Kanagawa. There, we are running an ad on the inside of the magazine, and starting from November 2021, our commercial will run on this railway, although it is only 15 seconds long.*

In particular, we are doing targeted advertising online on social networking services by region, age, and genre. There are also offline events in Tokyo at the end of this month and in December 2021, and hence, we will be setting up our store at those events. In addition, we are putting up flyers at our partner stores and universities. For the time being, we are working on that, but the strongest advertising measure is word-of-mouth. This is what we ask for the most, and I think it is probably the most effective and trustworthy to say, "We do this business, so please keep us in your mind"'.

3.2.3   Social impact and sustainability of the business

The main impacts expected from sharing economy are those on economic, environmental, and social sustainability. The company's understanding of the environmental effects of its services is described as follows.

A: *'When we started this business, we did not consider environmental sustainability much. However, after we started this business, we realised that we had to adapt it to the current times'.*

The company did not initially intend to run the business with environment sustainability in mind. According to Mr A's comment, this was not the most important issue for the company in the first place. It seems that they are becoming aware of it after starting the business. Now, what does Mr A think about the contribution of his company's business to the sustainability of society? The objective of this musical instrument-sharing business is to provide an environment in which people can easily and readily start playing music as mentioned above in Section 3.2.1. The impact on the industry due to the increased number of people using musical instruments through this sharing business was also mentioned.

A: *'The more people play musical instruments, the bigger the musical instrument industry will become. Moreover, the more people use a musical instrument, the more it needs to be maintained, and if that instrument needs to be repaired, of course it should be repaired. There is also a possibility that more and more people will want to use artisanal instruments. Of course, for the musical instrument industry, it is important to increase the number of people who play and use musical instruments; so, it would be great if we could provide an opportunity to increase the number of people who enjoy the instrument'.*

In addition, Mr A also intended to develop the relationship among the users on a C to C basis as follows.

A: *'Although it is somewhat unsuitable in this day and age considering the COVID-19 pandemic, we recommend hand delivery between lenders and borrowers. The reason why we recommend hand delivery is, of course, that hand delivery is better for musical instruments than other forms of delivery to maintain their condition. In addition, people who want to start playing music are scared to go to a music store they do not know anything about; so we are very conscious of the fact that we would like to connect people who do not know anything about music with people who know more about music and have known it for longer periods. So, we would like to revitalise the music community as a C to C service by connecting people who know about music with people who do not know about music'.*

This emphasis on building relationships and networks among the users is different from renting in a merely B to C or subscription model, and this was explained as follows.

A: 'It is not just about sharing things, but also about meeting people, and by meeting craftspeople and players in person, new connections can be made. This is something that is not possible with subscription or B to C rental services'.

Furthermore, the company intended to expand the scope of its sharing services to include space and skill sharing services for playing musical instruments to develop and strengthen the community as described below:

A: 'As the number of instruments increases, we will need more people to teach, have more people who want to learn to play music, as well as require more space to play instruments. For example, it is difficult to give away grand pianos, drums, etc., but if someone has an empty studio, they can rent it out for half an hour.

There is also a company that provides a platform for almost all kinds of skill sharing. Our goal is to provide a service where people who have no musical experience can easily and casually start playing by integrating our service with other kinds of sharing services. We would like to solve the problems of people who have no place to play, who are looking for students, and who want to learn to play an instrument'.

Finally, he also highlights the value of a musical performance itself, given the current situation where it is possible to play music using machines. In this manner, the mission of the company is to help people take up musical instruments and play them as well as facilitate connections with other like-minded individuals.

A: 'Machines (like music programming) have their advantages, such as constant tone and sound quality, but when a person plays or creates something, it has its own significance and advantages. There are times when the player's emotions come into play. I believe that a performance is truly unique and original because the amount of effort put into it changes depending on the emotion.

So, since there is a sound that only that person can produce, there is a value (value is created), such as the case wherein some person becomes famous.

Machines are certainly convenient and have their advantages, but I think the most important thing in this world for the future is to take the good parts from technology and do what people can do appropriately'.

4   Discussion

As observed in the previous section, the interviewee is highly conscious of the sustainability of social values, in terms of growing the music culture through his sharing business, rather than the economic or financial aspect of the business. Figure 11.1 provides an overview of the case. Mr A is concerned that unless the number of people who use musical instruments increases, the culture of music instrument artisans will be lost or altered unfavourably. This is the most critical concern in the operation of his business. Increasing the number of people who can enjoy musical instruments through this sharing business would lead to the maintenance and expansion of the work of musical instrument craftspeople, such as building and repairing instruments.

The interviewee also intended to create a large music community through the operation of his musical instrument-sharing business. Traditionally, the sale and rental of musical instruments have been centred around music stores and music classes closely related to the makers of musical instrument (See, Tanaka, 2021; Nishino, 2015; Tanaka, 2011; Iwabuchi, 1988). This inevitably has led to the customers being locked in and tied to specific music stores and sellers in Japan. As described by Mr A, who used the example of the relationship between a hairdresser and customer to explain the situation in a musical instrument shop, the established community of users who love musical instruments is relatively small. However,

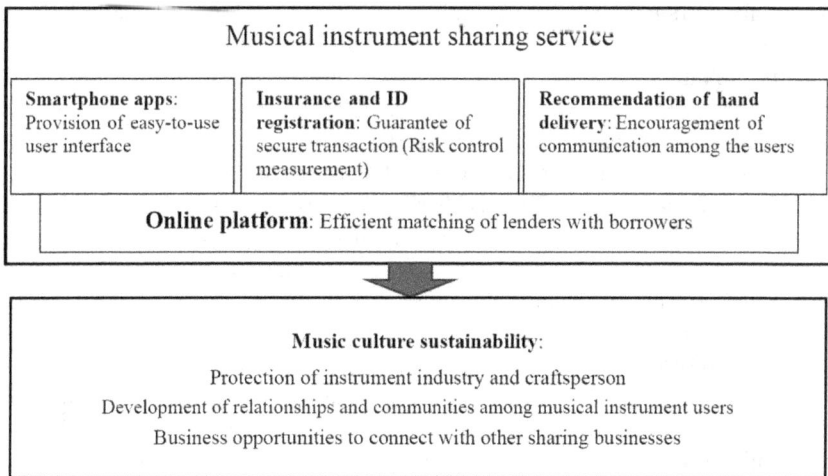

Figure 11.1 Model of the musical instrument-sharing service in this study.

their business, which developed an online platform enabling users to rent or borrow musical instruments from complete strangers, is expected to create a new community by bridging the gap between people beyond the existing community. It will also lower the barriers for people who want to start a new music career, thus expanding the community.

In addition, this business model encourages the lender to hand over the instrument to the borrower in person. Thus, the company also aims to create an interaction opportunity and consciously avoids becoming a mere facilitator of economic transactions. In many cases, anonymous economic transactions exclude individual subjectivity in the exchange and do not provide an opportunity for mutual recognition that leads to mutual identity construction (Yamauchi, 2015). Building a true community requires communication to understand and to approve each other more intensely than economic exchange. This musical instrument-sharing business intends to encourage its users, as individual subjects, to communicate with each other who has each identity and to make a community using the sharing service. In this sense, they are expected to create a true musical community. In fact, if the users have an opportunity to see each other in person, they can talk about, say, how to handle and play an instrument when it is hand-delivered. Moreover, if a lender can listen to music a borrower plays on a lent instrument live, a new form of interpersonal communication would emerge.

Furthermore, this interviewee desired to expand their business to include services that enabled the intermediation of music instructors or teachers with prospective students, as discussed in Section 3.2.3. The creation of teaching and learning relationships and opportunities to interact by playing musical instruments together will lead to a more intense exchange between consumers, one that is beyond mere economic exchange. Thus, it will promote growth in the musical community, as hoped for by the interviewee. The opportunities for interaction and communication among users may pave the way to developing the music community he aims to create.

While the interviewee does not seem to be strongly aware of Sustainable Development Goals (SDGs), his sharing business is aligned with SDGs to a certain extent, and he has recognised the need to be aware of SDGs. Certainly, the business model in this case can be expected to have the same environmental sustainability that is common across many goods-sharing businesses, that is, the utilisation of idle assets. This musical instrument-sharing business may also have a substantial positive effect on environmental sustainability, despite the managers being unaware of this advantage. In the future, the way in which this particular business model may change as people become more aware of and associate environmental sustainability with it should be considered. On the other hand, in a somewhat hyperbolic manner, the new approach toward sharing businesses as illustrated in this case may imply that the cultural aspects of the SDGs have not been highlighted sufficiently until now. Cultural sustainability may be the 'missing link' in the concept and achievement of the SDGs. This is because approaches to cultural sustainability or cultural value that people consider important may serve to link multiple aspects of sustainable development goals related to issues such as the conservation of the natural environment and artisan's working conditions as we discussed here. Aiming for cultural sustainability is regarded as important for people's well-being, it may help to achieve other social values across the spectrum, as it may also indirectly facilitate the realisation of other goals.

5 Concluding discussion

Based on the case of the musical instrument-sharing business, we argue that it is oriented toward social sustainability, one of the three expected effects or sustainability of a sharing economy, in terms of achieving cultural sustainability in the musical instrument industry. It seems that the Japanese society primarily did not expect this impact on cultural sustainability from a sharing economy obviously, and the interviewee's intentions of preserving the world of craftsmanship and music culture may not necessarily resonate with all of the customers. However, as mentioned above, this sharing business intends to ensure compatibility between the economic rationale of efficiency and convenience and the social value associated with sustaining the craftsmanship of musical instruments, which is more likely to engender the empathy of consumers. Their goal is to maintain cultural sustainability and gain cooperation from diverse actors, thus linking both their own and others' interests without any contradiction. In addition, it is also interesting to note that while the interviewee understood and acknowledged the impacts of the development of digital technology on music culture to some extent, they set up a sharing business using digital technology themselves to share real, non-digital, and musical instruments.

Thus, this musical instrument-sharing business attempts to show the 'convenience' and 'economic efficiency' of services to its customers by providing easy-to-use apps and tailor-made insurance policies. Then, when customers use their service, they will contribute to realising the goal of the business even if they are unaware of their contribution. That is, they will contribute to achieving cultural sustainability by increasing the number of music players and building a new music community that respects the craftspeople of musical instruments. One of the most insightful points of this case is that the founders are attempting to achieve their goals without their customers directly knowing about it. In other words, this case indicates that if the surface of the sharing economy, which is deemed to be Japanised and emphasising only economic efficiency, is peeled back, we can see a variety of other agendas and practices.

In this regard, the company may attempt to *translate* (Latour, 1987) or define a sharing economy based on their own understanding. This sharing business has attempted to integrate its interests with those of the consumers and other companies by linking itself with technologies, such as online platforms and smartphone apps, and networking with other companies, such as railroad and insurance companies. Additionally, it has brought a social change in the context of the COVID-19 pandemic, as discussed in Section 3. However, as we mentioned above, the relationship between anonymous economic exchange and the fostering of mutual recognition is somewhat contradictory, making it difficult to balance the pursuit of economic rationality and community building. Therefore, an understanding of how to achieve such a delicate balance is also required for the development of sharing businesses.

On the other hand, through our discussion of this case, we attempted to extend the understanding of a sharing economy in Japan, where economic effects, such as economic revitalisation, are expected first and foremost (Fors et al., 2021; Majima et al., 2021; Orito et al., 2021), and to imply the necessity to revise the concept of sharing economy in continuous manner. This understanding is shaped not only by policymakers and researchers, but also by those who engage in the clever translation strategies to realise other aims, such as the sharing company in this case. The concept of a sharing economy as a new economic tool is by no means monolithic and may in fact be the product of complicated translation or politics. Moreover, it is likely to be translated in different ways depending on contextual changes. For example, as we mentioned in Section 4, the recent contextual shift to an increased interest in SDGs and social values may change our understanding of the concept of a sharing economy.

Furthermore, these days, many prominent sharing businesses are facing intense criticism for their negative influence and monopolistic and dominating actions (Fickenscher, 2022; Guttentag, 2018). As a result, they are facing backlash worldwide, and even in Japan, they are perceived negatively in relation to cultural preservation and urban development. In contrast to the prominent large sharing business case, this musical instrument-sharing business case discussed here can embody the concept of sharing economy in its original sense and be more promising owing to its characteristics that are distinct from renowned sharing businesses.

There is no common understanding of the sharing economy in the business world since it is a complex interplay of intentions and understandings. The concept of sharing economy as a product of various conceptualisations is not limited to the company discussed here. Similar aspects can be found in other companies as well. Sharing businesses encompass not only the sharing of goods but also different categories of services, such as space, skills, and money, and these categories translate differently. This implies that the understanding of a sharing economy, particularly in the academia and governmental administration fields (Fors et al., 2021; Majima et al., 2021; Orito et al., 2021), must be investigated deeply. In the future, more interviews with relevant people and further analysis of interview results using text mining and other qualitative analysis tools should be conducted.

Acknowledgement

We really would like to thank the interviewee, Mr A. Thanks to his insightful talks, this research could be conducted. We also wish to thank the Ehime university students who supported this interview, Taizen Kasaoka, Choji Shido, Ayaka Shirakata, Akiho Takigawa, Yuu Hamano, Shunya Yoshimi, Suzuho Wakisaka, and research assistants who supported our survey, Haruka Suzuki and Kana Yoshimi, and the reviewers who provided insightful comments on our paper.

This study was supported by the JSPS/STINT Bilateral Joint Research Project, 'Information and Communication Technology for Sustainability and Ethics: Cross-national Studies between Japan and Sweden' (JPJSBP120185411) and Kakenhi(19K12528).

Notes

1 This does not means that they do not deal in music instruments other than those for classical or Western music.
2 Kataoka (2000) discussed that Japanese young men tended to enjoy not only classical music but also popular music such as rock, jazz, and other genres (the phenomeon called cultural omnivorous), and that they don't incline to listen to *enka* (traditional Japanese ballads) or *minyo* (Japanese folk songs).
3 Some points were answered by e-mail for confirmation after this semi-structured interview.

References

Akamatsu, Y. (2018). Product development strategies of Japanese domestic wind instrument manufacturers: A case study of flute and saxophone manufacturing, *Annals of the Society for Industrial Studies, 33*, 167–185 (in Japanese).

Belk, R. (2014). Sharing versus pseudo-sharing in Web 2.0. *The Anthropologist, 18*(1), 7–23. https://doi.org/10.1080/09720073.2014.11891518

Benjamin, W. (1968). The work of art in the age of mechanical reproduction. In H. Arendt (Ed.), *Illuminations* (pp. 217–251). Schoken Books.

Böcker, L., & Meelen, T. (2017). Sharing for people, planet or profit? Analysing motivations for intended sharing economy participation. *Environmental Innovation and Societal Transitions, 23*, 28–39. http://dx.doi.org/10.1016/j.eist.2016.09.004

Botsman, R., & Rogers, R. (2010). *What's mine is yours. The rise of collaborative consumption.* HarperCollins.

Ciulli, F., & Kolk, A. (2019). Incumbents and business model innovation for the sharing economy: Implications for sustainability. *Journal of Cleaner Production, 214*, 995–1010. https://doi.org/10.1016/j.jclepro.2018.12.295

Daunorienė, A., Drakšaitė, A., Snieška, V., & Valodkienė, G. (2015). Evaluating sustainability of sharing economy business models. *Procedia-Social and Behavioral Sciences, 213*, 836–841. https://doi.org/10.1016/j.sbspro.2015.11.486

du Gay, P., Hall, S., Janes, L., Mackay, H., & Negus, K. (1997). *Doing cultural studies: The story of the Sony Walkman.* Sage Publications.

Fickenscher, L. (2022). *NYC proposes strict Airbnb registration rules to take effect in January.* New York Post. https://nypost.com/2022/11/04/nyc-proposes-strict-airbnb-registration-rules-to-take-effect-in-january

Fors, P., Inutsuka, Y., Majima, T. & Orito, Y. (2021). Is the meaning of the "sharing economy" shared among us? Comparing the perspectives of Japanese and Swedish policymakers and politicians. *The Review of Socionetwork Strategies, 15*(1), 107–121. https://doi.org/10.1007/s12626-021-00070-z

Guttentag, D. (2018). *What Airbnb really does to a neighbourhood.* BBC. https://www.bbc.com/news/business-45083954

Henry, M., Schraven, D., Bocken, N., Frenken, K., Hekkert, M., & Kirchherr, J. (2021). The battle of the buzzwords: A comparative review of the circular economy and the sharing economy concepts. *Environmental Innovation and Societal Transitions, 38,* 1–21. https://doi.org/10.1016/j.eist.2020.10.008

Homma, C. (2013). Studies on classical music after World War II in Japan: classical music as cultural capital ad disintegration of modern audience. *Keio Journal of Economics, 106*(2), 111–133 (in Japanese).

Inoue, K. (2018). Transformation in the social meaning of piano ownership in Japan: Prestige, education, and luxury. In K. Furuta (Ed.), *SSM Research Report 2015, vol.4* (education Ⅰ) SSM research group, 77–102 (in Japanese).

Iwabuchi, A. (1988). *YAMAHA-New culture creation strategy,* TBS-Britannica (in Japanese).

Kakihara, M. (2003). The impact of digitization on music production: From a perspective of modularity. *Journal of Business Administration* (Kwansei Gakuin University), *51*(2), 87–108 (in Japanese).

Kataoka, E. (2002). Cultural tolerance and symbolic boundaries. In T. Imada (Ed.), *The postmodernism of social stratification in Japan,* Tokyo, 181–220 (in Japanese).

Kato, A. (2012). Two paths formed by digitization of sound recording technology and the fourth step in evolution of recording industry. *The Journal of the Japan Association for Social and Economic Systems Studies, 33,* 47–54. https://doi.org/10.20795/jasess.33.0_47 (in Japanese).

Kyodo. (2020). *Uber Eats sees 20% contract jump in Japan as restaurants lose guests.* The Japan Times. https://www.japantimes.co.jp/news/2020/04/21/business/uber-eats-contract-jump-eateries-japan-coronavirus/

Latour, B. (1987). *Science in action: How to follow scientists and engineers through society.* Harvard University Press.

Majima, T., Fors, P., Inutsuka, Y., & Orito, Y. (2021). Is the meaning of the "sharing economy" shared among us? Comparing the perspectives of Japanese and Swedish researchers. *Review of Socionetwork Strategy, 15*(1), 87–106. https://doi.org/10.1007/s12626-021-00068-7

Masuda, S. (2008). What will the 'digitalization of music' bring to music culture? In M. Toya (Ed.), *How should we capture the 'spreading music culture'?,* Keiso shobo, Tokyo, 3–24. (in Japanese).

Mi, Z., & Coffman, D. M. (2019). The sharing economy promotes sustainable societies. *Nature Communications, 10*(1), 1–3. https://doi.org/10.1038/s41467-019-09260-4

Ministry of Internal Affairs and Communications. (2022). The basic survey of social life 2021. https://www.stat.go.jp/data/shakai/2021/kekka.html (in Japanese).

Negoro, T. (2017). The essentials of sharing economy and its success factors. *Nextcom, 30,* 4–17 (in Japanese).

Nishijima, C. (2010), The emerging of listening t in modern Japan: Expression of the feelings of the 'audience'. *Human and Socio-Environmental Studies, 20,* 73–88 (in Japanese).

Nishino, K. (2015). The talent education (Suzuki method), Yamaha music school, and the development of the musical instrument manufacturing. *Management and Information, 28*(1), 33–43 (in Japanese).

Orito, Y., Majima, T., Inutsuka, Y., & Fors, P. (2021). Japanised "sharing Economy": In comparison with Sweden. *Proceedings of the Annual Conference of Japan Society for Management Information 2021,* 293–296 (in Japanese).

Paulauskaite, D., Powell, R., Coca-Stefaniak, J. A., & Morrison, A. M. (2017). Living like a local: Authentic tourism experiences and the sharing economy. *International Journal of Tourism Research, 19*(6), 619–628. https://doi.org/10.1002/jtr.2134

Riessman, C. K. (2008). *Narrative methods for the human sciences.* Sage Publications.

Rifkin, J. (2014). *The zero marginal cost society: The internet of things, the collaborative commons, and the eclipse of capitalism.* Palgrave Macmillan.

Schor, J. (2016). Debating the sharing economy. *Journal of Self-Governance & Management Economics, 4*(3), 7–22.

Sharing Economy Association Japan. (n.d.). *Vision of the association.* https://sharing-economy.jp/ja/about/ (in Japanese).

Siraj-Blatchford, J. (2008). The implications of early understandings of inequality, science and technology for the development of sustainable societies. In I. Pramling Samuelson & Y. Kaga (Eds.), *The contribution of early childhood education to a sustainable society.* UNESCO: Paris, France.

Sundararajan, A. (2016). *The sharing economy: The end of employment and the rise of crowd-based capitalism.* The MIT Press.

Tanaka, T. (2011). The competitive advantage of Nippon Gakki Co.'s marketing strategy: Hearing the voice of the piano & organ market during the high economic growth period in Japan. *Keiei Shigaku (Japan Business History Review), 45*(4), 4_52–4_76 (in Japanese).

Tanaka, T. (2021). *The Japanese history of piano: The formation of music instrument industry and consumers.* The University of Nagoya Press.

Tussyadiah, I. P. (2015). An exploratory study on drivers and deterrents of collaborative consumption in travel. In I. Tussyadiah, & A. Inversini (Eds.), *Information and communication technologies in tourism* 2015 (pp. 817–830). Springer: Cham. https://doi.org/10.1007/978-3-319-14343-9_59

Watanabe, H., Masuda, S., Shimizu, M., Tonoshita, T., Kato, Y., Wajima, Y., & Wakabayashi, M. (2005). *Politics of classical music.* Seikyusya: Tokyo (in Japanese).

Yamauchi, Y. (2015). *Service as "struggle".* Chuokeizaisya (in Japanese).

Yano Research Institute Ltd. (2018). *Sharing economy service market to grow at double-digit pace! Despite the impact of legal restrictions, the outlook is that the contraction in the market for home-stay services will be temporary.* https://www.yano.co.jp/press-release/show/press_id/1988 (in Japanese).

Appendix 1

Question items on semi-structured interview survey

a Questions on the sharing business
   - What is your background and education?
   - How did you come up with your business plan and what was your motivation?
   - What are the types of tasks you undertake in your business, and what type of problems do you solve through your business?
   - When you explain your business, do you use the term 'sharing economy'? Do you identify with 'sharing economy' or is it another word you use?
b Questions on the sustainability of the business
   - What is 'special' about being in the C2C or P2P market for you/your company?
   - How are sustainability and ethical values respected in your company?
   - Is there an established methodology or model for conducting business that maintains sustainability?
   - How do you prioritise sustainability over economic and commercial values such as sales and customer satisfaction?
   - What is your company doing to achieve Sustainable Development Goals (SDGs) and corporate social responsibility? Or does your company conduct business with an emphasis on social values such as achieving the SDGs? If so, what are you doing to achieve them?

12 A block in the chain of sustainability? On blockchain technology and its economic, social, and environmental impact

Matthew Davis, Rickard Grassman,
Vanessa Bracamonte and Maki Sato

1   Introduction

The concept of sustainable development was first popularised with the publication of the Brundtland Commission Report, which defines sustainability as 'meeting the needs of the present without compromising the ability of future generations to meet their own needs' (UN, 1987). This dominant discourse is, however, diluted by the breadth of its appeal to stakeholders with opposing aims (Fors, 2019). Various scholars have characterised the concept as a political compromise between growth and environmental sustainability (Castro, 2004), as well as a meta-fix that unites capitalists, socialists, and political elites (Lélé, 1991); there are even suggestions that the notion has been intentionally crafted to maintain a degree of ambiguity, thus allowing for discretionary flexibility and contextual adjustability (Gibson, 2001, p. 5). In line with this overarching view of sustainable development, academic proponents of blockchain technology have often focussed either on the economic benefits or attempted to counter mainstream environmentalist narratives about excessive electricity consumption. A common assertion, for example, is that compared to traditional financial industries, Bitcoin, as in the most famous blockchain powered alternative, is actually much more sustainable (Rybarczyk, Armstrong and Fabiano, 2021). There is also no shortage of voices pointing to the darker spectrum of socioeconomic effects of blockchain decentralisation, ranging from narcotics dealings and money laundering to facilitating terrorism (see, for example, Krishnan, 2020; Whyte, 2019). Many authors, however, choose to focus on a single aspect of sustainability in order to explore particular issues in depth. For example, Mora et al. (2021) assess how blockchain can be used to address certain social challenges of the UN Sustainable Development Goals (SDGs), and Parmentola et al. (2021) ask whether blockchain is able to enhance environmental sustainability.

Indeed, many scholars of sustainability tend to split their analyses into economic, environmental, and social sustainability discussions (see, for example, Gül Şenkardeş, 2021), and whilst this is a useful exercise and can help to analyse phenomena from different perspectives, it is rarely representative of the complexities of real life, nor does it acknowledge the interrelatedness of these dimensions. More recent perspectives are better equipped to address these limitations, and Kate

DOI: 10.4324/9781003367451-14

Raworth's Donut Economics (2017), for example, reimagines how societies and economies can configure themselves into 21st century versions that respect the environmental limitations of our planet whilst also meeting fundamental social needs. One aim of this chapter then is to expand on and critique prevailing sustainability narratives of blockchain, whilst also shining a light on alternative applications of the technology that bridge across sustainability related issues.

In addition to these overarching sustainability challenges, we will equally consider the technology's undeniable cultural appeal. In other words, we believe that part of the reason as to why there is a lack of similar scholarly engagements is because there seems to be a powerful counter cultural sway that this new technology holds in public discourse and its users' imagination. The 'origin story' of the first blockchain, Bitcoin, as the brainchild of a mysterious character, 'Satoshi Nakamoto', embodies the very idea of grass-root ingenuity and resistance; a challenge to existing financial institutions in the aftermath of their systemic excesses. Consequently, this 'underdog' narrative may have bestowed supporters of blockchain with powerful visions of a more equitable world and an overall cultural spirit of great innovative appeal. We thus consider blockchain both a 'socio-economic' and 'culturally imbued' technology, as it continues to create passionate communities and global followings across political spectra, which rarely appears to be considered in academic blockchain literature.

Furthermore, on a technical level, blockchain is essentially about decentralisation through incentivising a diverse range of third parties to verify and store an immutable record of transactions. However, recognising that technology often mobilises a series of values that can articulate certain visions of the world (Semenzin, 2021), it simultaneously embodies decentralisation from authority on a more cultural level as well, mirroring a so-called hacker ethic, cherishing anarchic undertones (cf. Himanen, 2001) and the broader ideals of financial inclusion and social justice. This dualism in significance is inescapable, and intrinsically two sides of the same coin. Therefore, merely considering the practical effects of the technology vis-a-vis sustainability would be incomplete if we fail to consider this additional cultural aspect, and what it might mean moving forward. Hence, we will not just try to problematise and make visible this somewhat under-represented relationship on a more holistic level but also consider how this cultural spirit materialises throughout our examples, and what its power of appeal might mean for the future of the technology.

For now, however, let us consider decentralisation in the narrower technical sense to make clear that this is the process whereby distributed but identical nodes make it rather difficult to distort records, or, as in the case of cryptocurrencies, to spend funds one does not have. In Bitcoin, the process of consensus between nodes is largely done by something called a 'proof-of-work' protocol, and forming a decentralised 'chain' of records has, in recent times, proven immensely energy-consuming due to the sheer volume of computing power required to maintain it. This is not so problematic if the energy used to secure the network comes from renewable sources that might otherwise be wasted; however, much of the bitcoin network is powered by 'dirty' coal power stations (Jiang et al., 2021). Newer and more experimental cryptocurrencies like Cardano and Solana utilise different consensus algorithms, such as 'proof-of-stake', that are far more energy efficient, to mitigate against this concern. If we problematise this example from a traditional sustainability perspective, we begin to see just how interrelated these core aspects are; perhaps, there is a cost to this alternative system of safe-guarding the immutability of the blockchain through proof-of-stake instead of proof-of-work? With significantly lower energy consumption, that cost does not seem to fall on the environmental side of the sustainability equation, but what about the economic and social spectrum? We will come to this later.

We therefore begin this chapter by considering sustainability in classically broad terms to demonstrate the varied and complex contexts in which blockchain technology is being applied today. We then question the underlying assumptions behind these sustainability discussions, to help orient where blockchain might fit in a more modern sustainability discourse, and with consideration to its cultural significance.

2   A block in the chain?

In recent years, the number of applications for blockchain technology has expanded to other fields beyond cryptocurrencies (Casino et al, 2019). Since the technology enables a virtual form of 'trust', or the capability for truly 'trustless' transactions, there is a potential for other forms of value transaction to be digitised and democratised. It can be useful to understand briefly that blockchain technologies are generally divided into three types (Ichikawa et al., 2017):

- *Public Blockchains*: Everyone can participate in the distributed network, check data transactions and verify them, and participate in the process of reaching consensus.
- *Consortium Blockchains*: The node that has authority may be selected in advance and usually has partnerships, such as business-to-business (peer-to-peer) partnerships. The data in the blockchain can be open or private and it is considered a partly decentralised network.
- *Private Blockchains*: Nodes will be restricted, and not everyone can participate in the blockchain, which has strict authority management of data access.

To start with, the easiest way to make a blockchain more energy efficient is to make it a private blockchain rather than public. This means that only authorised nodes such as an individual or a small number of agents will have access and authority over the network, and whoever would like to use or support its functionality in one way or other would require permission from this central authorising party to do so. There are of course several opportunities to use more centralised private blockchains (for example, see Galvanits, 2020), where the benefits of decentralisation such as data immutability, transparency, and information redundancy are required, but this has yet to achieve large-scale adoption. After all, why bother to make use of the technology of decentralisation if the use case demands a centralised model, and traditional technologies can achieve the same results with better scalability and lower environmental cost?

In any case, the ambition of making blockchains in general more energy efficient by considering a private blockchain infrastructure may risk undermining the very purpose of the blockchain itself, which is to make it autonomous from centralised control by openly incentivizing whoever wishes to participate, and thus exert a more decentralised form of governance. Consequently, it would seem as though the arguments rallied against blockchain for reasons of energy efficiency are somewhat disingenuous. As already indicated, there are alternative forms of public blockchain infrastructures such as the 'proof-of-stake' variety, with innovative new incentives and features that are significantly more energy efficient than their 'proof-of-work' counterparts (like Bitcoin). How might they affect the future of our society?

If we look briefly beyond the energy efficiency debate, there is growing disenchantment among people who are beginning to realise how thoroughly they are being monitored and monetised in what Shoshana Zuboff describes as Surveillance Capitalism (2019), and by extension, a cultural rejection of these predatory digital practices. This economic model that has displaced oil companies from the top of the Fortune 500 in the past decade extracts what she describes as behavioural surplus by way of predicting or even stimulating certain behaviours as opposed to others, which in turn is being commodified and sold off to a plethora of corporations that have all kinds of monetary interest in knowing and modifying what we do and how we do it online. Whilst this obviously has been extremely lucrative for a few companies, one could argue it being less sustainable in an economic and social sense considering the extreme centralisation of data extraction along with its concomitant concentration of wealth accumulation that generates hitherto unseen economic disparities. Not even the alienating symptoms of scientific management and Fordism at its height, strikingly ridiculed in Charlie Chaplin's 1920s *Modern Times*, are comparable to the wealth concentration brought forth by the onset of a Surveillance Capitalist oligarchy. Whereas Ford and other 20th century wonders of manufacturing relied on droves of workers, even having to make up for its tedious nature with comparatively competitive wages, there is virtually no such scalability of labour in the data-extraction complex of the likes of Facebook, Amazon, Netflix, and Google as most work is done by algorithms. Against this sobering backdrop, however, there are potentially areas within which blockchain-powered technologies could help alleviate some symptoms of extreme economic centralisation and resultant concentration of wealth, offering ways to better control one's data or the value it generates, and we will have reason to come back to this crucial point in the following sections of this chapter.

There are of course challenges of adoption that could be attributed to the norms of traditional technological transitions, but as blockchain usage continues to grow, network scalability may also become an issue. Consequently, this chapter discusses both the positive and negative aspects of blockchain usage, with the intention of highlighting debates which bridge traditional sustainability perspectives. Hence, we also contextualise this study with various real-world use cases to demonstrate how efficient blockchains can address some of these issues. Let us start off then by introducing aspects relating to the economic sustainability of this technology, which will then, in subsequent sections, allow us to better appreciate this multi-dimensional phenomenon in a more balanced way.

3   Beyond profits: Blockchain for economic sustainability?

Some may find it odd to speak about economic sustainability in the context of blockchain technology, since, more than anything, extreme volatility in the prices of various blockchain assets has characterised this emergent industry. In February of 2021, one Reddit user converted 8,000 dollars' worth of bitcoin into Shiba Inu, a so-called meme-coin such as its predecessor 'Dogecoin' (deriving its name from the combination of meme and dog). Three months later, this one user was a dollar billionaire on account of this one conversion, as the price of Shiba Inu had shot up more than 60 million percent in little more than a month. On the other side of this extreme volatility in cryptocurrency prices, there is undoubtedly a great many who have lost significant amounts to these very same breakneck price fluctuations, which certainly reek of anything but sustainability in the economic sense of the term. Indeed, the earliest adopters of blockchain were tech enthusiasts, but with the dawn of cryptocurrencies as the main use case, a plethora of overconfident 'get rich quick' schemers jumped on board (Sudzina et al., 2021). Nevertheless, let us take a moment to consider the context in which this technology was brought to the fore, one in which the financial crisis was already wreaking havoc throughout the financial markets of the world.

An unknown inventor, under the pseudonym Satoshi Nakamoto, launched this new blockchain technology at the precipice of the 2008 financial armageddon, along with the explicit declaration of making money more autonomous from a global banking system that had gambled the savings of millions on reckless credit schemes. Consequently, the system-wide time bomb resulting therefrom eventually blew up, and the financial institutions responsible had to be bailed out by the same taxpayers that had unknowingly been leveraging such toxic bets in the first place. Against this backdrop, Nakamoto (2008) certainly has a point in the way that the existing financial system failed its citizens and that it was able to do so because of the extreme centralisation of common people's money into the hands of a few financial institutions that could gamble with the value of it all. In other words, the financial crisis was brought about by a system in which the common person does not have full autonomy over his or her money. On the other hand, in launching his invention called Bitcoin, Nakamoto suggests that decentralisation provides people with 'the absolute autonomy of money'. Moreover, in this sense, one could argue that he is not only right that Bitcoin clearly offers its users greater autonomy as no central authority could do anything with the money that the individual user does not consent to, but also by this very same virtue of decentralisation, Bitcoin would have been much more economically sustainable than the legacy system in 2008 by making any such gambling with other people's money impossible.

Acknowledging the fact that blockchain-powered decentralisation would have been more economically sustainable than what turned out to be a devastating global financial meltdown is not to say that it would necessarily always be a more economically sustainable alternative. Centralisation has provided security from fraud, and the ability to override transactions and recover stolen funds for the eve-ryday user; true autonomy of money, however, forces the individual to carry the ultimate responsibility for their own funds, which can sometimes have disastrous effects. Hence, the main arguments for blockchain offering more economically sustainable solutions tend to focus on its potential applications around tracing and trust, not just in resilience to the occasional crisis, but in broader terms when com-pared to the centralised counterparts we have today. As consumers are evidently increasingly willing to pay more for products they are confident originates from certain places rather than others, being associated with trusted fair trade/labour processes or organic certification standards in production, blockchains are likely to significantly improve the transparency and traceability that make up such confi-dence. The four TRs (Traceability, Trust, Tracking, and Transparency), which cat-egorise the main value propositions for blockchain in this context, are the subject of much academic study and the application of blockchain technology is expected by many to provide a paradigm shift in the industrial landscape (Centobelli et al., 2021). Or so it is claimed.

Manufacturers with poor sustainability credentials, such as the typical sweat-shop variety strewn across large swathes of the global south, will be increasingly singled out, automatically disqualifying themselves from this type of transparency and traceability. The alternatives who do qualify may thus become more competi-tive in turn. Indeed, as Davis et al. (2021) point out, blockchain can be used by companies in their operations to mitigate and externalise the practice of corrup-tion in emerging markets, potentially leading to more trust in those economies and subsequent economic growth as foreign investment rises. Yet, this relies on the assumption that those emerging economies have access to sufficient infrastructure and reliable internet access to partake in the blockchain network, which is not always the case.

As for the predominantly service-based economies of the more post-industrial state of Western democracies, we increasingly hear the call for a 'sharing economy' opening up new frontiers, where some of the apparent deadlocks of contempo-rary capitalism, vis-a-vis sustainability, could be sidestepped. The various visions around a sharing economy are essentially about optimising the use of resources so that they can be shared and used collectively in smart new ways, rather than for each individual to own all required resources that are idle and unused most of the time (Pazaitis et al., 2017). If we look toward the similar drives for sustainability such as the circular economy, the lowest hanging fruit in terms of implementation may be to enable tracing the product life cycle, from mine, factory, to landfill (Leng et al., 2020; Liu et al., 2019; Lu & Xu, 2017). For managing the supply chain ecosystem of retail products, IBM has already been providing software using blockchain to keep transparency and traceability that in turn not only grasps the environmental impact of each product but also is expected to build customers' trust by certifying its provenance (such as how and where it has been processed). Furthermore, IBM and Energy Blockchain Lab[1] have been expanding blockchain technology to create a blockchain platform to trade carbon assets in China, enabling users to monitor their carbon footprints and supporting a peer-to-peer platform to trade renewable energy, using tokens or digitals assets.

Blockchain adoption then could enable considerable growth prospects for organisations and communities. But perhaps more importantly, it offers the technological infrastructure of a more sustainable economic model for a range of industries that rely on supply chain production processes, wherein the blockchain provides complete traceability of each component part. Putting concerns regarding remote infrastructure development aside, it could be a literal win-win effect which sees the value of production chains increase as a reflection of our propensity to pay more for products that are completely transparent about their respective trajectories of assembly. At the same time, it empowers labour processes to become more fair and sustainable as exploitative conditions jeopardise corporate valuations in light of this blockchain-powered holistic transparency. While blockchain is perhaps not what Marx envisioned as the catalyst of a post-capitalist economy, it could arguably be said to greatly improve one crucial problem he identified in the commodity form, and that is making visible the actual labour involved in the process of commodification (cf. Marx, 2008). Of course, as we have just elucidated, the reality of economic sustainability focused initiatives has a direct impact on social dimensions too; hence, we turn now to a more focussed view of how blockchain impacts social sustainability.

4   Empowering communities: The social side of blockchain

There are many blockchain-based solutions that have been proposed for the achievement of social sustainability goals; proposals related quality education (Raimundo & Rosário, 2021), identity (Wang & De Filippi, 2020), social governance processes like voting (Kewell et al., 2017), and cultural heritage (Vacchio & Bitulco, 2022), to name but a few. Since blockchain technology and blockchain-based solutions are relatively new and at an emerging stage, this section highlights recent research and solutions in terms of their impact on social issues.

In the last section, we discussed the applicability of blockchain for economic sustainability, and part of that discussion revolves around how blockchain technology can be used as a way to achieve the goal of financial inclusion (see, for example, Schuetz & Venkatesh, 2020). If blockchain-based solutions could help reduce the costs related to financial transactions and remove the need for intermediaries, then the use of these solutions may be a way in which rural populations could have access to global markets (ibid). Yet, this is not the only challenge. Part of the problem of obtaining access to financial services is due to lack of identification (Wang & De Filippi, 2020). As of 2018, the World Bank estimates that nearly 1 billion people lack access to a government-issued or legal identity (Marskell et al., 2018). This is problematic for many reasons, since legal identities are required for engaging with modern aspects of wider society. Opening a bank account, voting in elections, or registering land property, for example, all require legal credentials that prove identity. Blockchain technology has also been proposed as a solution for providing identity services that are decentralised and transparent, allowing people to have complete control over their identity (Dunphy & Petitcolas, 2018). One example is the project IAMX,[2] which has the stated goal to provide digital identifiers through a biometric identity gateway.

In emerging economies as well as in places with larger rural populations, the problem of lack of identity is more widespread, due in part to a lack of mature infrastructure to support registration. Being critical, one might argue that enforcing a legal identity is a reduction of basic human freedoms, even if the recognition of one's existence is foundational to the very idea of human rights and dignity (Sperfeldt, 2022). An important innovation in this context therefore is the decentralised nature of blockchain, since this removes reliance on centralised parties to act as trusted certifiers and places the individual as the holder of a self-sovereign identity. Consequently, it is no longer necessary for legal identities to be issued by governments or nation-states, who may not be even aware of the origins of an individual. Of course, the digital identity must still be legally acceptable to the government if the individual is to unlock access to state services, protections, and benefits, but the registration of a digital identity need not be tied to birth in a nation-state. It should be noted that new technologies always come with big promises, and blockchain is no different. For IAMX, such a large-scale project requires heavy institutional cooperation. The project claims to already be partnered with 49 telecommunications companies, which helps users connect to existing organisations to enable over 840 verified use cases, though it does not state what they are at this time. In terms of sustainability then, this particular blockchain use case strongly supports the social dimension but also bridges over to some measure of economic sustainability as well. Yet, with a digital identity tied to the real world comes ethical challenges.

Technology-based proposals for sustainability are for the most part data-driven. An important ethical consideration for these proposals therefore concerns safeguarding the right to privacy, alongside general issues of cybersecurity and data rights (Michael et al., 2019). Privacy violations can have serious consequences, in particular for vulnerable populations, and traditionally users have relied on central authorities to hold and secure their data. This is problematic for several reasons, most notably because it adds a single point of failure. Future industrial projects such as IoT assume that there will be detailed data collected from devices, which can be a concern for the tracking of user behaviour (Chanson et al., 2019; IIASA, 2019).[3] This is true at the individual user-level (for consumer IoT devices) and in the case of the general public (for smart-city IoT devices). In blockchain-based solutions, however, cryptographic techniques such as zero knowledge proofs (ZKPs) can be leveraged to preserve privacy, and here, the user is self-sovereign, which means they maintain control of the distribution of, and access to, their data. Proposals that deal with privacy sensitive information such as in health (Al Omar et al., 2017; Rupasinghe et al., 2019) and voting-related use cases (Hardwick et al., 2018) explicitly include techniques to protect the privacy of that data and keep the power in the hands of the individual. Proposals for technology-based sustainability solutions, therefore, should have privacy protection as a primary aspect and should be designed with consideration of privacy challenges (Cavoukian et al., 2013; Michael et al., 2019).

Bridging back to parts of economic sustainability, the use of blockchain technology has also been proposed as a way to preserve user privacy while retaining profitability, and one such example is the case of online advertising. In traditional web marketing, online ads are targeted to users who are profiled and tracked in order to be served ads. This type of advertising violates users' privacy by exposing their browsing habits to the highest bidder (Castelluccia et al., 2012). With growing distrust of the black box society (Pasquale, 2015), data security and veracity of data content in the complex web and layers of social networks become all the more important. From a social sustainability perspective, in perhaps a worst case scenario, targeted advertising has been used to promote misinformation, leading to social unrest (Silva et al., 2020), and population control (Pomerantsev, 2019). Brave is a privacy-focused browser that makes use of blockchain technology in order to reward users for viewing ads (Brave Software, 2021). They do so through the use of what they call the Basic Attention Token (BAT). The tokens are implemented as 'blind tokens', which means that the Brave browser can know certain performance data of the ads, but not which ads users viewed (Brave Software, 2019). This 'opt-in' approach means that users maintain their privacy, as well as digital autonomy, and it is this preservation of user privacy that is crucial if users are to use blockchain-based solutions and technology with peace of mind.

In the healthcare industry, the use of blockchain technology has been suggested to assist in exchanging patients' electronic health records, and here perhaps is where data privacy is most imperative. The challenge is to implement sharing and storage of medical records by providing standardised, yet secure, mechanisms for information exchange of medical data within the industry. Sharing of medical records enables doctors to collect, store, and analyse patient data in a more comprehensive and integrated manner, which leads to better service, eliminating unnecessary treatments, and helps in longitudinal analysis and validation of the effects of a particular procedure. With rapid technological advancements, there are already various literature review papers that discuss the possibilities and anticipation in broadening and expanding beyond the exchange of patients' medical records, to improving the ecosystem of healthcare industries in general (see, for example, Akbar et al., 2021; Hasselgren et al., 2020; Mayer et al., 2020).

Another area of social sustainability concerns the education industry. Prominent universities such as MIT in the US (as of 2017) and Keio University in Japan (as of 2020) are seeking to utilise blockchain to secure digital diplomas, as a digital identity for personnel who have received higher education at their university.[4] With more people receiving higher education and moving towards the international job market, as well as a more general fluidity in career development under global businesses, faking educational background has increasingly been recognised as a problem (Børresen, 2020; Chen et al., 2018; Grolleau, 2008). In 2018, the Malaysian Ministry of Education (MoE) announced, with a consortium of six universities, that they would be using the NEM blockchain to address a rampant use of fake diploma certificates (their application is named Tasdiq Al-Syahadah, TAS for short).[5] Not all communities are receptive to blockchain-based technologies, however, and there is still a hesitation with its use as a platform to secure learning and certification of compulsory education. During the Covid-19 pandemic in Japan, one solution that had been discussed was to provide a non-fungible token (NFT) through online classes that proves the learning (Proof of Learning, PoL) of children taking remote classes. Interestingly, there was social unrest among parents whose children were at the stage of compulsory education, and it was not widely accepted due to the fear of excluding children from low-income households.

For students seeking higher education with credits where university education is not accessible, credits gained from massive open online courses (MOOCs) would be crucial for those in need of PoL. In this instance, blockchain could underpin a sustainable learning platform for recurrent education and for gaining new skills for post-graduates with positive learning performances that may lead to job skill improvements (Kuleto et al., 2022). The use of blockchain as a platform for PoL, however, has a few critical challenges in how to attract technological maintenance through blockchain consensus (inward sustainability), coupled with a feeble orientation to collective development to ensure the quality of education and to certify the degree (outward sustainability) (Park, 2021). Simultaneously, there is a discussion around to what extent education should be covered by a blockchain platform at both a macro and micro level, to enable social inclusion and access to education (Kwok & Treiblmaier, 2022). Further exploration of how blockchain can impact education management and development to improve educational quality, accessibility, social inclusion, and equality is yet to be analysed. In terms of securing data privacy with regards to education, the World Wide Web Consortium (W3C), an international community that develops open standards for the long-term growth of the Web society, has been discussing Decentralised Identity (DID) and published its proposed recommend architecture on securing DIDs in 2021.[6] As for the impact on wider social sustainability, in a transitioning world overwhelmed with new digital technologies, it is important to note that life-long learning becomes increasingly necessary.

Reflecting on the social sustainability dimension, we can see that there are a wide variety of potential applications and indeed areas that are already actively being explored. Many of the benefits discussed may also materialise through standard digitalisation and transformation from legacy systems. However, the value of blockchain in this context is fundamental to its trustless modus operandi, to be able to transact without intermediation, and to retain ownership of one's own digital footprint. Of course, digitalisation comes at a cost in terms of new servers and computing hardware, and decentralisation exponentially more so. How then, does blockchain fare when we analyse its environmental sustainability aspects?

5   The chain reaction: Blockchain's eco-impact

The concept of environmental sustainability is perhaps the most familiar to an everyday reader, invoking connotations of global warming, and humanity's ongoing struggle to avoid catastrophic climate change. Hence, it is unsurprising that much of the literature regarding blockchain and environmental sustainability takes issue with the energy usage of the Bitcoin network, often quoting it as being comparable to the consumption of a small country such as Ireland (O'Dwyer & Malone, 2014) or Denmark (Küfeoğlu & Özkuran, 2019; Vranken, 2017). Most recently, the Cambridge Bitcoin Electricity Consumption Index[7] puts the estimated power demand at an average of 14.27 gigawatts, or 125.13 Terawatt hours annually, which is more comparable to Norway or Argentina.[8] This figure is likely to continue to rise as long as there is profit to be gained.

To provide additional context, it was mentioned briefly in the introduction that Bitcoin achieves consensus among validators through a proof-of-work (PoW) algorithm, which requires computational energy to be spent, solving a cryptographic puzzle. The difficulty of this puzzle is proportional to the number of users who are engaged in validation of the network, and this difficulty factor changes programmatically in order to ensure that transactions are validated and a new block generated every 10 minutes. To incentivise users to spend energy solving the puzzle, the validator is rewarded with newly minted bitcoins. Through clever design, and perhaps as a way to stimulate adoption, the minting schedule follows a deflationary economic model and has a fixed supply; there will only ever be 21 million coins available. Thus psychological and economic factors align to give bitcoins a measure of scarcity and therefore potential value. Since energy is not free, and validators are in competition with one another, they are incentivised to increase their computational power to improve their chances of receiving the block reward. In the early days of bitcoin, this was not a big problem since the number of users was relatively small, which meant that the difficulty was also low and the puzzle easy to solve; most validators were able to mine bitcoin using their home PCs. However, as more users realised the potential utility of bitcoin and blockchain, the network began to scale up in size, and the difficulty level increased substantially. New developments in mining hardware continue to increase the difficulty, lowering the ROI for miners and forcing them to upgrade their equipment to stay profitable. Whilst new generations of mining hardware often improve energy efficiency, the fact that the hardware (in this case, application-specific integrated circuits) can only mine bitcoin means that older generations essentially become useless e-waste (de Vries & Stoll, 2021).

Some proponents of Bitcoin are attempting to change this negative sustainability narrative by demonstrating alternative use cases for the perceived waste. Kryptovault, for instance, is a Norwegian data centre company that also mines bitcoin and uses the excess heat generated by the mining equipment to dry out wood (Boffey, 2022). Arguably, this is not quite the bastion of sustainability that advocates might like to proclaim, but it does represent a fundamental shift in the ways in which bitcoin miners are approaching the problem, moving from secret practices, to a more transparent and open discussion about how energy is used. In any case, Kryptovault uses only renewable energy to power its machines, and perhaps, this is the community's best accolade, even if one concedes that the energy might be better spent elsewhere. El Salvador, which controversially made bitcoin legal tender in 2021 (Lopez & Livni, 2021), has plans to use surplus geothermal energy generated from its volcanoes to power bitcoin mining equipment, in order to improve its local economy (Murray, 2021). Hence, whilst it may be true that the bitcoin network requires ever increasing amounts of energy as users vie for computational supremacy and thus financial reward, the widespread criticism based purely on energy consumption of cryptocurrencies like bitcoin, is rather one dimensional.

Indeed, McCook (2015, 2018) takes a much more nuanced view than most critics comparing the energy usage of the bitcoin network and production of mining equipment, to the entire supply chain for the global financial macroeconomy. His first review (2015) provides a relative sustainability assessment of bitcoin and the legacy banking system, finding that the bitcoin network barely even registers when compared to the energy required to run the banking system as a whole and mint paper currency. The comparison may seem a little uneven; however, since Bitcoin is not a precise substitute for any one legacy system (Rybarczyk et al., 2021). In a revisit, McCook (2018) notes that the economic activity required to mine and recycle gold outstrips bitcoin mining as well, both in terms of energy used (kWh) and tonnes of CO_2 produced, though one must accept that gold has additional value in the electronics and jewellery industries. More recently, Rybarczyk et al. (2021) attempt to calculate the same use case and find that the bitcoin network uses around half that of the banking and gold industries. Even if we take these somewhat reasonable estimations with a pinch of salt, it is obviously not quite the environmental disaster that some may have considered it in the past (e.g. Carstens, 2018) and, in effect, showcases the role of digitalisation in significantly reducing the environmental externalities of legacy economic systems.

There are of course many factors which affect energy usage, such as which consensus mechanism is used to secure the network. More recent papers, for example Saleh (2021), note a community shift away from energy-hungry consensus algorithms like proof-of-work (PoW), towards others such as proof-of-stake (PoS). PoS removes the incentive for validators to engage in the aforementioned computational arms race, by assigning block rewards randomly across holders of a particular token. Consequently, users are incentivised to simply hold tokens to secure the network, which requires very little energy in comparison.

In any case, the energy efficiency of the network is only one half of the proverbial coin. As a technology, blockchain has many relative advantages which are already being used to improve aspects of traditional industries, with a noticeable impact on environmental issues. For instance, the growth of renewable energy generation has changed the way that energy networks can be configured. Micro-generation has opened opportunities for increased efficiency of delivery, but the complexity of implementing the necessary infrastructure has made scaled adoption difficult. Wörner et al. (2019) present the concept of decentralised energy networks, where blockchain is used to arbitrate new market structures for peer-to-peer energy trading, and improve the security of supply. By creating a prototype market, the team were able to connect the micro-generation capabilities of 37 local households in Switzerland and allow them to buy and sell from one another, thus ensuring that energy is used at the point of generation, and not wasted through transformation and transmission across the country.

Another example is provided by Ahl et al. (2020), who study a blockchain-based smart-energy pilot system in Urawa Misono, Japan, consisting of ten consumers, five prosumers, and one shopping mall. They find that blockchain has the potential to incentivise adoption of clean energy micro-generation, since transparency and traceability of the supply metrics simplifies the process for all stakeholders, and enables instantaneous monetisation of generation for prosumers through tokenomics (Schinckus, 2020). It can be important to note, however, that although the technology is mature enough to support the complexity of managing an infrastructure for mixed grids, the implementation of such schemes on a broader scale is likely to require challenging multi-stakeholder collaboration and regulatory flexibility.

In any case, even if the above examples are only prototypes, other, more straightforward opportunities exist. For example, the Green Assets Wallet[9] is a blockchain-driven investment platform for trusted green bonds, which aims to unlock greater volumes of credible investment opportunities that deliver environmental impact. Another organisation, Poseidon,[10] is using blockchain to improve transparency in supply chains and reduce transaction fees for carbon emission trading, allowing organisations to make more informed decisions about their carbon footprint and make it easier to offset their production (Pan et al., 2019). Lin et al. (2018) suggest that blockchain can be used alongside other IoT technologies to immutably record water transactions, thereby improving trust and data security and offering the potential to improve water resource management in real time.

Peer-to-peer financing, which is facilitated by blockchain, can also speed up development of clean technologies through easier access to investment opportunities (Dorfleitner & Braun, 2019). In a more recent review, Dorfleitner et al. (2021) showcase 85 green applications across a variety of blockchains that contribute to positive environmental action; Gainforest,[11] for instance, uses blockchain's inherent transparency to provide green investors with assurance that their funds are being properly utilised to protect and restore natural environments. They use a combination of smart contracts and the so-called 'NFTrees' (their label for non-fungible tokens) to trace the impact of individual contributions, giving investors a sense of ownership over their conservation or restoration project. Clearly, the technology itself has much to offer in terms of improving the environmental impact of many aspects of modern society, even if it is the Bitcoin network's energy consumption that still dominates the mainstream narrative.

6   Navigating the ethos of blockchain culture: revolutionary sustainability?

As we continue our discourse on the convergence and interconnectedness of these sustainability aspects, we now present an example of how blockchain could fundamentally revolutionise the approach to tackling the social and economic disparities of our era. By challenging the prevailing power structures through novel methods of resource organisation, it may also offer the opportunity to reshape our engagement with the environment. This transformative potential is not merely grounded in the technical capabilities of the technology, but in the cultural and ideological ethos of its proponents, who see it as a means to realise a vision of revolutionary sustainability that transcends the boundaries of conventional thinking, opening up new horizons of possibility. In this regard, blockchain appears to represent a new frontier of innovation, where the visionary and the practical intersect to create a platform for systemic change and sustainable development. Nowhere is this fresh paradigm more clear than with proof-of-stake blockchains.

In the particular case of Cardano, users with an ADA balance in their digital wallet are rewarded for delegating their stake to a pool, in order to support the network consensus protocol.[12] While anyone may operate a pool, there is a reasonably high barrier to entry which requires operators to have specific technical knowledge, such as how to configure a server node. By offering the ability to delegate a user's stake, this ensures everyone can participate in securing the network, irrespective of technical experience, and arguably, this is a much more sustainable model. Pool operators are able to charge a fee, as a percentage of the pool's mining rewards to, for example, support maintenance and improve reliability. However, given the competitive open market dynamic of the pool system, these fees are usually kept to a minimum. Consequently, pool operators have turned to other ways of attracting stakers. Some donate part of their staking profits to charities or environmentally friendly organisations such as Veritree,[13] in an attempt to appeal to the core values of their potential delegators. Others entice new stakers with zero fees, and then slyly increase the fee at a later date, hoping that stakers don't notice.

In many ways, running a mining pool for proof-of-stake operators becomes an enterprise and one where varying aspects of sustainability can be prioritised. Some upcoming projects that utilise the Cardano blockchain, and indeed many that have already been implemented, adopt the novel business model of an Initial Stake Pool Offering (ISPO or ISO) to fund project development. This phenomena is a very much understudied aspect of the blockchain ecosystem, and thus, it deserves some discussion here. In plain terms, the idea is that instead of traditional financing routes, through venture capital and such, projects pitch their idea to the community of token stakers, who, when delegating their stake to the project pool, forgo part or all of their staking rewards in exchange for a share of the promised project value. This could be other tokens which the staker hopes might afford some future economic value, or perhaps tokens which will eventually have some kind of utility, such as governance, within the project.

The pool staking rewards are then used or sold by the project team to finance continued development. For Cardano, staking rewards are calculated based on a rolling five-day epoch, and stakers are able to redelegate their stake every epoch, as and how they wish, hence there is a lot of competition between new projects for potential stakers. Obviously, when it comes to ISPOs, there is some risk involved that a project will never deliver. Since stakers retain ownership of their tokens, however, if a project is unsuccessful, or worse, a scam, the only losses are from unearned income in staking rewards. The process of an ISPO is thus inherently more economically sustainable for users of the cryptocurrency than the preceding ICO (initial coin offering) business model, where investors would invest upfront, thereby losing ownership of their funds in advance of project delivery.

If we contrast this ISPO process against the typical funding models available to the everyday investor, we find that most early-stage investment opportunities are only open to accredited venture capitalists, or those already with significant financial capital. And here we begin to see how such opportunities gather cultural appeal. The average retail investor struggles to invest in start-up companies pre-initial public offering (IPO) and hence does not get the same opportunities for return on their investment as the financial elite. The ISPO model thereby serves as both a socially and economically sustainable democratising force on a traditionally unequal power distribution, allowing anyone holding ADA tokens to invest in any project they see as valuable, and reap the benefits (and risks) of early investment.

Looking towards the developers of other blockchain solutions, Semezin and Gandini (2021) note that entrepreneurs and tech experts who are interested in the potential of blockchain for social change are motivated by broadly libertarian values; hence the rhetoric of 'revolution' used by many projects is unsurprising and reflects the underlying culture of 'technosolutionism' (Morozov, 2013) at the heart of these communities. There are equally a great many users of public blockchains who, despite getting involved for profit seeking activities, consider the reigning socio-economic systems unsustainable and are more than happy to, in their eyes at least, contribute to a new decentralised alternative (Grassman et al., 2021).

7   The sustainable blockchain paradox

In this book chapter, we have embarked on a rather broad and overarching view concerning the different sustainability aspects of blockchain technology. Indeed, teasing out a theoretical or applied discussion on the ethics of blockchains requires such multidimensional undertakings, but it is certainly not unproblematic, and we note an overwhelming technophilia among the literature. If we are to stake out some emergent strands for our discussion, we might as well start with signalling the broader philosophical questions on how blockchain can make an impact, for example, to gauge present-day financial inclusion and autonomy of money. Decentralisation may create resilience against the occasional financial crisis, but we must then again balance all this against a colossal increase in electricity and e-waste, contributing in no small measure to the perils of a warming planet and the impact this has on future generations.

What adds further nuance to this already rather complicated ethical balancing act is the previously mentioned comparison between the sustainability of current financial systems as they stand today, along with that of the gold industry that historically has been the guarantor of last resort for financial stability, considering that each respectively consumes more energy than Bitcoin (Rybarczyk et al., 2021). Moreover, one might need to also consider the likelihood of less energy-consuming alternatives, and innovations such as the proof-of-stake protocols, scaling up to become the credible and dominant standard of potential future financial systems.

Which brings us to an equally puzzling issue already hinted at above, in terms of economic and/or social sustainability, is there something of a concession made when a transition to a more energy efficient proof-of-stake alternative occurs? To what extent might inequalities in the number of tokens held exacerbate economic and social inequalities among users, given that larger stakes essentially translate into more power and potential returns? On the other hand, there is virtually no non-state financial practice within capitalist societies where the volume in financial holdings does not translate into potential for earning proportionally more profit. Needless to say this is the very DNA of our financial markets, where the scale of the economy harshly works. And at one level, it is what powers what some may argue to be the more egalitarian version of its PoW counterpart, because control of data-processing power is proportional to the likelihood of earning more in the mining process.

The main difference, however, is that in the PoW model, it is still random whether the reward will go to a big holder in data-mining power, as opposed to a small-scale single miner who once in a while happens to luck out against the odds of proportional inferiority. In the PoS protocol, there is no randomness left; those with the greater stake are always the ones who are rewarded with proportionally more. Whether or not one would argue this to be more unfair or perhaps honest about a power imbalance that is also very much present in the PoW variety, one thing is clear. The silver lining to such unfairness and or realist honesty, depending on how you see it, is that the PoS model is considerably more energy efficient and consequently sparing of our environment.

In any case, to make an outright ethical judgement on which consensus mechanism is preferable, if we were to propose such a thing would be counter-productive. A more sensible discussion should revolve around the ways in which blockchain can be made sustainable from a holistic perspective, and the first contentious debate that covers all of the above is perhaps the dichotomy of centralisation versus decentralisation. Both PoW and PoS need decentralisation to function appropriately, but the implications for sustainability are different for each type. In PoW, centralisation of miners can lead to potential security threats, such as a 51% attack (Sayeed & Marco-Gisbert, 2019), which can negatively affect the trust of users, and thus its utility. For networks like Bitcoin, however, where the current network hash power makes this type of exploit extremely expensive in terms of required computing power, groups of miners centralise their hash power to share rewards in order to maintain a steady income and finance their operations regardless of mining luck.

Centralised consensus is also required by communities to achieve network upgrades in the form of a 'hard fork', which, as the name suggests, continues building the blockchain along a new tree branch, with different parameters (Jiang et al., 2022). For PoS blockchains, a 51% attack is much less likely; stakers are incentivised not to disrupt or fraudulently attack the network since negative trust would lead to lower valuations of their own stake. In some blockchains, centralising the node validators can speed up network processing times, though it can also lead to a less secure network in terms of stability. A major criticism of PoS is that many node operators choose to run their node on centralised services, such as Google Cloud or Amazon Web Services (AWS), which can negatively affect the perception of decentralisation,[14] leaving the blockchain exposed to central points of failure and regulatory pressures. In both PoS and PoW, decentralisation of users and validators/miners is beneficial for adoption of the network into the wider socio-economic systems. Taking this dimension along a more cultural divide, we note amongst proponents a significant, almost dogmatic belief in decentralisation as the key value proposition of blockchain, with many projects keen to advertise themselves as a solution to the 'problem' of centralisation. And yet, to maintain a purely decentralised operating model is perhaps to isolate oneself from mainstream adoption and limit its potential to supersede the very institutions that they aim to disrupt. To have any utility in the real world, decentralised identities, for example, must be decentralised, except that for now at least they need to integrate seamlessly with centralised authorities and services. The sustainability of blockchain is therefore paradoxically contingent on a 'goldilocks' measure of both centralised and decentralised traits to ensure the longevity and usability of a given token; how much of each, however, is quite context-specific.

Another complementary dichotomy relating to the discussion above, and touched upon briefly in prior sections, is that of privacy versus transparency. In the autonomy of money debate, privacy and transparency are often considered contradicting terms in the use of digital or conventional currency for transactions; cash allows some measure of user privacy, while digital cash allows traceability and transparency for authorities, because transactions are processed by a central intermediary. Blockchain, as an infrastructure, has the possibility to add value to transacted data by transparently guaranteeing or certifying data content at the time of data creation and transaction, without a central authority or intermediary. Without this transparency, one of the main value propositions of blockchain would be significantly diminished. Indeed, whilst making public the entire transaction history of a blockchain could potentially open up new opportunities for value creation through open data analysis, more importantly perhaps, it democratises access for all, reducing the power of global corporations by normalising their ability to manipulate access to and profit from information. Personal privacy, however, is widely recognised as a fundamental human right (Noam & Rottenstreich, 2021) and some of the use cases where blockchain could have a beneficial impact, such as healthcare, legally require this capability. A key question therefore is, with all records published on a blockchain and available for anyone to peruse, how can the privacy of individual users be protected?

Putting private permissioned blockchains to one side for a moment, the decentralised and immutable nature of blockchain is particularly problematic where data protection laws are strong. Within the EU, for example, GDPR[15] legislates the right to be forgotten. On public blockchains, there is some measure of pseudo-anonymity since a lack of regulation allows anyone to open a wallet without their 'real' identity attached and begin to transact. As soon as a user wants to convert their cryptocurrency into fiat or make purchases in the real world, however, this digital footprint can be traced and their identity compromised. Of course, researchers are working hard to find technical solutions, such as the aforementioned zero-knowledge proofs, to solve these issues, but their implementation is complex and can affect network performance (Noam & Rottenstreich, 2021).

A future blockchain infrastructure that can accommodate these requirements may even remove the necessity for data protection legislation such as GDPR, as it will operate in a different way than today's internet. Trust is no longer required, since blockchain facilitates the automation of trust, or removal of the concept entirely (Semenzin & Gandini, 2021), meaning users engage at will, on their own terms, with verifiable consequences. The fundamental question, we suppose, is that of data sovereignty; who should own the data that is stored on the blockchain, and who should benefit from its exploitation? Taking a sustainability-focussed position in this regard we find that again the answer is context-specific but perhaps ethically related to personal autonomy, as blockchain offers the opportunity to repatriate the intangible parts of ourselves that have been colonised by capitalist forces online.

Moreover, this is where the cultural impact of blockchain technology becomes all the more salient. As we have demonstrated, blockchain could be used in a range of different ways to optimise the functioning of governmentally or commercially centralised authorities, through sovereign control and autonomy over one's personal data. However, a powerfully subversive culture of a cyberpunk hacker ethic that has accompanied the technology from its very invention is problematically embodied by the dominant use cases. Smart contracts, for example, are mostly coded by developers rather than lawyers and hence lack the necessary social nuance in implementation that would be required from more sustainability-focussed projects. Indeed, despite many of the examples we discussed having promised to disrupt entrenched institutions through a more 'morally acceptable' modus operandi, they often embody 'neoliberal cultural concepts of entrepreneurialism, individualism and meritocracy, characterised by mainstream tech culture' (Semenzin & Gandini, 2021). In other words, whilst this technology could be used in a range of different ways in a purely technical sense, it is anything but culturally neutral, and public blockchain projects tend to attain an almost cult-like following of profit seekers. There is extraordinary potential in the types of use case where aim and purpose align with the underlying cyberpunk ethos it undeniably carries, but in order to realise this utility and apply the technology in a sustainable way, proponents should also be sensitive to the cultural diversity of their audience, as well as the ethical implications imposed by their vision.

8   Conclusion

Much of the public and academic discourse on blockchain sustainability remains around environmental concerns and energy efficiency. Taking a more nuanced view, we have uncovered several dichotomies which may initially seem contradictory but will better help future researchers to explore the realm of blockchain with a more sustainable eye. These competing dimensions are somewhat described at a technical level by the so-called blockchain trilemma; a proposition attributed to Vitalik Buterin in 2016 that posits every blockchain implementation trades off between the scalability of the system, its decentralisation, and its ability to be secure or tolerate faults. However, this simplistic view does not accommodate that adopters of blockchain in the future will need to consider that the effects of these dimensions bridge across all areas of sustainability and thus require deeper consideration. In the same way that Raworth (2017) sets a hypothetical floor for social needs, and ceiling for environmental costs, we suggest that the dimensions of decentralisation and privacy also have a context-specific 'goldilocks' zone and should be considered of vital importance for a sustainable future.

Any technology can be appropriated by good or bad actors, and here, we have discussed the potential of blockchain to redress or exacerbate the resource limitations of our planet, alongside its cultural subtext. In fact, returning full circle now to the beginning of this chapter, if one is to build an infrastructure using blockchain that meets the needs of the present, without compromising the needs of future generations, developers, entrepreneurs and policy makers must all expand their thinking to bridge across the different aspects of sustainability, to reflect on the cultural divides and socio-economic conditions that stimulate the evocations and trepidations of a decentralised economy, and strive to ensure that blockchain is used appropriately, rather than as a knee-jerk response to the failings of capitalism or nefarious intentions of bad actors.

Acknowledgement

This study was supported by the JSPS/STINT Bilateral Joint Research Project, 'Information and Communication Technology for Sustainability and Ethics: Cross-national Studies between Japan and Sweden' (JPJSBP120185411) and Kakenhi (19K12528).

Notes

1 As described by https://www.ibm.com/case-studies/energy-blockchain-labs-inc.
2 https://iamx.id/.
3 The International Institute for Applied Systems Analysis.
4 For more details regarding MIT announcement and Keio University announcement in using blockchain to certify their issued diploma see the following website: https://news.mit.edu/2017/mit-debuts-secure-digital-diploma-using-bitcoin-blockchain-technology-1017.
   https://www.keio.ac.jp/ja/press-releases/2020/10/26/28–75892/. Accessed 31/03/2022.
5 https://www.asiablockchainreview.com/malaysia-uses-blockchain-for-certificate-verification-to-tackle-degree-fraud/. Accessed 31/03/2022.
6 https://www.w3.org/TR/did-core/. Accessed 31/03/2022.
7 Run by the University of Cambridge Judge Business School, Cambridge Centre for Alternative finance. Available online, here: https://ccaf.io/cbeci/index. Accessed 11/02/2022.
8 The most recent data available is based on 2018/2019 estimations from the US. Energy Information Administration (EIA).
9 https://greenassetswallet.io/.
10 https://poseidon.eco/solution.html.
11 https://www.gainforest.app/documentation.
12 For Cardano, the PoS consensus protocol is called Ouroboros. https://docs.cardano.org/new-to-cardano/proof-of-stake.
13 Another organisation that utilises blockchain to provide transparency and verifiable information of environmentally restorative practices. https://www.veritree.com/.

14 As of the 3rd of October 2022, 65% of the ethereum network is hosted on cloud services, with Amazon.com hosting nearly 36% of nodes. Source: https://ethernodes.org/networkType/Hosting.
15 Under Article 17 of the EU GDPR, individuals have the right to have personal data erased. This is also known as the 'right to be forgotten'. Source: https://gdpr-info.eu/art-17-gdpr/.

References

Ahl, A., Yarime, M., Goto, M., Chopra, S. S., Kumar, N. M., Tanaka, K., & Sagawa, D. (2020). Exploring blockchain for the energy transition: Opportunities and challenges based on a case study in Japan. *Renewable and Sustainable Energy Reviews, 117*, 109488. https://doi.org/10.1016/j.rser.2019.109488

Akbar, I. M., Bhawiyuga, A., & Siregar, R. (2021). An ethereum blockchain based electronic health record system for inter-hospital secure data sharing. In *6th International conference on sustainable information engineering and technology 2021*, Universitas Brawijaya, Malang, Jawa Timur (pp. 226–230). https://doi.org/10.1145/3479645.3479699

Al Omar, A., Rahman, M. S., Basu, A., & Kiyomoto, S. (2017). Medibchain: A blockchain based privacy preserving platform for healthcare data. In G. Wang, M. Atiquzzaman, Z. Yan, & K.-K. R. Choo (Eds.), *Security, privacy, and anonymity in computation, communication, and storage,* SpaCCS 2017 International Workshops, Guangzhou, China (pp. 534–543). Springer International Publishing. https://doi.org/10.1007/978-3-319-72395-2_49

Boffey, D. (2022, February 9). Can bitcoin be sustainable? Inside the Norwegian mine that also dries wood. *The Guardian.* Retrieved February 11, 2022, from https://www.theguardian.com/technology/2022/feb/09/can-bitcoin-be-sustainable-inside-the-norwegian-mine-that-also-dries-wood

Børresen, L. J., Meier, E., & Skjerven, S. A. (2020). Detecting fake university degrees in a digital world. In *Corruption in higher education* (pp. 102–107). Leiden: Brill. https://doi.org/10.1163/9789004433885_016

Brave. (2019). Security and privacy model for ad confirmations brave/brave-browser Wiki. *Brave Browser Wiki.* Retrieved January 16, 2022, from https://github.com/brave/brave-browser/wiki/Security-and-privacy-model-for-ad-confirmations

Brave. (2021, February 10). Basic attention token (BAT) blockchain based digital advertising. https://basicattentiontoken.org/static-assets/documents/BasicAttentionTokenWhitePaper-4.pdf

Carstens, A. (2018). Money in the digital age: What role for central banks? *Lecture at the House of Finance, Goethe University, Frankfurt, 6.* Available at: https://www.bis.org/speeches/sp180206.pdf. Accessed 10 November 2021.

Casino, F., Dasaklis, T. K., & Patsakis, C. (2019). A systematic literature review of blockchain-based applications: Current status, classification and open issues. *Telematics and Informatics, 36*, 55–81.

Castelluccia, C., Kaafar, M. A., & Tran, M. D. (2012, July). Betrayed by your ads! Reconstructing user profiles from targeted ads. In Fischer-Hübner, S., Wright, M. (eds.) *International symposium on privacy enhancing technologies symposium* (pp. 1–17). Springer, Berlin.

Castro, C. J. (2004). Sustainable development: Mainstream and critical perspectives. *Organization & Environment, 17*(2), 195–225.

Cavoukian, A., Polonetsky, J., & Winn, C. (2013). Privacy by design and third party access to customer energy usage data. Accessed May 2022. Available online: https://www.ipc.on.ca/wp-content/uploads/Resources/pbd-thirdparty-CEUD.pdf

Centobelli, P., Cerchione, R., Esposito, E., & Oropallo, E. (2021). Surfing blockchain wave, or drowning? Shaping the future of distributed ledgers and decentralized technologies. *Technological Forecasting and Social Change, 165*, 120463. https://doi-org.ezproxy.its. uu.se/10.1016/j.techfore.2020.120463

Chanson, M., Bogner, A., Bilgeri, D., Fleisch, E., & Wortmann, F. (2019). Blockchain for the IoT: Privacy-preserving protection of sensor data. *Journal of the Association for Information Systems, 20*(9), 1274–1309.

Chen, G., Xu, B., Lu, M., & Chen, N. S. (2018). Exploring blockchain technology and its potential applications for education. *Smart Learning Environments, 5*(1), 1–10. https:// doi.org/10.1186/s40561-017-0050-x

Davis, M., Lennerfors, T. T., & Tolstoy, D. (2021). Can blockchain-technology fight corruption in MNEs' operations in emerging markets?. *Review of International Business and Strategy*. https://doi.org/10.1108/RIBS-12-2020-0155

de Vries, A., & Stoll, C. (2021). Bitcoin's growing e-waste problem. *Resources, Conservation and Recycling, 175*, 105901. https://doi.org/10.1016/j.resconrec.2021.105901

Dorfleitner, G., & Braun, D. (2019). Fintech, digitalization and blockchain: Possible applications for green finance. In Migliorelli, M., Dessertine, P. (eds.), *The rise of green finance in Europe* (pp. 207–237). Palgrave Macmillan, Cham. https://doi. org/10.1007/978-3-030-22510-0_9

Dorfleitner, G., Muck, F., & Scheckenbach, I. (2021). Blockchain applications for climate protection: A global empirical investigation. *Renewable and Sustainable Energy Reviews, 149*, 111378. https://doi.org/10.1016/j.rser.2021.111378

Dunphy, P., & Petitcolas, F. A. (2018). A first look at identity management schemes on the blockchain. *IEEE Security & Privacy, 16*(4), 20–29. https://doi.org/10.1109/ MSP.2018.3111247

Fors, P. (2019). *Problematizing sustainable ICT* (Doctoral dissertation, Acta Universitatis Upsaliensis).

Glavanits, J. (2020). Sustainable public spending through blockchain. *European Journal of Sustainable Development, 9*(4), 317. https://doi.org/10.14207/ejsd.2020.v9n4p317

Gibson, R. B. (2001). *Specification of sustainability-based environmental assessment decision criteria and implications for determining" significance" in environmental assessment*. Ottawa: Canadian Environmental Assessment Agency.

Goldfeder, S., Kalodner, H., Reisman, D., & Narayanan, A. (2017). When the cookie meets the blockchain: Privacy risks of web payments via cryptocurrencies. arXiv preprint arXiv:1708.04748.

Grassman, R., Bracamonte, V., Davis, M., & Sato, M. (2021). Attitudes to cryptocurrencies: A comparative study between Sweden and Japan. *The Review of Socionetwork Strategies, 15*, 169–194. https://doi.org/10.1007/s12626-021-00069-6

Grolleau, G., Lakhal, T., & Mzoughi, N. (2008). An introduction to the economics of fake degrees. *Journal of Economic Issues, 42*(3), 673–693. https://doi.org/10.1080/00213624 .2008.11507173

Gül Şenkardeş, C. (2021). A discussion on the effects of blockchain technology within the context of sustainable development. *Journal of Information and Communication Technologies, 3*(2), 243–262. https://doi.org/10.53694/bited.1021926

Hardwick, F. S., Gioulis, A., Akram, R. N., & Markantonakis, K. (2018, July). E-voting with blockchain: An e-voting protocol with decentralisation and voter privacy. In *2018 IEEE International Conference on Internet of Things (iThings) and IEEE Green Computing and Communications (GreenCom) and IEEE Cyber, Physical and Social Computing (CPSCom) and IEEE Smart Data (SmartData)* (pp. 1561–1567). IEEE. https://doi. org/10.1109/Cybermatics_2018.2018.00262

Hasselgren, A., Kralevska, K., Gligoroski, D., Pedersen, S. A., & Faxvaag, A. (2020). Blockchain in healthcare and health sciences—A scoping review. *International Journal of Medical Informatics, 134*, 104040. https://doi.org/10.1016/j.ijmedinf.2019.104040

Himanen, P. (2001). *The hacker ethic and the spirit of the information age.* New York: Random House.

Ichikawa, D., Kashiyama, M., & Ueno, T. (2017). Tamper-resistant mobile health using blockchain technology. *JMIR mHealth and uHealth, 5*(7), e7938.

Jiang, S., Li, Y., Lu, Q., Hong, Y., Guan, D., Xiong, Y., and Wang, S. (2021). Policy assessments for the carbon emission flows and sustainability of Bitcoin blockchain operation in China. *Nature Communications, 12*(1). https://doi.org/10.1038/s41467-021-22256-3

Jiang, S., Li, Y., Wang, S., & Zhao, L. (2022). Blockchain competition: The tradeoff between platform stability and efficiency. *European Journal of Operational Research, 296*(3), 1084–1097. https://doi.org/10.1016/j.ejor.2021.05.031

Kewell, B., Adams, R., & Parry, G. (2017). Blockchain for good? *Strategic Change, 26*(5), 429–437.

Krishnan, A. (2020). Blockchain empowers social resistance and terrorism through decentralized autonomous organizations. *Journal of Strategic Security, 13*(1), 41–58. https://www.jstor.org/stable/26907412

Küfeoğlu, S., & Özkuran, M. (2019). Bitcoin mining: A global review of energy and power demand. *Energy Research & Social Science, 58*, 101273. https://doi.org/10.1016/j.erss.2019.101273

Kuleto, V., Bucea-Manea-Ţoniş, R., Bucea-Manea-Ţoniş, R., Ilić, M. P., Martins, O. M., Ranković, M., & Coelho, A. S. (2022). The potential of blockchain technology in higher education as perceived by students in Serbia, Romania, and Portugal. *Sustainability, 14*(2), 749. https://doi.org/10.3390/su14020749

Kwok, A. O., & Treiblmaier, H. (2022). No one left behind in education: Blockchain-based transformation and its potential for social inclusion. *Asia Pacific Education Review*, 1–11. https://doi.org/10.1007/s12564-021-09735-4

Lélé, S. M. (1991). Sustainable development: A critical review. *World Development, 19*(6), 607–621.

Leng, J., Ruan, G., Jiang, P., Xu, K., Liu, Q., Zhou, X., & Liu, C. (2020). Blockchain-empowered sustainable manufacturing and product lifecycle management in industry 4.0: A survey. *Renewable and Sustainable Energy Reviews, 132*, 110112. https://doi.org/10.1016/j.rser.2020.110112

Lin, Y. P., Petway, J. R., Lien, W. Y., & Settele, J. (2018). Blockchain with artificial intelligence to efficiently manage water use under climate change. *Environments, 5*(3), 34. https://doi.org/10.3390/environments5030034

Liu, K. H., Chang, S. F., Huang, W. H., & Lu, I. (2019). The framework of the integration of carbon footprint and blockchain: Using blockchain as a carbon emission management tool. In Hu, A., Matsumoto, M., Kuo, T., Smith, S. (eds.), *Technologies and Eco-Innovation towards Sustainability I* (pp. 15–22). Springer, Singapore. https://doi.org/10.1007/978-981-13-1181-9_2

Lopez, O., & Livni, E. (2021, October 7). In Global First, El Salvador Adopts Bitcoin as Currency. *The New York Times.* Retrieved February 11, 2022, from https://www.nytimes.com/2021/09/07/world/americas/el-salvador-bitcoin.html

Lu, Q., & Xu, X. (2017). Adaptable blockchain-based systems: A case study for product traceability. *Ieee Software, 34*(6), 21–27. https://doi.org/10.1109/MS.2017.4121227

Marskell, J., Metz, A., & Lu, J. (2018). Identification for Development (ID4D) global dataset. *The World Bank Group*. Retrieved February 9, 2022, from https://datacatalog.worldbank. org/search/dataset/0040787/Identification-for-Development--ID4D--Global-Dataset

Marx, K. (2008). *Capital.* New York: Oxford University Press.

Mayer, A. H., da Costa, C. A., & Righi, R. da R. (2020). Electronic health records in a Blockchain: A systematic review. *Health Informatics Journal,* 1273–1288. https://doi. org/10.1177/1460458219866350

McCook, H. (2015). *An order-of-magnitude estimate of the relative sustainability of the Bitcoin network.* 3rd Edition, Academia.edu. Available at: https://www.academia. edu/7666373/An_Order-of-Magnitude_Estimate_of_the_Relative_Sustainability_of_ the_Bitcoin_Network_-_3rd_Edition. Accessed 10 November 2021.

McCook, H. (2018). The cost & sustainability of Bitcoin. *Unpublished Working Paper.* Retrieved November 10, 2021, from https://www.academia.edu/37178295/ The_Cost_and_Sustainability_of_Bitcoin_August_2018_

Michael, K., Kobran, S., Abbas, R., & Hamdoun, S. (2019, November). Privacy, data rights and cybersecurity: Technology for good in the achievement of sustainable development goals. In *2019 IEEE International Symposium on Technology and Society (ISTAS)* (pp. 1–13). IEEE. https://doi.org/10.1109/ISTAS48451.2019.8937956

Mora, H., Mendoza-Tello, J. C., Varela-Guzmán, E. G., & Szymanski, J. (2021). Blockchain technologies to address smart city and society challenges. *Computers in Human Behavior, 122,* 106854. https://doi-org.ezproxy.its.uu.se/10.1016/j.chb.2021.106854

Morozov, E. (2013). *To save everything, click here: The folly of technological solutionism.* Public Affairs.

Murray, C. (2021). El Salvador plans 'bitcoin city' powered by volcano. Retrieved February 11, 2022, from [online] Financial Times. Available at https://www.ft.com/ content/67515f23-ccdc-4dbc-a184-70848e183ac3

Nakamoto, S. (2008) *Bitcoin: A peer-to-peer electronic cash system.* Available at: https:// bitcoin.org/bitcoin.pdf

Noam, O., & Rottenstreich, O. (2021). Realizing privacy aspects in blockchain networks. *Annals of Telecommunications, 77*(1), 3–12. https://doi.org/10.1007/s12243-021-00861-z

O'Dwyer, K. J., & Malone, D. (2014). Bitcoin mining and its energy footprint. In *Proceedings of the 25th joint IET Irish signals & systems conference 2014 and 2014 China-Ireland international conference on information and communications technologies (ISSC 2014/CIICT 2014)* (pp. 280–285). IET. https://doi.org/10.1049/cp.2014.0699

Pan, Y., Zhang, X., Wang, Y., Yan, J., Zhou, S., Li, G., & Bao, J. (2019). Application of blockchain in carbon trading. *Energy Procedia, 158,* 4286–4291. https://doi.org/10.1016/j. egypro.2019.01.509

Park, J. (2021). Promises and challenges of Blockchain in education. *Smart Learning Environments, 8*(1), 1–13. https://doi.org/10.1186/s40561-021-00179-2

Parmentola, A., Petrillo, A., Tutore, I., & De Felice, F. (2021). Is blockchain able to enhance environmental sustainability? A systematic review and research agenda from the perspective of Sustainable Development Goals (SDGs). *Business Strategy and the Environment, 31*(1), 194–217. https://doi-org.ezproxy.its.uu.se/10.1002/bse.2882

Pasquale, F. (2015). *The black box society: The secret algorithms that control money and information.* Boston, MA: Harvard University Press.

Pazaitis, A., De Filippi, P., & Kostakis, V. (2017). Blockchain and value systems in the sharing economy: The illustrative case of Backfeed. *Technological Forecasting and Social Change, 125,* 105–115. https://doi-org.ezproxy.its.uu.se/10.1016/j.techfore.2017.05.025

Pomerantsev, P. (2019). *This is not propaganda: Adventures in the war against reality.* London: Faber & Faber.

Raimundo, R., & Rosário, A. (2021). Blockchain system in the higher education. *European Journal of Investigation in Health, Psychology and Education, 11*(1), 276–293.

Raworth, K. (2017). *Doughnut economics.* London: Random House Business Books.

Rupasinghe, T., Burstein, F., & Rudolph, C. (2019). Blockchain based dynamic patient consent: A privacy-preserving data acquisition architecture for clinical data analytics. *ICIS 2019 Proceedings.* 14. Available online: https://aisel.aisnet.org/icis2019/blockchain_fintech/blockchain_fintech/14.

Rybarczyk, R., Armstrong, D., & Fabiano, A. (2021). On Bitcoin's energy consumption: A quantitative approach to a subjective question. *Galaxy Digital Mining,* 1–13. Available online: https://github.com/GalaxyDigitalLLC/Financial-Industry-Electricity-Balance

Sayeed, S., & Marco-Gisbert, H. (2019). Assessing blockchain consensus and security mechanisms against the 51% attack. *Applied Sciences, 9*(9), 1788. MDPI AG. http://dx.doi.org/10.3390/app9091788

Saleh, F. (2021). Blockchain without waste: Proof-of-stake. *The Review of Financial Studies, 34*(3), 1156–1190. https://doi.org/10.1093/rfs/hhaa075

Schinckus, C. (2020). The good, the bad and the ugly: An overview of the sustainability of blockchain technology. *Energy Research & Social Science, 69,* 101614. https://doi.org/10.1016/j.erss.2020.101614

Schuetz, S., & Venkatesh, V. (2020). Blockchain, adoption, and financial inclusion in India: Research opportunities. *International Journal of Information Management, 52,* 101936.

Semenzin, S. (2021). *Blockchain & Data Justice; The political culture of technology.* Università degli Studi di Milano; Università degli Studi di Torino.

Semenzin, S., & Gandini, A. (2021). Automating trust with the Blockchain? A critical investigation of 'Blockchain 2.0' cultures. *Global Perspectives, 2*(1). https://doi.org/10.1525/gp.2021.24912

Sudzina, F., Dobes, M., & Pavlicek, A. (2021). Towards the psychological profile of cryptocurrency early adopters: Overconfidence and self-control as predictors of cryptocurrency use. *Current Psychology.* https://doi.org/10.1007/s12144-021-02225-1

Silva, M., Santos de Oliveira, L., Andreou, A., Vaz de Melo, P. O., Goga, O., & Benevenuto, F. (2020, April). Facebook ads monitor: An independent auditing system for political ads on facebook. In *Proceedings of the web conference 2020* (WWW'20), Taiwan (pp. 224–234). New York: Association for Computing Machinery. https://dl.acm.org/doi/proceedings/10.1145/3366423

Sperfeldt, C. (2022). Legal identity in the sustainable development agenda: Actors, perspectives and trends in an emerging field of research. *The International Journal of Human Rights, 26*(2), 217–238. https://doi.org/10.1007/978-3-642-31680-7_1

The International Institute for Applied Systems Analysis. (2019). Policy brief #16: Digitalization and the future of energy systems. Available online: https://sdgs.un.org/documents/policy-brief-16-digitalization-and-future-e-25347

Brundtland Commission. (1987). *Our common future* [*Brundtland* report]. New York: Oxford University Press.

Vacchio, E., & Francesco, B. (2022). Blockchain in cultural heritage: Insights from literature review. *Sustainability, 14*(4), 2324. https://doi.org/10.3390/su14042324

Vranken, H. (2017). Sustainability of bitcoin and blockchains. *Current Opinion in Environmental Sustainability, 28,* 1–9. https://doi.org/10.1016/j.cosust.2017.04.011

Wang, F., & De Filippi, P. (2020). Self-sovereign identity in a globalized world: Credentials-based identity systems as a driver for economic inclusion. *Frontiers in Blockchain*, 28. https://doi.org/10.3389/fbloc.2019.00028

Whyte, C. (2019). Cryptoterrorism: Assessing the utility of blockchain technologies for terrorist enterprise. *Studies in Conflict & Terrorism*, 1–24. https://doi.org/10.1080/10576 10X.2018.1531565

Wörner, A., Meeuw, A., Ableitner, L., Wortmann, F., Schopfer, S., & Tiefenbeck, V. (2019). Trading solar energy within the neighborhood: Field implementation of a blockchain-based electricity market. *Energy Informatics*, 2(1), 1–12.

Zuboff, S. (2019). *The age of surveillance capitalism*. New York: Public Affairs.

13 Using bits to consume less – Consuming less when using bits

A European perspective

Norberto Patrignani

1 ICT as a *phármakon*

The Greek term φάρμακον (*phármakon*) has many meanings, but in this context, it is used to denote any kind of drug, and as medical science explains, drugs may have a beneficial effect on organisms but they can also have very dangerous effects. The same applies to digital technologies; they can be considered as the *phármakon* of the 21st century and humans have to learn how to use them wisely.

A similar debate happened, in a completely different scenario, many centuries ago, as Plato describes in his dialogue *Phaedrus* between the great inventor Theuth and the king of Egypt Thamus. When Theuth describes his discovery about the use of letters, he says that it will make the Egyptians wiser and give them better memories. Then Thamus replies that the discovery will create forgetfulness in the learners' souls, because they will not use their memories (Plato, 1952).

At that time, the introduction of letters and writing that can be considered one of the first "information technologies" is already seen as a *phármakon* with a potential negative impact on society. Nowadays, in a hyperconnected world, human beings are flooded with data and information, but they have very little time to reflect, to transform this information into real knowledge, into autonomous cognitive structures. Humans will be hearers of many things and will have learned nothing is the warning of Thamus in Plato's dialogue. ICT are in a good position to be considered the last version of this kind of *phármakon*.

The climate emergency is pushing the entire humanity towards a careful analysis also about the digital world. What are the conditions for a wise use of digital technologies?

The study of the relationship between ICT and energy and, in general, the dilemma information energy is one of the most intriguing in the history of science (Patrignani and Kavathatzopoulos, 2018). James Clerk Maxwell, in 1867, opens this debate with his famous "Maxwell's devil" – a thought experiment where a very high speed creature, by discriminating between slow and fast molecules, is able to create a temperature difference against the law of entropy (Leff and Rex, 1990). But the acquisition of information on the molecules requires energy; information (collection, storage, processing) is not free (Szilard, 1929; Brillouin, 1953).

DOI: 10.4324/9781003367451-15

Digital technologies promise to help humans in facing climate change, the key challenge of the Anthropocene era. For example, exchanging a paper letter produces about 29 grams of CO_2 while an email produces only about 4 grams of CO_2. At the same time, the number of emails is exploding to a staggering 400 billion emails daily, and the majority is spam! (McGovern, 2020). Nevertheless, it is important to investigate these two faces of this 21st century's *phármakon*.

Looking for only market or technological solutions to this complex issue is not enough. It is necessary to investigate the difficult possibilities of ICT to develop within the limits of the planet, and while this requires a different vision about ICT, it needs also a new vision of the role of policymakers and of the relationship with the planet Earth. This is the core of a European perspective for ethical and sustainable computing, which will be developed during the paper.

The search for a new vision in the relationship with the planet Earth has a long history. March 2022 marks the 50th anniversary of the publication of "The Limits to Growth" (Meadows et al., 1972). This fundamental book, for the first time, puts together the complex interconnections of the dynamics of population, food, energy and economies, and clarifies that, on a finite planet, an exponential growth is impossible and unsustainable. Humans have to realize that it is wiser to live "inside" these limits.

In 1987, Gro Brundtland wrote "Our common future" where the term "Sustainable Development" was coined: a development that satisfies the needs of the present without compromising the possibility of future generations to satisfy their own needs (Brundtland, 1987).

Despite these alarming voices, and a lot of international conferences, the global emissions from human activities still continue to grow. After a drop of about 5.4% due to the pandemic in 2020, global fossil CO_2 emissions are set to rebound, and in 2021, the global total emissions reached 33.0 GtonCO$_2$ (Giga tonnes CO_2) (from 31.5 GtonCO$_2$ in 2020) (source: *IEA org*) but, according to other sources, can set a new record of 36.4 GtonCO$_2$ (GCP, 2022).

As a result, extreme weather events are becoming frequent: storms, heat waves, melting glaciers, and rising sea levels. Now, it is time to act.

The scientific community has made many efforts to raise awareness, to inform, and to make society in general aware of the effects of climate change. Now humans, apparently the animals best at making long-term predictions, have important and life-critical decisions in front of themselves. Today humans have a large amount of information; it is possible to quantify the impact on the planet, to forecast in the long term and to predict in the short term what the environmental consequences will be, and to take important decisions consequently. The scientific community, represented by the IPCC (Intergovernmental Panel on Climate Change), has provided all the basic knowledge for designing a transition from a "termo-industrial" society based on fossils, towards a decarbonized society based on renewable energies and a circular economy and, most importantly, consuming less energy and materials.

Humans need a new vision for their relationship with the planet. Only with a clear systemic vision, it will be possible to define the necessary policies, norms, projects, and infrastructures, which will influence behaviours, lifestyles, and daily habits.

If the first alarm of "The Limits to Growth" (1972) did not help, if the concept of sustainable development (1987) proved to be insufficient, the vision of a Global Stationary State is now the most promising. Humans must abandon the narratives based on exponentials, and no entity can grow exponentially on a finite planet; there can be only cycles, areas where the quantities involved can vary but within certain limits. For example, in a two-dimensional plane, a circular trajectory shows the periodic variation of two quantities. In a three-dimensional space, the trajectories must remain contained in a "donut", a useful model used to visualize four quantities that must be kept within sustainable limits of the planet. The most important indicators are climate change, consumption of soils, water, biodiversity, air pollution, ocean acidification, chemical pollution, etc. (Raworth, 2017; Scalia et al., 2020).

How can ICT be the phármakon of the 21st century and at the same time to help in remaining within the limits of a global stationary state? How can a digital sufficiency principle be implemented?

2   Using bits for consuming less

Digital technologies could promote sustainable lifestyles and support people in the transition towards a global stationary state: in the last IPCC report, among the dimensions of potential feasibility, digital technologies play an important role in many climate responses and adaptation options, the major ones are sustainable forest management, agroforestry, biodiversity management and ecosystem connectivity, improved cropland management, green infrastructure and ecosystem services, sustainable urban water management, improve water use efficiency, resilient power systems, energy reliability, risk spreading and sharing (IPCC, 2022, p. 24).

All these areas have in common the smart use of information for using natural resources more wisely. By using sensors for collecting data and data-analysis software, many processes can be optimized; it is possible to significantly decrease the environmental impact of many human activities, to move from "plentiness" (and waste) to "enoughness". The most important indicator is represented by the CO_2 emissions, and it is clear that digital technologies can play an important role for decreasing these emissions and so in addressing many of the climate risks. It is important to focus on "using bits to consume less", also called "Greening by IT" (Lennerfors et al., 2015), as a central value when designing new digital systems.

For example, before carrying out a project, some "dashboards" powered by Big Data analysis could be created for policymakers, allowing for an in-depth evaluation of all aspects: benefits for society, impact on ecosystems, and erosion of natural resources; in short, a deep investigation of the stakeholders' network. Up until now, the positive aspects of projects were measured only in terms of GDP and this is no longer acceptable; the evaluations will have to adopt much more advanced indicators such as the BLI, the Better Life Index by OECD. BLI measures housing, income, jobs, community, education, environment, civic engagement, health, life satisfaction, safety, and work-life balance (OECD, 2020).

In general, digital innovations could improve the ability to make important decisions on environmental issues by developing high-precision software models and using the large amount of data and the processing power provided by digital platforms. Systems for preventing and managing ecological disasters with monitoring tools able to detect pollution and other concerning patterns of data are becoming feasible.

ICTs are also enabling technologies for the circular economy, an epochal transformation of all industrial sectors and their respective supply chains and value chains. It requires keeping objects under control throughout their life cycle. Since it is released on the market, a physical product is assigned with an "electronic passport" with the origin, composition, status of maintenance, repairability, and recyclability. The sensors and the related software applications (IoT, Internet-of-Things) will make it possible to know the position, functionality, lifetime, maintenance interventions carried out, hardware diagrams (for facilitating on-site repairing), and software version, allowing each product to be reused, and easily repaired, extending its lifetime. An interesting pioneering project is Recirce.de funded by the Federal Ministry for the Environment in Germany. Of course also vendors have to update their business models, by proposing to users moving from possession to the use of objects: the dawn of the service-based economy, where companies need to maximize the life of products.

Digital technologies will be essential to design such scenarios and to manage the amount of information needed. The result should be less consumption of materials and energy (Rejeb et al., 2022).

Digital innovations based on sensors are also important for industry for improving the efficiency of electric motors (present in all sectors) and for decreasing consumption in agriculture like careful and precise water management.

Some recent studies estimate that it would be enough to invest 1.2% of global GDP in climate-friendly technologies to keep global warming below 1.5°C and to reach zero emissions by 2050 (De Weerdt, 2020).

Digital technologies can play a central role in this long-term strategy (decarbonization through digitalization); they can help to reduce power consumption and CO_2 emissions in many ways.

Society should invest in increasing the "digital literacy" by educating about the wise use of ICT: there are many examples of apps (like carbon calculators) that help in reducing the individual footprint, in reflecting on personal lifestyles, in providing multimodal sustainable mobility options, etc.

By speeding up many processes, digital technologies could help in streamlining many activities. This productivity growth induced by ICT should be invested in time for society; people could have more time for caring activities and local communities' support.

One example of decarbonization and productivity enabled by mobile devices is proposed by an important stakeholder of the ICT global scenario: the GSM Association (GSMA), representing the main mobile operators. Of course they are an "interested" party in this debate; nevertheless, a recent GSMA report claims that in one year (2018) mobile technologies avoided about 2.13 $GtCO_2$ by increasing connectivity, improving efficiency, and impacting behaviour change. This should be compared to the total annual emissions from the mobile sector itself: about 0.22 $GtCO_2$: apparently, digital mobile technologies have a positive impact ten times higher than their negative one (GSMA, 2022).

On another front, for example, at community level, the environmental impact of food waste is estimated in 0.17 $GtonCO_2$ per year (EPA, 2021). A simple scenario can be envisaged: the optimization of a university canteen. The vision could be an artificial intelligence application (software that fine-tunes itself with much data) that manages the food in the pantry with the goal to reduce food waste. It can check that all food supplies are there and good for the health, fruit and vegetables are in season and locally produced, most of them are legumes and nuts. It can check also the preferences of students and academic personnel, the expiration date of products, and then recommend the best menu for the day matching the expectation of all stakeholders.

Of course it is possible to develop many applications in many contexts, at a social and community level and personal level. Many processes could be dematerialized saving matter and energy. Many systems for advanced logistics management could be developed: for example, with the creation of temporary warehouses shared between many actors – the same paradigm of Internet routers that move bits – to move goods more efficiently would allow to optimize the use of resources: a good example of moving bits instead of moving atoms.

One of the key applications of "using bits to consume less" is software for smart grids. For the efficient management of supply and demand in renewable energy communities, these applications represent a strategic development with huge potential savings and for improving energy availability and "security". The shift to renewable energies implies more electricity produced by a large number of small decentralized units and the alignment of demand with supply of electricity; as a consequence, this implies significant information exchange between different market participants: the electricity grid and the electricity market must be controlled digitally (Ipakchi and Albuyeh, 2009).

Other examples of using bits to consume less could be

- computerized management of mobility, where people and goods movements are optimized for minimizing the energy consumption;
- software for optimizing agendas for the reduction of commuter traffic through a balanced mix between work in presence and remote work;
- energy management software in buildings, one key application for all future scenarios for saving energy.

It is difficult to estimate the exact contribution of ICT in decreasing CO_2, also because all projects' results depend on the starting date. Nevertheless, according to a now dated report, the estimated total contribution of ICT could be the reduction of 7.8 $GtonCO_2$ emissions (GeSI, 2015), several times its own footprint. But this report needs to be updated and, in this area, it is strongly needed a more independent and peer-reviewed research. Another recent study focused on applications based on "artificial intelligence" in sectors such as energy, transport, water management, and agriculture; the results are globally interesting: up to 2.4 $GtCO_2$ could be saved by 2030 (PWC, 2020).

It is also important to be aware that, while ICT can help in improving cost, time, and energy efficiencies, this usually generates rebound effects that counterbalance the gains; ICT opens also new areas of production and consumption, and these induction effects must be carefully taken into account (Santarius et al., 2022).

3   Consuming less when using bits

In the last 50 years, the world population multiplied by two, but the global consumption of digital devices on the planet multiplied by six. The environmental impact of ICT, consuming less when using bits, also called "Greening of IT" (Lennerfors et al., 2015), is a relatively new area of study.

At the global level, according to some recent research, in 2019 the ICT sector consumed about 2,500 TWh of electricity (10% of total electricity consumed globally in a year) corresponding to 3.9% of total CO_2 emissions (Andrae and Edler, 2015; Green IT, 2019). According to one of the more comprehensive and detailed report, the entire ICT sector (including devices, networks, and data centres) could emit 0.74 $GtCO_2$ in 2040 (Belkhir and Elmeligi, 2018). In this report, Belkhir and Elmeligi show a very clear trend towards the so-called cloud computing architectures. Indeed, the share of total ICT emissions due to smartphones was 4% in 2010, while in 2020, it was 11%. The scenario is clear: digital technologies are shifting towards a centralized architecture with billions of users "tethered" with their smartphones to data centres "on the other side" of the network.

The power (not only the computational one: memory and CPU) is in the hands of the Big Tech industry controlling the data centres. This is the result of "cloud computing" era started around the year 2005: users start having only a touchscreen in their hands (input and output), while the core computing power and algorithms are "on the other side": heteronomy is the result. Just the opposite of the autonomy promised by the "personal computing" era started in 1965 with the first personal computer sold by Olivetti (WSJ, 1965).

Not surprisingly, among the top ten companies in the world, seven out of ten are now the so-called Big Tech: Apple, Amazon, Facebook (now Meta), Google, Microsoft, Tencent, and Alibaba (Forbes, 2021).

This also complicates the environmental accountability of companies in general: cloud service providers do not disclose data related to their emissions as they are considered "competitive secrets", so when the user of the cloud is a company (that rely on external ICT services provided by Big Tech), for this company it is difficult to display its environmental impact. This can be a problem for serious companies that publish every year their environmental balance. While consistent and transparent financial accounting rules are central for a healthy economy since it enables the analysis and comparison of companies, the same should apply to environmental impact reporting. Unfortunately, as more workloads move to the cloud, for companies become more difficult to show their environmental credentials (Sissa, 2021b).

It is interesting to show a complete picture of the ICT energy consumption: the "cloud architecture" requires devices that are "always on" and connected via the network to data centres.

The latest breakdown of the energy consumed (Santarius et al., 2022) is

- 14% devices (in the hands of users, smartphones, computers, laptops, smart TVs);
- 13% networks (from WiFi to fixed and mobile networks);
- 58% data centres (for powering and cooling servers in gigantic data centres, one third of the energy is used only for air conditioning);
- 15% energy needed to produce all these physical components.

The higher consumption is due to data centres; they are the higher "weight" of ICT, 24/24 always-on, and, most importantly, many of them are still powered by non-renewable sources (Lucivero et al., 2020).

The worst trend is that their power need is increasing at a rate of 10% every year. Without drastic measures, by 2025 they will consume a large amount of the electricity produced on the entire planet, resulting in an exponential growth in CO_2 global emissions (Andrae, 2017; Uptime, 2019; Frazelle, 2020; Jirotka, 2020).

There are also other factors to be included in this analysis: the billions of IoT sensors (Internet of Things) that will be produced, powered, and disposed of (IDC, 2019); the new ("artificial intelligence") applications that are turning out to be "carbon hungry" (e.g. "training" neural networks); particular applications such as bitcoins which, in order to be generated, consume about 45.8 TWh per year (Stoll et al., 2019) (as a reference, in 2019, the entire Italy consumed approximately 319 TWh).

This exponential growth of the speed of consumption cycles of ICT (in particular the smartphone devices and the use of online videos) risks to overcome the contributions of the digital world for consuming less and represents one of the best examples of an "unsustainable growth" on a finite planet: the total contribution of ICT to greenhouse gases is projected to reach 15% of CO_2 in 2040, the same impact of the transport sector (Belkhir and Elmeligi, 2018). This study for the first time includes smartphones into the evaluation of the environmental impact of digital technologies and describes that about 90% of the total energy adsorbed by a smartphone in its lifetime is required just for its production (up to 80 $KgCO_2$), while on the other side its average life is about two years. As a result, the CO_2 emission due to a single smartphone is about 45.3 $kgCO_2$ per year. The smartphone is becoming a daily tool for billions of people, so maybe interesting to know how much CO_2 is produced per year by our digital companions. Taking into account that in many countries, people have more than one smartphone, it can be estimated that 10 billion smartphones can produce 0.453 $GtonCO_2$ per year, about 1.4% of global CO_2 emissions, a significant impact considered that is due just to smartphones. The conclusions of the report are a strong warning about the business model that, while highly profitable for the smart phone manufacturers and the telecom industry, it is unsustainable and quite detrimental to the global efforts in emissions reductions (Belkhir and Elmeligi, 2018, p. 458).

During the weeks of the lockdown for the 2020 pandemic, many people appreciated the beneficial effects of reduced traffic and of the possibility to continue to work remotely. In these situations, of course ICT is critically essential and becomes clear the fundamental role of digital infrastructures for society. On the other hand, when billions of people are connected for remote work, videoconferencing, streaming videos, and social networks, then the networks become saturated with zettabytes of data (Moss, 2020). While it is true that the CO_2 emissions decreased, in general, most of these savings were adsorbed by the growth of digital equipment's energy consumption: as a result, the CO_2 reductions in Europe were just 5% even in the presence of a historical pandemic (Taylor, 2020).

Software and terms like "cloud" generated the myth of digital technologies as "clean" by definition, completely "detached" from the real world of ecosystems. But software without hardware does not exist. An important hint for consuming less when using bits is to think this complex task as divided into three dimensions: producing, powering (using), and managing the lifecycle (in particular the e-waste) of digital technologies.

3.1   *Producing ICT*

For analyzing the impact of ICT, it is fundamental to take into account the need of materials: the production of hardware requires the extraction of rare minerals, and their reuse and recycling is a great challenge for the nascent circular economy. Since the main part of CO_2 emissions are due to the production of new devices, this is one more reason to extend the life of digital equipment and to minimize the extraction of new minerals, which implies well-known human and environmental costs (Vazquez-Figueroa, 2012). It becomes important to understand the need to produce fewer devices that consume less energy and that last for more time.

The main issue on the hardware side is the short lifetime of digital devices (due to software unreliability, fashion of new smartphones, planned obsolescence, etc.). As a result, the growth of the number of digital devices is exponential: from 18 billion in 2017 to over 27 billion in 2022 (Santarius et al., 2022).

Indeed, producing digital devices requires many elements and rare earths, and their mining is very dangerous and often take place in very poor working conditions. Taking into account that the large part of the energy required by devices is related to the production phase, ICT vendors and manufacturers have the main responsibility for offering hardware that is repairable and upgradeable. On the software side, upgrades and maintenance should ensure the support of the physical devices for their entire lifetime.

Applying the principles of circular economy in ICT is very challenging: devices are far from being designed with a repairable-by-design approach and made of recyclable-by-design materials. As a first step, policymakers could require ICT vendors to inform customers with clear labels describing the materials contained in the devices, their country of origin, and the compliance with the human rights of workers along the entire supply chain. It is also important to provide information about the energy required for the production of the device, for its use, and how to repair and update it. Also, these labels should improve the capability of users to get the appropriate device for the right use, avoiding pointless functions and waste of matter and energy.

On another side, also the business models of ICT vendors could change: from possession to use. Introducing new Device-As-A-Service (DAAS) business models, of course, will stimulate companies to produce fewer devices but that will last longer.

3.2   *Using and powering ICT*

Of course, this is the most interesting part of this analysis: why humans build and use these digital technologies? Their operations require energy for moving bits around the planet (networks), for storing and processing data, and for using this information from the users' point of view. As a consequence, software development is becoming an important domain for consuming less.

The new "cloud" ICT architecture is based on billions of devices connected to very large data centres. Indeed about half of the electricity demand of ICT is consumed by data centres and networks. Of course, the CO_2 emissions of this energy depend on the electricity mix, but in many countries, this mix is still based mainly on fossils. These large data centres require energy for storing data, processing data, cooling servers, and energy and materials for building them.

In software development, some organizations concentrate their attentions to the power consumption and some interesting tools are becoming available. An interesting, open, and freely available tool is "CodeCarbon.io". It helps software developers to estimate the CO_2 produced by digital resources to execute their software code and provide interesting suggestions to decrease those emissions.

Software developers should start a reflection of the power consumption of their applications: among the several coding options what are the ones that minimize the absolute energy required? That minimizes the hardware obsolescence? That ensures backward compatibility via open standards? When designing a web site, what are the video resolutions that are really needed?

On the "vendors" side, an important contribution to decreasing the environmental impact of ICT (including the software layer) is the standardization of processes. Several international bodies are currently working in this new area like IEEE (Institute of Electrical and Electronics Engineers) with its "sustainable ICT" project (IEEE, 2022), European Telecommunication Standards Institute (ETSI), and the International Telecommunication Union (ITU). These efforts produce standards that can be used by companies as "certification" of their attempts and attention when developing new digital technologies (Sissa, 2021a).

In simple words, ICT emissions will decrease (or will be as low as possible) if the applications are designed considering environmental sustainability as a requirement.

Recent advances in "artificial intelligence" are based on fine-tuning complex algorithms by feeding them with very large "data lakes", but this tuning requires a huge amount of energy. Looking at the consequent amount of CO_2 emissions, training one big neural network could require up to 284.0 tonCO_2. As a reference, one car, on average (including fuel), releases 57.1 tonCO_2 in its lifetime (Worldbank, 2022). This means that advanced techniques for training one neural network produce about the same amount of CO_2 produced by five cars in their entire lifetime. The conclusions of the authors of a reference paper in this domain are interesting; they recommend a concerted effort by industry and academia to promote research of more computationally efficient hardware and software (Crawford et al., 2019; Strubell et al., 2019).

On the software side, it is becoming necessary paying more attention also in the design of algorithms.

It could be interesting that the development and application of sustainable software design principles that can ensure that software are programmed for minimizing the energy demand (e.g. default settings towards minimal energy demand, limiting connectivity requirements, using open standards, mitigating hardware obsolescence by ensuring backward compatibility, sustainable interaction design of web sites including accessibility via slow network connections, helping users to reduce data traffic by strict privacy settings, providing appropriate minimal resolution for online videos and images, deleting unused data generated during operations, avoiding autoplay, avoiding software-induced obsolescence, etc.).

It is also important to reduce the number of software updates by separating the security updates and evolutive updates, and giving priority to existing software maintenance instead of constantly releasing new software (Santarius et al., 2022).

A special attention must be dedicated to data centres responsible for the largest environmental impact of ICT. Some examples of initiatives in this directions could be to manage appropriately the software applications by setting them for minimal resource requirements, to introduce a price for online services for raising awareness of users about the energy and resource consumption, to look for strategies that minimize the data traffic, to concentrate the run of computation-intensive applications when it is most appropriate for renewable energy supply, to manage the infrastructures by putting them into "sleep-mode" when unused, etc.

3.3   *Managing e-waste*

It is now widely recognized the need to manage the entire lifecycle of digital technologies; in particular, the problem of electronic waste (e-waste) needs to be addressed with utmost urgency. In 2014, 42 million tonnes of e-waste were generated and it is no longer acceptable to export to the southern hemisphere increasing quantities of materials that contain toxic substances for human health and the environment. At the rate of 215,000 tonnes/year of e-waste, the location of Agbogbloshie, near Accra in Ghana, has become one of the most polluted sites on the planet (Bernhardt and Gysi, 2013; Baldé et al., 2014). According to United Nations e-waste monitor, more that 80% of digital devices end up in a landfill site (EWM, 2020).

In order to reduce the mountain of e-waste, it is becoming mandatory to develop reusable, repairable, recyclable, and low energy consumption digital technologies.

For example, several ICT vendors are starting to use plastic obtained by recycling fishing nets recovered from the oceans and are planning to become "carbon neutral" in a few years. Also, the disassembling of old smartphones is becoming a new business taking into account that the resulting product contains copper, gold, aluminium, cobalt, gold, lithium, rare earths, tantalum, tin, tungsten, and zinc that do not need to be extracted from the mines (Ruffilli, 2022).

It is also true that for some devices, the collection at their end of life is a challenge: personal computer, tablets, smartphones, servers from data centres, routers, antennas, etc., all of them require specific actions like removing toxic materials, extracting rare minerals, etc. Nevertheless, take-back programs should be mandatory now. Policymakers can provide norms and incentives to ICT vendors in order to move the entire industry in these directions.

On the other hand, we must not ignore one of the paradoxes brought out by the most attentive scholars of the environmental impact of ICT: the famous "rebound effect" (or "Jevons paradox"). More efficient devices risk increasing the purchases of the devices themselves, thus increasing the quantity of materials consumed, waste to be disposed of, and energy required. The paradox: better efficiency turns into an increased number of users and therefore in an increase in consumption (Hilty, 2012). From this point of view, digital technology represents a textbook example for the rebound effect: while the amount of energy for a certain processing operation halved every 18 months, the processing power doubled in the same period (Moore' law); efficiency increases result in rising consumption (Coroama and Friedemann, 2019).

Indeed, ICT sometimes accelerates the consumption cycles with the illusion of "instant gratification", with digital technologies people find themselves trapped in a world of short-term thinkers relentlessly peddling superficial desires (McGovern, 2020).

As a consequence, recycling should not be seen as a magic panacea because the exponential growth of the number of cycles to be closed itself requires increasing quantities of energy (Butera, 2020): it is time to start thinking about "slowing down" the cycles, implicit in the concept of Slow Tech that underlines the emerging need to slow down ICT consumption cycles (Patrignani and Whitehouse, 2018).

It is time to live in harmony with natural cycles and to build a more supportive society; it is time to slow down, to make a more desirable and resilient human community.

A precise assessment of the global contribution of ICT to the climate change is strongly needed, but on the other side, it is anyway urgent to introduce wise norms and standards (by policy makers), to innovate the business models and market strategies of ICT vendors, and to increase the awareness of the environmental impact of digital technologies (by the users and society in general). The need of norms and education in this domain is precisely the kernel of a European perspective.

4   A European perspective to ICT sustainability

In Europe, many researchers are investigating the ICT sustainability dilemma, for example: one of the first research centres in Europe is the Informatics and Sustainability Research at the University of Zurich (Hilty, 2022); The conferences ICT for Sustainability (ICT4S) (conf.researchr.org/series/ict4s) and the Workshops on Computing Within Limits (computingwithinlimits.org) that gather the research community every year; the Shift Project (theshiftproject.org); and many others.

The most provoking idea around these studies is that the current evolution of ICT, with its impact on ecosystems, is creating the conditions for its own failure. For example, in 2020, Steffen Lange and Tilman Santarius claim that only a strict adoption of principles for sustainable digitalization can ensure that digitalization makes a positive contribution to a social and ecological transformation; these principles are digital sufficiency (as much digitalization as necessary and as little as possible), strict data protection, and focus on the common good (Lange and Santarius, 2020).

Other researchers, at a deeper level of investigation, claim that digital technologies, with their "philosophy" based on individualism, positivism approach, and free market-fundamentalism, are inherently "against nature". For addressing the limits of ICT, it is not enough to look at information systems in their current historical context. For a more radical view of the relationship between ICT and the environment, it is required a "counter-philosophy" based on a collective view of society and on ecological principles and to recognize the climate change as the main existential threat in the Anthropocene era (Kreps, 2018).

But probably the most interesting and unique European perspective is the important role of policymakers, the need for norms, and education in ICT.

If one wants to define a starting date for the exponential growth of ICT and, as a consequence, also of its environmental impact, this date can be set to January 3rd, 1996. On that date, the United States Congress approved the "Telecommunications Act of 1996", one of the most important changes in the history of norms (de)regulating ICT (FCC, 1996):

The main points contained in this deregulation act were

- enabled competition via deregulation;
- removed restrictions on media ownership and resulted in immediate consolidation within that segment of the industry;
- protected Internet service providers from liability for content of third parties on their service.

This historical decision with its blending of low taxes, relaxing anti-monopolistic norms, de-responsibility about content represents the main enabler of the exponential growth of the Big Tech platforms in the last 30 years.

The myth of self-regulating market and of the "digital technologies" was immediately accepted and celebrated all over the world, including Europe. On those days, the world was completely immersed into the mantra "there is no alternative" (to market and technologies). Indeed, the myth of the frontier, where there are no rules, just competition, created the Digital Wild West as recently defined by Christel Schaldemose, a member of European Parliament, that affirms that for too long that tech giants have been benefited from an absence of rules, the digital world has developed into a Wild West, with the biggest and strongest setting the rules (EU, 2022a).

If there is a European perspective in this domain, it can be defined in this position:

As democratic societies, it is unacceptable to delegate only to market and technologies the governance of complex systems with a huge social and environmental impact like ICT, it is strongly needed to find the right blending also with norms and education.

This position represents a main deviation from the US approach and nowadays informs the most important European policies in ICT.

The first fundamental watershed between the US approach and a European perspective is the general regulation on data protection (GDPR, General Data Protection Regulation) introduced in Europe since May 2018. Indeed, in the US, personal data are tradable, and in Europe, it is a human right. In Europe, the protection of personal data is now strengthened and privacy regulation within Europe is homogeneous. A clear move for facing the power of Big Tech is the strict control introduced on exporting personal data outside Europe.

This is also related to ICT sustainability: data protection and privacy-by-design software can help in reducing data traffic (and then of energy consumption and CO_2 emissions) instead of stimulating connectivity and data extraction of users.

Another example is the European new vision of "Responsible Research and Innovation" implying that the different actors (researchers, users, policymakers, companies, associations) work together to better align the innovation process and its results with the values, needs, and expectations of society (RRI, 2020).

It represents a strong push towards a more complete stakeholders' network of ICT. And of course it is mitigating the strong market powers by including important social actors.

In the area of long-term sustainability, one of the main programs launched in Europe in December 2019 is the European Green Deal. It covers all sectors of the economy, including ICT (EU, 2019). The European Green Deal represents the largest strategic project in the EU: by 2050 to become the first carbon-neutral continent in the world. The plan includes the transition to renewable energy and an epochal turning point towards the circular economy (reaching zero emissions of the so-called greenhouse gases in 2050); sustainable and intelligent mobility (smarter transport); respect for ecosystems and biodiversity; and great attention to the entire food chain (high-quality food) (EU, 2019).

In general, the achievement of the European objectives in 2050 requires a major breakthrough in terms of divestments in fossil sources and investments for the development of renewable sources, for energy saving and for the closing of cycles in compliance with ecosystems. For ICT sustainability, this has important consequences on the hardware side and on the powering of large data centres.

In line with the European Green Deal, in 2019, was launched also the clean energy for all European acts where the concept of renewable energy communities was introduced and boosted. It represents a big switch from a centralized energy architecture based on few big power plants to a decentralized architecture where many small renewable energy sources are connected. It enables citizen-driven energy actions based on the optimization of the balance between the offer and the demand for energy based on renewable sources (solar, wind) and energy storage. In order to manage this complexity, renewable energy communities require also energy management software systems. It represents the most advanced "nudge" of European policymakers towards a wise use of ICT.

Probably, the most advanced step towards ICT sustainability is the European "right to repair". It represents one of the key measures to reduce e-waste by means of repairability. This "revolutionary" measure represents an historic turning point since it requires companies to use a repairable-by-design approach and a recyclable-by-design approach. Finally, we see the end of the madness of the linear economy, of the disposable devices, of the planned obsolescence, of the lack of repair manuals and spare parts for ICT. New business models based on repairability, expandability, modularity, and recyclability will finally have to unfold (EU, 2019). The new EU norm introduced in 2021 requires manufacturers of electronic device to make their products repairable for at least ten years after first coming to market (Loughran, 2021). This will create a whole new market: the sale of spare parts has been an integral part of the history of many industries (think of the automotive industry); now it is the time for ICT (Minter, 2016).

A best practice is the Dutch social enterprise Fairphone; it develops a modular smartphone, in which the individual components can be replaced or updated. It offers a five-year warranty and the return of old devices for recycling in exchange for a discount; it also carefully checks the working conditions of people along the entire supply chain, the first "fair trade" smartphone (Fairphone, 2022). It can be considered a flagship example of a European perspective.

Also related to hardware design is the European Ecodesign Directive that regulates the energy consumption of products including computers and ICT devices. It could be extended also to software products, and labelling environmental-friendly hardware and software applications could help users and public authorities in the selection phase and promote awareness of ICT impact (EU, 2009).

Policymakers have also the responsibility of supporting standards that require vendors to provide repairable devices, to increase the use of recycled materials, and to provide modular and standard interfaces. Providing incentives for recovering materials and low-energy consumption devices is also a useful policy.

Indeed, looking at the recent European policy decisions like the "right-to-repair" on the hardware side, or the DMA, Digital Markets Act (EU, 2022) on the software side, it looks like there is a growing attention of European policymakers to regain control of technologies and markets by means of education and legal norms.

In particular, the DMA, approved in 2022, for the first time in the era of cloud computing is regulating the so-called gatekeepers, the Big Tech platforms that have a too strong economic position (many of them have a market value greater than the GDP of major European countries), and a strong social impact.

The main rules in DMA enforce gatekeepers to allow interoperability with third parties, users to access data generated using the platform, and business users to promote their offer and conclude contracts outside the platform.

Also gatekeepers may no longer treat their services and products more favourably, or prevent users from linking to business outside the platform, or prevent users from un-installing any software or app, and finally, but most importantly, gatekeepers cannot track end users outside the platform service for targeted advertising.

Related to DMA is the Digital Service Act (DSA), also introduced in 2022, it contains several measures to empower users and civil society; in particular, it has introduced the protection of users' fundamental rights online. One example: it includes procedures for faster removal of illegal content.

These moves "from the society" point of view by the European institutions probably are connected with the unique position of Europe in the global ICT market: there are no more large hardware and software vendors, and most of the large online platforms (Big Tech) are based in the US or in Asia. Nevertheless, Europe is a very important rich market with several hundred millions of users.

The most recent example of a European perspective on ICT sustainability is the Corporate Sustainability Reporting Directive, introduced in 2022, where EU rules require companies to publish reports on the social and environmental impacts of their activities. This will help the civil society and policymakers to evaluate the non-financial performance of large public-interest companies with more than 500 employees, approximately 11,700 organizations in EU (2022b).

One of the most controversial areas of ICT sustainability is the transparency of its supply chain. In particular, there are minerals coming from areas where child labour is involved, and the local mining business is involved in financing wars. Since 2021, Europe has defined the so-called Conflict Minerals Regulations that state that EU importers must ensure that what they are purchasing has not been produced in a way that funds conflicts or other illegal practices (EP, 2021).

The European Parliament is also working on a more comprehensive and mandatory due diligence for environmental and human rights protection (e.g. child labour or other practices in extractive industries in developing countries). The goal is to achieve a consistent and more complete supply-chains transparency.

In Europe, one of the more interesting case studies of ethical and sustainable computing is the city of Barcelona, Spain. The project "Barcelona digital city – Putting technology at the service of people" is one of the best examples of digital platforms based on citizens' participation, cooperative and collaborative economic models, non-profit-oriented, and enabling new business models like Mobility-As-A-Service (MAAS), peer-to-peer sharing platforms, e-commerce for local companies, open data and open software, knowledge considered as a commons (BDCP, 2019).

Another good example of a European perspective is the need to embed sustainability into design practice: there are many new ideas in the software domain like "software sustainability design" principles of the Karlskrona Manifesto for Sustainable Design. Originated in Karlskrona, Sweden in 2015, it represents a milestone in the reflection on digital technologies and their environmental impact by showing the need for a cross-disciplinary approach when designing software, including the environmental, social, economic, technical, and personal dimensions (Becker et al., 2015).

Of course, policymakers can define very clear policies for ICT like requiring all "cloud providers" to run their data centres 100% on renewable energies, to define standards for companies to describe their environmental impact, to elaborate standards for the software development impact assessment (e.g. in terms of CO_2), to define clear incentives for keeping the same smartphone for a much longer period (five years and more), and to support the growing attention among users about the ICT potential benefits of a wise use of ICT. For example, users can learn how to use different (low-tech) tools for consuming less energy and resources or just decide to use ICT for less time.

The roles of policymakers and ICT vendors are fundamental: norms and digital technologies should support a transition to an economy not based on growth but on what is called "digital sufficiency": "production and consumption sufficient to serve existing societal and individual needs" (Santarius et al., 2022). Despite these promising horizons, the majority of digital applications are still based on a "growth" vision, leading to unsustainable production and consumption patterns. This leads to an exponential growth of Internet traffic triggering a reinforcing cycle: the growth rate of bit flow over the net is increasing by 25% per year: in 2017, it was 1.5 zettabyte per year; in 2022, it will be 4.8 zettabytes per year (CVN, 2022). The majority of these bits are due to the addiction-by-design capability of many popular apps: the business model of free online services is based on advertising, and it requires to keep users tethered online (Zuboff, 2019).

Will the European position be able to counterbalance the power of the Big Tech? Policymakers should investigate the possible areas of "de-commercialization" of the Internet like a ban on online advertising (Cyphers and Schwartz, 2022). This could be the evolution of a European perspective on ethical and sustainable computing.

5   Conclusions

The environmental dilemma of ICT is not just a "technology" problem. Addressing this complexity requires a deep reflection on the models and values that have driven the economy up to now and to creatively imagine different ones. What are the new ways of thinking about consumption lifestyles, about relationships, and about quality of life? ICT have accelerated dramatically the speed of many processes and multiplied exponentially the daily number messages to deal with. But quantity of information does not mean to be able to go in deep; or to construct complex sentences (*hypotaxis*), there is a growing risk of navigating over many pages, but with little depth, to construct only very simple and plain sentences without connections between them (*parataxis*).

Of course at personal level, ICT can provide sustainability assistants that help humans in decreasing their impact and learning more sustainable daily habits. ICT could help also for better balancing the time dedicated to work and the time dedicated to leisure: a wise use of technologies based on a "digital sobriety" approach. But this requires a complete different set of values: slowing down instead of accelerating, concentrating on a few things but in deep, taking time to reflect, and for living real experiences with all the senses, not just stimulating just the sight and hearing navigating the Web.

The vision of "digital sufficiency" is in line with the goal of a global reduction in energy and material consumption and with the vision of a global stationary state that is mandatory for respecting the planetary boundaries and limits. Indeed, a transition is needed from a "digital consistency" concept ("doing things better", like circular economy), to "digital efficiency" ("doing more with less"), and to a true "digital sufficiency" based on an absolute reduction of resource and energy demand while maintaining, or even improving, immaterial living conditions and a "good life" for all (Santarius et al., 2022).

In particular, this will require favouring lifestyles no longer based on consumption and objects, but centred on the quality of relationships and processes, a shift from "possession" to "use", a "digital sobriety", a "frugal abundance", frugal in consumption and abundant in relationships (Latouche, 2011).

In conclusion, digital technologies offer an interesting scenario: on one hand, they allow humans using bits to consume less; on the other hand, it has become urgent consuming less when using bits. The digital world itself requires careful evaluation from the point of view of the energy and materials needed.

This deep ethical dilemma is in front of policymakers, designers (computer professionals), ICT companies, users, and society in general, and this will require deep ethical skills and competences. These ethical skills and competences will be a requisite for the designers of future complex digital systems. In this direction, they can be supported by the Slow Tech's holistic compass, an approach that proposes the set of important questions that will be unavoidable for a wise use of ICT (the phármakon of 21st century). The typical designers of ICT systems are engineers coming from years of technical education; today, they need to enlarge their horizon and be able to depict the most complete stakeholders' network implied by their projects and innovations.

They need to design a good ICT (socially desirable), starting from human needs by using participatory design methods and design-for-all criteria. They need to design reliable systems, but also helping humans in finding the right balancing otium-negotium (leisure and working). Respecting the habeas data and the privacy-by-design principles. Using technologies that are based on open standards (open hardware, software, and data).

They need to design a clean ICT (environmentally sustainable), since the most important stakeholder is the planet Earth. Designing systems that are repairable and recyclable by design, consistent with the circular economy principles. Applying the best efforts for lowering the material and power consumption. Powering machines with renewable energies, and most importantly, these innovations must help in decreasing the CO_2 emissions.

They need to design fair ICT (ethically acceptable) with a focus on the working environments: all the actors of the supply-chain must respect the workers' safety and human rights. A fair ICT should also help address the social conflicts by creating and sharing value within and for the communities (Patrignani, 2020).

Indeed, the future designers need to develop the skills and competencies for thinking and acting in complex situations, to be able to understand the relationships of the entire ICT stakeholders' network and take decisions (the virtue of "phronesis"). But even if the next generation of engineers will be well prepared in using this Slow Tech compass, it is not enough: policymakers need to enter into the play for counterbalancing the power of Big Tech.

Of course, ICT can help in producing renewable energies for powering all digital devices; this is necessary but not enough. It is also needed a deep reflection of the values driving humans' lives: if all the reports provided by IPCC scientists describe scenarios and their horizon is very close, the risk of catastrophic changes in climate and the environment in general is becoming high. This can be a big failure for individuals, for society, and for the species (Drake, 2015). Of course, one can blame the lobbies and the strong powers of fossil industry on governments, but in reality, it is true that also individual behaviours should change. The transition towards a sustainable living requires also different visions, infrastructures, behavioural patterns, and a different systemic view.

A systemic view cannot emerge from market and technology only; it requires a vision as a starting point. Here, Europe can play an important role in front of the giants of the global scenario: in front of the "big business" approach of the USA, of the "big state" approach of China, the European contribution could be a "big democracy" approach based on its unique combination of history (the home of Magna Charta, 1215) and cultures (Kant's homeland).

This is a European perspective on ethical and sustainable computing, based on the fundamental roles of policymakers, defining norms and education for the governance of these complex systems.

Acknowledgements

This study was supported by the JSPS/STINT Bilateral Joint Research Project, "Information and Communication Technology for Sustainability and Ethics: Cross-national Studies between Japan and Sweden" (JPJSBP120185411) and Kakenhi (19K12528).

References

Andrae, A. (2017). *Total Consumer Power Consumption Forecast. Nordic Digital Business Summit*, Helsinki, Finland, October 5, 2017. www.researchgate.net/publication/320225452.

Andrae, A.S.G., & Edler, T. (2015). On global electricity usage of communication technology: Trends to 2030. *Challenges*, 6(1), 117–157.

Baldé, C.P., Wang, F., Kuehr, R., & Huisman, J. (2014). *The Global e-Waste Monitor.* United Nations University.

BDCP. (2019). *Barcelona Digital City Plan.* https://ajuntament.barcelona.cat.

Becker, C., Chitchyan, R., Duboc, L., Easterbrook, S., Penzenstadler, B., Seyff, N., & Venters, C. (2015). Sustainability design and software: The Karlskrona manifesto. In: *37th International Conference on Software Engineering*, 16th–24th May 2015, Florence, Italy.

Belkhir, L., & Elmeligi, A. (2018). Assessing ICT global emissions footprint: Trends to 2040 & recommendations. *Journal of Cleaner Production*, 177, 448–463.

Bernhardt, A., & Gysi, N. (eds). (2013). *The Worlds Worst 2013: The Top Ten Toxic Threats.* Blacksmith Institute, Green Cross Switzerland.

Brillouin, L. (1953). Negentropy principle of information. *Journal of Applied Physics*, V24(9), 1152–1163.

Brundtland, G.H. (1987). *Our Common Future, Report of the World Commission on Environment and Development.* UN Nations.

Butera, F. (2020, August 10). *The More Waste We Recycle, the Better?* connettere.org (retrieved March 23 2022).

Coroama, V., & Friedemann, M. (2019). Digital rebound – Why digitalization will not redeem our environmental sins. In: *Proceedings of the 6th International Conference on ICT for Sustainability* (ICT4S.2019). Laapeenranta, Finland, June, 2019.

Crawford, K., Dobbe, R., Dryer, T., Fried, G., Green, B., Kaziunas, E., Kak, A., Mathur, V., McElroy, E., Nill Sánchez, A., Raji, D., Lisi Rankin, J., Richardson, R., Schultz, J., Myers West, S., & Whittaker, M. (2019). *AI Now 2019 Report.* New York: AI Now Institute. https://ainowinstitute.org/AI_Now_2019_Report.html.

CVN. (2022). *Cisco Visual Networking Index: Forecast and Trends, 2017–2022.* cisco.com (retrieved December 23 2022).

Cyphers, B., & Schwartz, A. (2022). *Ban Online Behavioral Advertising.* eff.org.

De Weerdt, S. (2020, August 11). Post-coronavirus economic recovery policies could make or break the effort to limit global warming to 1.5 degrees. *Anthropocene.*

Drake, N. (2015). *Will Humans Survive the Sixth Great Extinction?* National Geographic.

EP. (2021). *Critical Raw Materials in EU External Policies.* European Parliament, May 2021.

EPA. (2021). *From Farm to Kitchen: The Environmental Impacts of U.S. Food Waste.* US Environmental Protection Agency. https://www.epa.gov/land-research/farm-kitchen-environmental-impacts-us-food-waste.

EU. (2009). *Ecodesign and Energy Labelling.* Directive 2009/125/EC and Regulation (EU) 2017/1369. https://single-market-economy.ec.europa.eu/single-market/european-standards/harmonised-standards/ecodesign_en (retrieved March 23 2022).

EU. (2019). *The European Green Deal.* eur-lex.europa.eu (retrieved December 23 2022).

EU. (2022). *The Digital Markets Act.* https://ec.europa.eu/info/strategy/priorities-2019-2024/europe-fit-digital-age/digital-markets-act-ensuring-fair-and-open-digital-markets_en (retrieved September 23 2022).

EU. (2022a, July 5). *MEPs Approve Two Laws That Aim to Make the Internet Safer and Fairer.* Euronews.

EU. (2022b). *Corporate Sustainability Reporting.* https://finance.ec.europa.eu (retrieved December 23 2022).

EWM. (2020). *E-Waste Monitor.* United Nations Institute for Training and Research (UNITAR). https://ewastemonitor.info/ (retrieved March 23 2022).

Fairphone. (2022). *How Can We Create a Fairer Smartphone?* https://www.fairphone.com/it/impact (retrieved March 23 2022).

FCC. (1996). *Telecommunications Act of 1996*. Federal Communications Commission. https://www.fcc.gov/general/telecommunications-act-1996 (retrieved December 23 2022).

Forbes. (2021). *The 10 Largest Companies in the World by Market Capitalization*. forbes.com (retrieved March 23 2022).

Frazelle, J. (2020). Power to the people. Reducing data center carbon footprints. *ACM Queue*, 18(2), 5–18.

GCP. (2022). *Global Carbon Project*. https://www.globalcarbonproject.org/ (retrieved December 23 2022).

GeSI. (2015). *Global eSustainability Initiative*. gesi.org (retrieved December 23 2022).

GreenIT. (2019). *Environmental Footprint of the Digital World*. https://www.greenit.fr/ (retrieved March 23 2022).

GSMA. (2022). *Enablement Effect. The Impact of Mobile Communications Technologies on Carbon Emission Reductions*. www.gsma.com (retrieved March 23 2022).

Hilty, L. (2012). Why energy efficiency is not sufficient – Some remarks on "Green by IT". In: Arndt, H.K. (ed.), *EnviroInfo 2012, 26th International Conference Informatics for Environmental Protection*. Shaker Verlag.

Hilty, L. (2022). *University of Zurich, Department of Informatics Informatics and Sustainability Research*. https://www.ifi.uzh.ch/en/isr/people/people/hilty.html (retrieved March 23 2022).

IDC. (2019). *The Growth in Connected IoT Devices Is Expected to Generate 79.4 Zettabytes of Data in 2025*. International Data Corporation.

IEEE. (2022). *Sustainable ICT*. https://sustainableict.ieee.org/ (retrieved March 23 2022).

Ipakchi, A., & Albuyeh, F. (2009). Grid of the future. *IEEE Power and Energy Magazine*, 7(2), 52–62.

IPCC. (2022). *Climate Change 2022. Impacts, Adaptation and Vulnerability*, Working Group II Contribution to the Sixth Assessment Report of the *Intergovernmental Panel on Climate Change*, March 2022. www.ipcc.ch (retrieved March 23 2022).

Jirotka, M. (2020). *Is It Time for a 'Responsible' Revolution?* Keynote ICT4S-2020.

Kreps, D. (2018). *Against Nature: The Metaphysics of Information Systems*. Routledge.

Lange, S., & Santarius, T. (2020). *Smart Green World? Making Digitalization Work for Sustainability*. Routledge.

Latouche, S. (2011). *Vers une société d'abundance frugale. Controsens et controverses sur la décroissance*. Mille et une nuits.

Leff, H.S., & Rex, A.F. (1990). *Maxwell's Demon, Entropy, Information, Computing*. CRC Press.

Lennerfors, T.T., Fors, P., & van Rooijen, J. (2015). ICT and environmental sustainability in a changing society: The view of ecological World Systems Theory. *Information Technology & People*, 28(4), 758–774. https://doi.org/10.1108/ITP-09-2014-0219.

Loughran, J. (2021). *EU Introduces 'Right to Repair' Rules for Electrical Goods*. Engineering & Technology.

Lucivero, F., Samuel, G., Gordon, B., Darby, S.J., Fawcett, T., Hazas, M., Ten Holter, C., Jirotka, M., Parker, M., Webb, H., & Yuan, H. (2020, June 19). *Data-Driven Unsustainability? An Interdisciplinary Perspective on Governing the Environmental Impacts of a Data-Driven Society*. http://dx.doi.org/10.2139/ssrn.3631331 (retrieved March 23 2022).

McGovern, G. (2020). *World Wide Waste*. Silver Beach.

Meadows, D.H., Meadows, D.L., Randers, J., & Behrens, W.W., III. (1972). *The Limits to Growth*. Universe Books.

Minter, A. (2016). How we think about e-waste is in need of repair. *Anthropocene*.

Moss, S. (2020, March 12). *Italy's Coronavirus Lockdown: The View From SuperNAP*. data-centerdynamics.com (retrieved March 23 2022).

OECD. (2020). *Better Life Initiative: Measuring Well-Being and Progress.* www.oecd.org/statistics/better-life-initiative.htm (retrieved March 23 2022).

Patrignani, N., & Kavathatzopoulos, I. (2018). On the complex relationship between ICT systems and the planet. In: Kreps, D., Ess, C., Kimppa, K., Leenen, L. (eds.), *13th IFIP-TC 9, Human Choice and Computers: This Changes Everything.* Springer.

Patrignani, N., & Whitehouse, D. (2018). *Slow Tech and ICT, A Responsible, Sustainable and Ethical Approach.* Palgrave Macmillan.

Patrignani, N. (2020). *Teaching Computer Ethics: Steps towards Slow Tech, a Good, Clean, and Fair ICT.* Uppsala University Press.

Plato. (1952). *Plato's Phaedrus.* Cambridge University Press.

PWC. (2020). *How AI Can Enable a Sustainable Future.* www.pwc.co.uk (retrieved March 23 2022).

Raworth, K. (2017). *Doughnut Economics: Seven Ways to Think Like a 21st-Century Economist.* Random House.

Rejeb, A., Suhaiza, Z., Rejeb, K., Seuring, S., & Treiblmaier, H. (2022). The Internet of Things and the circular economy: A systematic literature review and research agenda. *Journal of Cleaner Production,* 350, 131439.

RRI. (2020). *Responsible Research and Innovation.* ec.europa.eu/programmes/horizon2020/en/h2020-section/responsible-research-innovation (retrieved March 23 2022).

Ruffilli, B. (2022, March 19). Cosa c'è di ecosostenibile dentro uno smartphone, *Green & Blue – LaRepubblica.*

Santarius, T., Bieser, J.C.T., Frick, V., Höjer, M., Gossen, M., Hilty, L.M., Kern, E., Pohl, J., Rohde, F., & Lange, S. (2022). Digital sufficiency: Conceptual considerations for ICTs on a finite planet. *Annals of Telecommunications.* https://doi.org/10.1007/s12243-022-00914-x

Scalia, M., Angelini, A., Farioli, F., Mattioli, G., Ragnisco, O., & Saviano, M. (2020). An Ecology and Economy Coupling Model. A global stationary state model for a sustainable economy in the Hamiltonian formalism. *Ecological Economics,* 172, 1–9.

Sissa, G. (2021a). *Sostenibilità ICT requisito "by design": così ci lavorano gli enti internazionali.* agendadigitale.eu (retrieved March 23 2022).

Sissa, G. (2021b). *Quanto inquina il Cloud? Ecco perché l'opacità dei vendor non è più sostenibile.* agendadigitale.eu (retrieved March 23 2022).

Stoll, C., Klaaßen, L., & Gallersdörfer, U. (2019, July 17). The Carbon Footprint of Bitcoin. *Joule,* 3(7), 1647–1661.

Strubell, E., Ganesh, A., & McCallum, A. (2019). *Energy and Policy Considerations for Deep Learning in NLP,* arXiv.org.

Szilard, L. (1929). On the reduction of entropy in a thermodynamic system by the intervention of intelligent beings. *Zeitschrift für Physik,* 53(11–12), 840–856.

Taylor, A. (2020). *Is Going Digital in a Pandemic as Green as We Think?* Cambridge University.

Uptime. (2019). *Data Center Industry Survey.* uptimeinstitute.com (retrieved March 23 2022).

Vazquez-Figueroa, A. (2012). *Coltan.* New worlds.

Worldbank. (2022). CO_2 *Emission per Capita.* https://data.worldbank.org/indicator/EN.ATM.CO2E.PC?locations=US (retrieved December 23 2022).

WSJ. (1965, October 15). Desk-top size computer is being sold by Olivetti for first time in US. *Wall Street Journal.*

Zuboff, S. (2019). *The Age of Surveillance Capitalism: The Fight for a Human Future at the New Frontier of Power.* Profile Books.

Index

For Product Safety Concerns and Information please contact our EU
representative GPSR@taylorandfrancis.com
Taylor & Francis Verlag GmbH, Kaufingerstraße 24, 80331 München, Germany

www.ingramcontent.com/pod-product-compliance
Lightning Source LLC
Chambersburg PA
CBHW052121230326
41598CB00080B/3931